CURRENT THEMES IN ENGINEERING TECHNOLOGIES

CURRENT THEMES IN ENGINEERING TECHNOLOGIES

Selected Papers of the World Congress on Engineering and Computer Science

San Francisco, California 24 – 26 October 2007

EDITORS

Sio-long Ao
Oxford University, United Kingdom

Mahyar A. Amouzegar
California State University, Long Beach

Su-Shing Chen
University of Florida, Gainesville

All papers have been peer reviewed

SPONSORING ORGANIZATIONS

International Association of Engineers
IAENG Society of Artifical Intelligence
IAENG Society of Bioinformatics
IAENG Society of Computer Science
IAENG Society of Data Mining
IAENG Society of Electrical Engineering
IAENG Society of Imaging Engineering
IAENG Society of Industrial Engineering
IAENG Society of Internet Computing and Web Services
IAENG Society of Operations Research
IAENG Society of Scientific Computing
IAENG Society of Software Engineering
IAENG Society of Wireless Networks

AMERICAN INSTITUTE OF PHYSICS

Melville, New York, 2008

AIP CONFERENCE PROCEEDINGS ■ VOLUME 1007

Editors:

Sio-Iong Ao
Oxford University Computing Laboratory
University of Oxford
Wolfson Building, Parks Road
Oxford OX1 3QD
United Kingdom
E-mail: siao@graduate.hku.hk

Mahyar A. Amouzegar
California State University, Long Beach
J. Brotman Hall
1250 Bellflower Blvd.
Long Beach, CA 90840
USA
E-mail: mahyar@csulb.edu

Su-Shing Chen
Department of Computer Science and Engineering
University of Florida
Gainesville, FL 32611
USA
E-mail: suchen@cise.edu.fl

L.C. Catalog Card No. 2008925211
ISBN 978-0-7354-0526-4
ISSN 0094-243X
Printed in the United States of America

CONTENTS

Preface

A large international conference on engineering and computer science was held in UC Berkeley in San Francisco Bay Area, USA, 24-26 October 2007, under the World Congress on Engineering and Computer Science (WCECS) 2007. The WCECS 2007 is organized by the International Association of Engineers (IAENG), a non-profit international association found by the engineering and the computer science pioneers in 1968. The WCECS conferences serve as good platforms for the engineering community to meet with each other and to exchange ideas. The conferences have also stroke a balance between theoretical and application development. The conference committees have been formed with over two hundred committee members who are mainly research center heads, faculty deans, department heads, professors, and research scientists from over 30 countries. The conferences are truly international meetings with a high level of participation from many countries. The response that we have received for the congress is excellent. There have been more than five hundred manuscript submissions for the WCECS 2007. All submitted papers have gone through the peer review process and the overall acceptance rate is 56.78%.

This volume contains revised and extended research articles on the engineering and computer science written by prominent researchers participating in the congress WCECS 2007. Topics covered include computational biology, chemical engineering, circuits, communications systems, electrical engineering, education and information technology, modeling health advances, systems engineering, engineering management, and industrial applications. The papers are representative in these subjects sitting on the top-end-high technologies in the fields. The book offers the states of arts of tremendous advances in engineering and computer science and also serves as an excellent reference work for researchers and graduate students working with engineering and computer science.

Constructing Linkage Disequilibrium Map with Iterative Approach

S. I. Ao

Oxford University Computing Laboratory, University of Oxford, Wolfson Building, Parks Road, Oxford OX1 3QD, UK

Abstract. With recent advance of the genotyping single nucleotide polymorphisms (SNPs) in mass scale of high density in a candidate region of the human genome, the linkage disequilibrium analysis can offer a much higher resolution of the biological samples than the traditional linkage maps. We have formulated this LD mapping problem as a constrained unidimensional scaling problem. Our method, which is directly based on the measurement of LD among SNPs, is non-parametric. Therefore it is different from LD maps derived from the given Malecot model. We have formulated with the quadratic programming approach for solving this constrained unidimensional scaling problem. Different from the classical metric unidimensional scaling problem, the constrained problem is not an NP-hard combinatorial problem. The optimal solution is determined by using the quadratic programming solver. Nevertheless, because of the large requirement for memory during the running time that may cause the out of memory problems, and the high computational time of the quadratic programming algorithm, the iterative algorithm has been developed for solving this LD constrained unidimensional scaling problem.

Keywords: Single Nucleotide Polymorphism (SNP), Unidimensional Scaling, Constrained, Iterative Approach, Linkage Disequilibrium (LD), LD Map.
PACS: 87.85.Tu

ADVANTAGES AND APPLICATIONS OF THE LD MAPS

The advantages of LD maps include the revealing of the fine scale recombination patterns, the facilitating of the optimal SNP/marker spacing, and the increasing of the power for localizing disease genes etc. In the LD maps, different regions of medical interests can be shown conveniently and visually. A cold spot for LD is an interval in the LD map where LD declines rapidly with distance. When the LD maps are used to mirror the recombination, the cold spot for LD is a hot spot for recombination and vice versa. The hot spot for LD is the region in the LD map that appears as a flat region, which LD declines slowly against distance. It should be pointed out that the corresponding relationship between the LD and the recombination is not always so well defined, as there are other unpredictable genetic factors that may influence this corresponding. Besides showing the recombination patterns in the LD map, the LD map can facilitate the optimal spacing of the tag SNPs. In order to have enough coverage of a genomic segment, it is required that there will be tag SNPs within each LD unit.

The first LD maps were proposed by Maniatis and colleagues [1]. The derivation of this LD map is parametric and requires the estimation of three coefficient parameters.

CP1007, *Current Themes in Engineering Technologies*, edited by S.-I. Ao, M. A. Amouzegar, and S.-S. Chen

In the above approach, these estimated parameters are found to have large variances among different populations. We have formulated this LD mapping problem as a constrained unidimensional scaling problem with the quadratic programming approach [2, 3, 4]. Our method, which is directly based on the measurement of LD among SNPs, is non-parametric. Therefore it is different from LD maps derived from the given Malecot model. Nevertheless, because of the large requirement for memory during the running time that may cause the out of memory problems, and the high computational time of the quadratic programming algorithm, the iterative algorithm has been proposed for solving this LD constrained unidimensional scaling problem.

BACKGROUNDS FOR NON-PARAMETRIC LD MAPS

When two diallelic SNPs A and B are said to be linkage equilibrium, it means that their distributions are the same as the table shown below:

TABLE 1. Distribution table for the two alleles A and B in linkage equilibrium.

	Allele B	Allele b
Allele A	pq	p(1-q)
Allele a	(1-p)q	(1-p)(1-q)
Column sum	q	(1-q)

where the p is the probability of obtaining an allele A from the pool of alleles A and a. Similarly, q is that of getting an allele B from the pool of alleles B and b. Usually, the probability p is used to denote the allele with the lowest probability and q for the second lowest one, so we have the following relationship: $p \le q, p \le 0.5, q \le 0.5, p \le 1-q$.

When the two SNPs are in linkage disequilibrium (LD), it can be represented by the parameter d, which is defined as followed [5]:

TABLE 2. Distribution table for the two alleles A and B in linkage disequilibrium.

	Allele B	Allele b
Allele A	pq + d	p(1-q) - d
Allele a	(1-p)q -d	(1-p)(1-q) + d
Column sum	q	(1-q)

Here, d is defined in such a way that it remains non-negative all the time. When d is equal to zero, then the above table is reduced to that for the linkage equilibrium case.

The unit for measuring the scaled linkage disequilibrium is d' and is defined by:

$$d' = \frac{d}{p(1-q)} \tag{1}$$

As an example, let $d = 0.1$, $p = 0.2$ and $q = 0.3$, then $d' = 0.1 / 0.2(1-0.3) = 0.7143$. From the theoretical population biology, it is known that the above d' decays by a factor of 1 – θ per generation, where θ is the recombination fraction. Thus, we can have: $-\ln d'^2 \propto -\ln(1-\theta)$. For small values of θ, it is approximate true that $-\ln d'^2 \propto \theta$, and, therefore, $-\ln d'^2$ is approximately close to the genetic map distance measured in units of Morgan [5].

Because of the above property of the $-\ln d'^2$, we have used it to measure the LD distance between each pair of SNPs. Consider the case of n SNPs, then, we can have the n-by-n distance matrix for representing the distances between each pair of these n SNPs, with the (i, j) element defined as $[-\ln d'^2_{ij}]$. It is noted that this distance matrix is symmetric and has the diagonal elements equal to zeros. When using in the LD maps, it is desirable to have the representation of these distances in one dimension. Also, the biological knowledge tells us that it is desirable to have the preservation of original ordering of the SNPs in the chromosome. This is because physically closer SNPs have smaller genetic map distances between them too. The objective of this scaling into one dimension is such that this scaled distance values can be in agreement with the original distances in the n-by-n distance matrix as much as possible. This objective is very similar with the classical unidimensional scaling problem.

In the classical unidimensional scaling problem, it is to place n objects on the real line, so that the interpoint distances can best approximate the observed dissimilarities between pairs of objects. It is well known that this problem is equivalent to an NP-hard combinatorial problem. Our constrained unidimensional scaling problem here is to place *n* objects in a given order on the real line, so that the interpoint distances best approximate the observed dissimilarities between pairs of objects. In the literature, researchers are interested in constrained multidimensional scaling problems. Lee [6] used least squares scaling to allow not only for equality constraints but also inequality constraints.

Mathematical Formulation of the LD Maps

Mathematically, the above constrained unidimensional scaling problem can be expressed with the objective function J defined as:

$$J(z_1, z_2, \ldots, z_{n-1}) = \sum_{i=1}^{n} \sum_{j=1}^{i-1} w_{ij} \left(d_{ij} - \sum_{k=j}^{i-1} z_k \right)^2 \tag{2}$$

where d_{ij} denotes the original distance between the *i*th SNP and *j*th SNP in our distance matrix, and z_k denotes the scaled distance between the *k*th SNP and the (*k+1*)th SNP. The weighting coefficients w_{ij} can be defined as, for example, the length of the 95% confidence interval of the distance $-\ln d'^2_{ij}$. Mathematically,

$$w_{ij} = \frac{1}{-\ln CIL^2_{ij} + \ln CIH^2_{ij}} \tag{3}$$

where ($CIL_{ij} > 0.05$ or $CIL_{ij} < d'_{ij}$) and ($CIH_{ij} > CIL_{ij}$); otherwise, $w_{ij} = 0$. CIL_{ij} represents the lower 95% confidence interval, and CIH_{ij} represents the upper 95% confidence interval of the estimated d'_{ij}. When the 95% confidence interval is narrower, it means that we know quite sure about the distance between the SNP pair. Thus, their weighing should be larger. It is also noted that w_{ij} is non-negative.

We also have the following constraints:

$$z_k \geq 0, \quad k = 1, 2, ..., n-1 \tag{4}$$

This is to ensure that the scaled distances are in consistence with the orderings of SNPs in the chromosome. The scaled distance between two non-consecutive SNPs is defined as the summation of the distances between all the consecutive SNP pairs that can be found between these two SNPs. Mathematically, $z_{j(i-1)} = \sum_{k=j}^{i-1} z_k$, where $i > j$, and the new symbol $z_{j(i-1)}$ is for representation of the scaled distance between SNP j and SNP (i-1).

MATHEMATICAL FORMULATION OF THE PROPOSED ITERATIVE ALGORITHMS

Because of the large requirement for memory during the running time that may cause the out of memory problems, and the high computational time of the quadratic programming algorithm, the iterative algorithm has been developed for solving this LD constrained unidimensional scaling problem. We start the iteration with defining the initial values as

$$d^0_{i,i+1} = -\ln d'^2_{i,i+1} \tag{5}$$

for i = $1,...,n$-1. Then, we are going to update these values with the recursive algorithm:

$$d^t_{i,i+1} = \frac{\sum_{j=1}^{i-1} \sum_{k=i+1}^{n} w_{j,k} \left(-\ln d'^{\,2}_{j,k} - \left(\sum_{u=j}^{k-1} d^{t-1}_{u,u+1} - d^{t-1}_{i,i+1} \right) \right)}{\sum_{j=1}^{i} \sum_{k=i+1}^{n} w_{j,k}} \tag{6}$$

where the parameter t denotes the current number of iteration.

TABLE 3. Computational time (seconds) for chromosomes with the iterative algorithm, compared with the quadratic programming algorithm.

	chr7p15	chr8q24
SNPs No.	466	533
Q.P. Time	2720	8993
All Nearby SNPs Time	5662	9066

From the above table or from our updating algorithm, it can be observed that the computational complexity of the above iterative equation (1) is $O(n^4)$, and the memory requirement for the iterative algorithm is $O(n^2)$.

In order to simplify the computational process, the approximate iterative algorithms are developed as followed:

$$d^{t}_{i,i+1} = \frac{\sum_{j=j\_near}^{i-1} \sum_{k=i+1}^{k\_near} w_{j,k}\left(-\ln d'^{\,2}_{j,k} - \left(\sum_{u=j}^{k-1} d^{t-1}_{u,u+1} - d^{t-1}_{i,i+1}\right)\right)}{\sum_{j=j\_near}^{i} \sum_{k=i+1}^{k\_near} w_{j,k}} \tag{7}$$

where the variable j_near is the defined as $\max(1, i - near)$, k_near as $\min(SNP\ no, i + near)$. The variable $near$ is the number of the near-by SNPs that are going to be used in the approximation. The complexity for the approximation algorithm is $O(n)$. We will call this approximate algorithm "iterative algorithm with *near* nearby SNPs', and the original iterative algorithm "iterative algorithm with all nearby SNPs'.

EXPERIMENTAL GENOMIC RESULTS OF LD MAP CONSTRUCTIONS WITH THE ITERATIVE ALGORITHMS

The figures 1 and 2 show us the scaled SNPs position against their original position with the iterative algorithm with all nearby SNPs for the chromosome regions chr7p15 and chr8q24. We have observed that the scaled positions are almost exactly the same as those from the quadratic programming algorithm. Thus, it suggests that both algorithms can find the optimal solutions for the constrained unidimensional scaling problem.

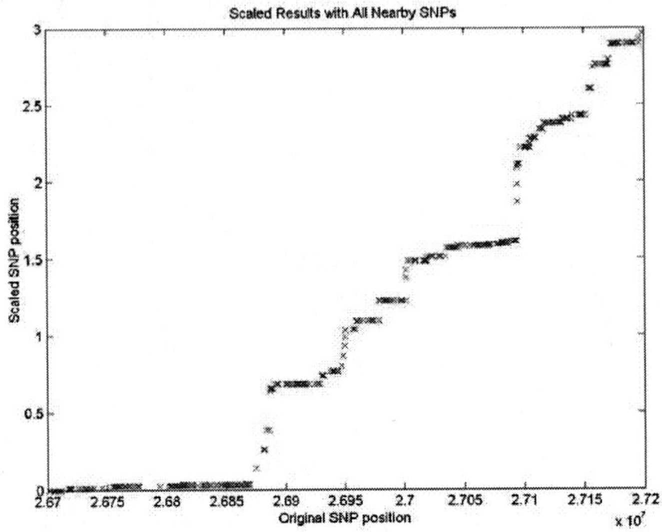

FIGURE 1. Scaled SNP position vs. original SNP position for chr7p15 with the iterative algorithm of all nearby SNPs.

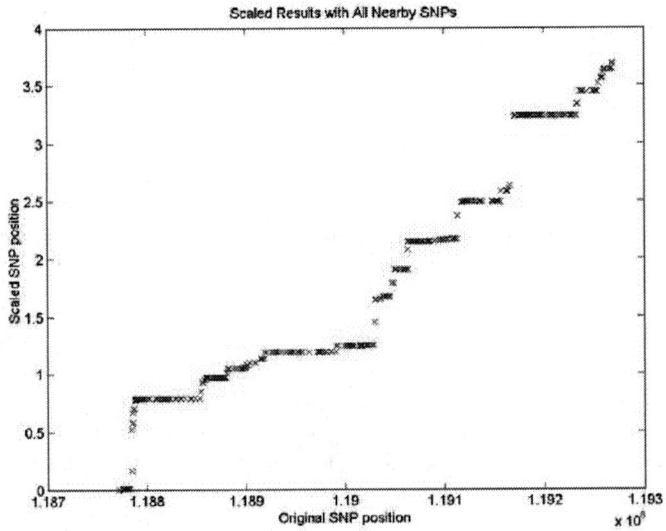

FIGURE 2. Scaled SNP position vs. original SNP position for chr8q24 with the iterative algorithm of all nearby SNPs.

Computational Time Results for the Iterative Algorithms

As we have discussed in the previous section, the approximate iterative algorithms have been developed because the computational time for the iterative algorithm with all nearby SNPs are still of order $O(n4)$. From the two tables below and from the counting of steps of the approximate iterative algorithms, we can find that the computational time is now of $O(n)$. The memory requirement is still the same as that of the iterative algorithm with all nearby SNPs.

TABLE 4. Computational time (seconds) for chromosome regions (with SNP number) by the iterative algorithms with the number of nearby SNPs set at different values.

	5 nearby	10 nearby
chr7p15 (466)	11.86	63.16
chr8q24 (533)	13.97	71.94
chr9q34 (426)	10.73	56.98

Convergence Rates of the Iterative Algorithm

The following two tables show us the convergence properties of the iterative algorithms with different nearby SNPs for the chromosome regions chr7p15 and chr8q24. We can observe that all the iterative algorithms converge fast. Generally, it only needs about 10 iterations for the relative changes to drop to a level close to zero. It is noted that, with a larger number of nearby SNPs, it usually has a larger initial relative changes.

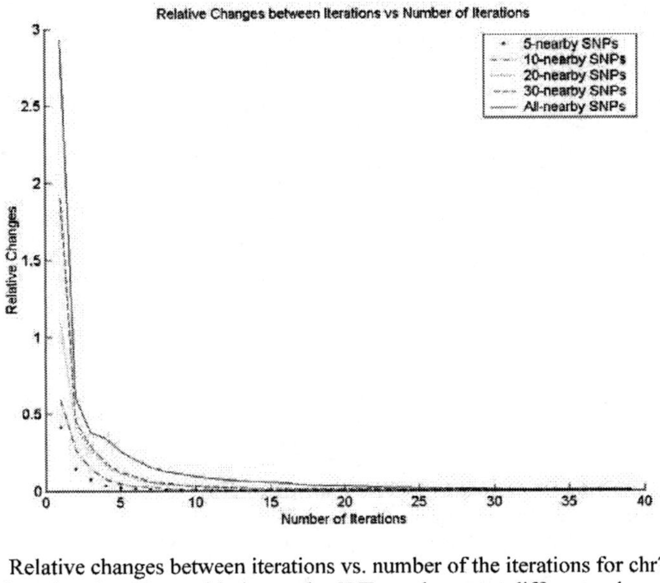

FIGURE 3. Relative changes between iterations vs. number of the iterations for chr7p15 with the iterative algorithms with the nearby SNP number set at different values.

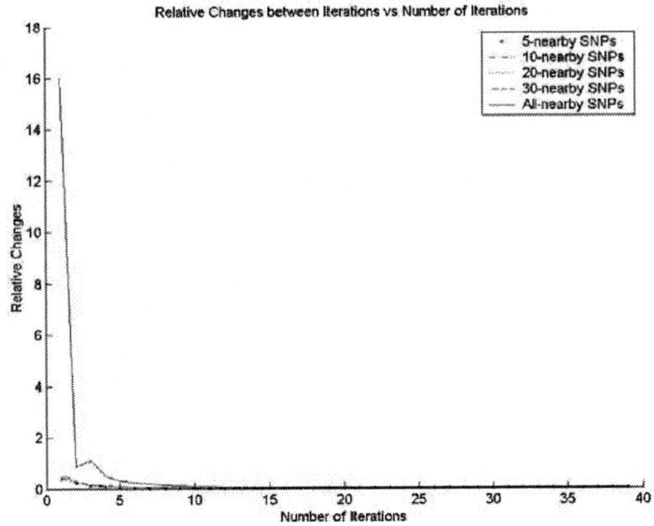

FIGURE 4. Relative changes between iterations vs. number of the iterations for chr8p24 with the iterative algorithms with the nearby SNP number set at different values.

Experimental Scaled Results for the Iterative Algorithms with Different Nearby SNPs

The figures below show us the scaled SNPs position results with the approximate iterative algorithms of different nearby SNPs. It can be observed that the iterative algorithms with different nearby SNP numbers and the quadratic programming algorithms are producing more or less similar patterns of hot and cold recombination regions. As we have seen previously, the iterative algorithm with all nearby SNPs is observed to produce the nearly the same results. The iterative algorithms with less nearby SNPs are producing results where the hot recombination regions are slightly flatter than those of the all-nearby SNP iterative algorithm and the quadratic programming algorithm.

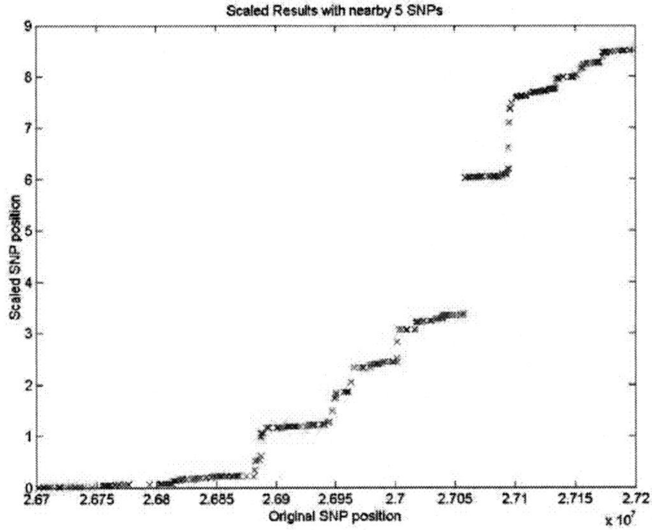

FIGURE 5. Scaled SNP position vs. original SNP position for chr7p15 with the iterative algorithm of 5 nearby SNPs.

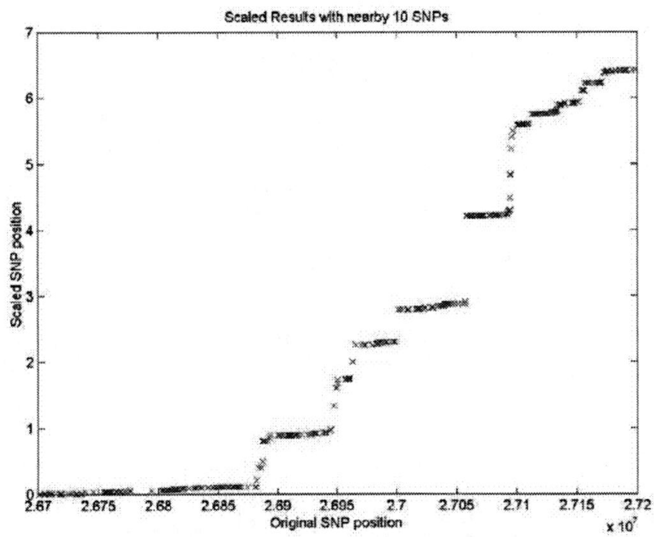

FIGURE 6. Scaled SNP position vs. original SNP position for chr7p15 with the iterative algorithm of 10 nearby SNPs.

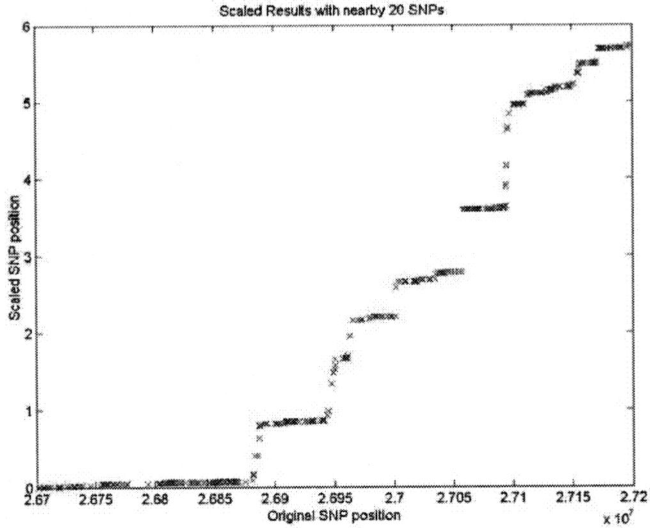

FIGURE 7. Scaled SNP position vs. original SNP position for chr7p15 with the iterative algorithm of 20 nearby SNPs.

REMARKS AND DISCUSSIONS

The characters of the linkage disequilibrium patterns can be helpful for the inferences of recombination. Combining them together, this will facilitate the search of signatures of recent selective sweeps across the human genome, for example, the regions that show more extensive linkage disequilibrium than predicted by the underlying recombination rate and which exhibit unusually low nucleotide diversity [7].

In this chapter, we have formulated and studied constrained unidimensional scaling models for constructing this linkage disequilibrium map. The model requirement is to place the objects in a given order on the real line according to the available information about the SNPs. Numerical results are presented to demonstrate the model for the application in linkage disequilibrium maps.

ACKNOWLEDGMENTS

The author would like to acknowledge the fellowship funding from the International Association of Engineers (IAENG), which supports his academics research in Oxford University Computing Laboratory, University of Oxford, UK.

REFERENCES

1. N. Maniatis, A. Collins, C. Xu, L. McCarthy, D. Hewett, W. Tapper, S. Ennis, X. Ke and N. Morton, "The First Linkage Disequilibrium (LD) Maps: Delineation of Hot and Cold Blocks by Diplotype Analysis", *PNAS*, 2002, 99:4, pp. 2228-2233.
2. S. I. Ao, M. Ng and P. Sham, "Constrained Unidimensional Scaling", *Programme and Abstracts, 3rd World Conference on Computational Statistics & Data Analysis*, International Association for Statistical Computing, 2005, pp. 49.
3. S. I. Ao, M. Ng and P. Sham, "Constrained Unidimensional Scaling with Application to Genomics", *Computational Statistics & Data Analysis*, 52:1, 2007 (advance online access: Dec. 1, 2006), pp 201-210.
4. S. I. Ao, "Data Mining Algorithms for Genomic Analysis", Ph.D. Thesis, The University of Hong Kong, May, 2007.
5. P. Sham, *Statistics in Human Genetics*, Arnold, 1998.
6. S. Lee, "Multidimensional Scaling Models with Inequality and Equality Constraints", *Commun. Statist. Simula. Computa.*, 1984, 13, 127-140.
7. M. Przwworski, "The Signature of Positive Selection at Randomly Chosen Loci", *Genetics*, 2002, pp162. 2053.

Suspension of Fine Solid Particles through the Transversal Section of Horizontal Cylindrical Channels

Herbert Loría Molina[a], Pedro Pereira-Almao[a] and Carlos Scott[a,b]

[a]*Chemical and Petroleum Engineering Department, University of Calgary, 2500 University Drive NW, Calgary. Alberta. Canada. T2N 1N4*
[b]*Facultad de Ciencias, Escuela de Química, Universidad Central de Venezuela, Los Chaguaramos, Apartado Postal 47102, Caracas, Venezuela*

Abstract. Deposition of dispersed and ultradispersed solid particles in cylindrical channels has received considerable attention due to its practical significance and direct application in industry. However, an adequate mathematical expression that studies the separation and suspension of dispersed and ultradispersed particles present in a horizontal cylindrical channel in unsteady state is still missing. In this paper we developed and solved a time-dependent, two-dimensional convective/dispersive model. This model simulates the transient deposition and suspension of dispersed and ultradispersed particles immersed in a stagnant fluid medium inside the cross section of a horizontal cylindrical channel. The results of the modeling are compared with images taken from a series of experiments especially designed for this work. The conditions which permit to control the suspension of the solid particles inside a liquid medium with cylindrical geometry are unveiled by the presented model. The experiments conducted in this work can be a simple way for obtaining values of the dispersion coefficient, via measuring the time required by the particles to reach the steady state. The model so far developed can now be further expanded to consider flow movement.

Keywords: Dispersed and Ultradispersed Particles, Suspension, Horizontal Cylindrical Channels, Modeling.
PACS: 47.55.Kf

INTRODUCTION

Mass transfer and deposition of fine particles in cylindrical channels has received considerable attention for a long time due to its practical significance and direct application in industry. For example, this knowledge is helpful in aerosol classification and its deposition under electrical fields, formation of deposits in heat exchangers and pipelines, hydrodynamic field chromatography, thrombus formation in organs, ultradispersed catalysis, etc. [1]. A theoretical prediction of the particle deposition in such systems would be very useful for the design and optimization of these processes.

The study of the particles behavior immersed in fluids began hundreds of years ago, when Archimedes introduced the concepts that made it possible to determine the forces acting on objects immersed in liquids. The next significant advance on the mechanics and dynamics of immersed objects was not accomplished until the late 17th century with Newton's publication *Principia* [2]. However, it was not until the last

CP1007, *Current Themes in Engineering Technologies*, edited by S.-I. Ao, M. A. Amouzegar, and S.-S. Chen

few decades when significant advancement was made in the understanding about particle sedimentation in vertical and horizontal channels.

These advances began with the study of the motion and settling of a single rigid particle in various media and also under boundary layer effect [3]. Forney and Spielman [4] investigated the phenomenon of the sedimentation of particles in vertical flow and gave expressions for their sedimentation velocity for a wide range of diameters. Around the same time, Sehmel [5] and Yoshioka et al. [6] reported their study on the settling of large particles (with diameters larger than 200 μm) in horizontal flow. Laurinat and Hanratty [7] took into account the motion of the particles in different directions and proposed an empirical fit to a representative deposition flux profile as a function of the angle around the cylindrical channel cross-section in horizontal flow. The determination of the driving forces present in the settling of particles immersed in fluids has been recently studied by Molls and Oliemans [8]. They have modeled the dispersion and deposition as a combined process of turbulent diffusion and gravitational settling fluxes of particles in a one-dimensional problem between two horizontal plates.

To summarize, we now realize that the sedimentation phenomenon in horizontal channels is more complex as compared to those in vertical channels due to the additional effect of the gravitational force. Investigations have shown that due to the gravitational pull the sedimentation velocity of the particles in horizontal cylindrical channel is nonsymmetrical along the cross-section of channels [9].

It is generally believed that the sedimentation velocity in a horizontal cylindrical channel essentially depends on the fluid properties and the position of the particles around the channel cross-section. Also the effect of the gravitational force cannot be neglected. However, an adequate comprehensive physical model that evaluates the effect of the separation and suspension of the particles present in a horizontal cylindrical channel is still missing in literature.

The aim of this paper is to present a study on the mechanical separation and suspension of dispersed (particle diameters: 1 μm – 100 μm) and ultradispersed particles (particle diameters: < 1 μm) based on their motion through a medium in a horizontal cylindrical channel. The objective of this work is to develop and solve a time-dependent, two-dimensional convective/dispersive model to simulate the deposition and suspension of dispersed and ultradispersed particles immersed in a fluid inside the cross section of a cylindrical channel. The model is combined with an expression that represents the velocity of these particles. This velocity is dependent on the size and density of the particle as well as on the viscosity and density of the medium. The results of the modeling are compared with images taken from a series of experiments especially designed for this work.

In some systems, it is important to ensure that the particles present in the fluid medium remain suspended in it. Because of this; one aspect that is studied in this work is to obtain the conditions (particle diameter and dispersion coefficient) that are necessary to maintain particle suspension in a fluid enclosed by the cross-section of a cylindrical geometry.

DEVELOPMENT OF THE MATHEMATICAL MODEL

Sedimentation velocity of solid particles

When a particle settles down in a liquid medium, it accelerates until the forces that cause the sedimentation equilibrate with the resistance or drag forces offered by the medium. Once this equilibrium is achieved, the particle has a constant sedimentation velocity called terminal sedimentation velocity, v_{pT}, which can be represented by [10]

$$v_{pT} = \frac{g(\rho_p - \rho_L)d_p^2}{18\mu_L}, \tag{1}$$

where g is the acceleration due to gravity, ρ_p is the density of the particle, ρ_L is the density of the liquid, d_p is the diameter of the particle and μ_L is the viscosity of liquid medium.

An expression for the time that the particle needs to reach certain sedimentation velocity can be calculated by

$$t_s = \frac{1}{36}\frac{\rho_p d_p^2}{\mu_L}\ln\left(\frac{v_{pT} + v_p}{v_{pT} - v_p}\right), \tag{2}$$

where t_s is the necessary time to reach certain sedimentation velocity, v_p.

The Stokes law applicability is well accepted for predicting sedimentation velocity of a particle in a fluid medium; several limitations start to arise when having several particles in the system and the equation becomes non-applicable when high concentrations of particles are present [11]. At substantial concentration, particles start to interact with each other and the sedimentation velocity changes.

Analysis of the particle distribution across the cylindrical channel circumference

A continuity equation for the particle concentration, C_P, in a fluid medium can be obtained by carrying out a mass balance on a differential element of mass [12]. Considering that the effective dispersion coefficient, D_E, is constant (it is called effective because includes the turbulence effects caused by the particles and the gravity force in all directions) and that the particles have a velocity field **v**; then the continuity equation can be written as:

$$\frac{\partial C_P}{\partial t} + (\mathbf{v} \cdot \nabla C_P) = D_E \nabla^2 C_P. \tag{3}$$

In order to develop a mathematical convection-dispersion model for the ultradispersed particles present in a fluid medium, consider a laminar incompressible flow which contains ultradispersed particles that flow inside a horizontal cylinder of radius R and length L. Therefore, cylindrical coordinates will be used in order to develop the model. The continuity equation in cylindrical coordinates is the following:

$$\frac{\partial C_P}{\partial t}+\left(v_r\frac{\partial C_P}{\partial r}+v_\theta\frac{1}{r}\frac{\partial C_P}{\partial \theta}+v_z\frac{\partial C_P}{\partial z}\right)= D_E\left[\frac{1}{r}\frac{\partial}{\partial r}\left(r\frac{\partial C_P}{\partial r}\right)+\frac{1}{r^2}\frac{\partial^2 C_P}{\partial \theta^2}+\frac{\partial^2 C_P}{\partial z^2}\right]. \tag{4}$$

The sedimentation of the ultradispersed particles affects the distribution of the particle concentration at each point of the horizontal cylinder. However, in this model only fully developed flow is considered, meaning that the model is taking on account a cross-section far enough from the entrance of the channel and consequently, the concentration profile does not change with the axial direction (z direction).

Thus, the particle concentration inside the horizontal channel will be modeled as a function of the position of the particles in the cross-section of the channel and the time, $C_P = C_P\,(r, \theta, t)$.

Equation 4 can be further simplified. This simplification would involve the assumption that the particle concentration is only a function of the radius r, and the angle θ. Also the angular velocity of the particle, v_θ, is small as compared to the radial velocity, v_r; therefore, it can be neglected. Equation 4 can be written as:

$$\frac{\partial C_P}{\partial t}+v_r\frac{\partial C_P}{\partial r}= D_E\left[\frac{1}{r}\frac{\partial}{\partial r}\left(r\frac{\partial C_P}{\partial r}\right)+\frac{1}{r^2}\frac{\partial^2 C_P}{\partial \theta^2}\right]. \tag{5}$$

The expression for the radial particle sedimentation velocity, v_r, is considered to be the projection of the vertical terminal velocity of the particle into the radial direction, that is, $v_r = v_{pT}\cos\theta$. Equation 5 can now be expressed as:

$$\frac{\partial C_P}{\partial t}+v_{pT}\cos\theta\frac{\partial C_P}{\partial r}= D_E\left[\frac{1}{r}\frac{\partial}{\partial r}\left(r\frac{\partial C_P}{\partial r}\right)+\frac{1}{r^2}\frac{\partial^2 C_P}{\partial \theta^2}\right]. \tag{6}$$

In the beginning of the process ($t = 0$) all the particles are well dispersed inside the cross section of the cylindrical channel. This means that the particle concentration is uniform everywhere inside the cross-section of the cylindrical channel. This concentration is the initial concentration of the particles, C_{P0}.

$$\text{At } t = 0,\ C_P(r, \theta, 0) = C_{P0} \text{ for } 0 \le r \le R \text{ and } 0 \le \theta \le 2\pi. \tag{7}$$

In the vertical axis there are symmetry boundary conditions that can be represented as:

$$-D_E\left(\frac{1}{r}\frac{\partial C_P}{\partial \theta}\right)+v_\theta C_P=0. \tag{8}$$

Previously it was assumed that the angular velocity of the particle, v_θ, is neglected, therefore:

$$\text{At } \theta=0,\ \frac{\partial C_P}{\partial \theta}=0 \text{ for } 0\leq r\leq R. \tag{9}$$

$$\text{At } \theta=\pi,\ \frac{\partial C_P}{\partial \theta}=0 \text{ for } 0\leq r\leq R. \tag{10}$$

In the center of the cylindrical channel the particle concentration does not vary with respect to the radius around all the angles:

$$\text{At } r=0,\ \frac{\partial C_P}{\partial r}=0 \text{ for } 0\leq \theta\leq 2\pi. \tag{11}$$

The walls of the channel represent a physical boundary where there is no mass exchange between the interior and exterior of the cylinder. This is an insulation boundary that will prevent any particle from leaving the channel and gather all them at the bottom. The insulation boundary means that there is no convective or dispersive flux across that boundary. This can be represented by:

$$-D_E\left(\frac{\partial C_P}{\partial r}\right)+v_r C_P=-D_E\left(\frac{\partial C_P}{\partial r}\right)+\left(v_{pT}\cos\theta\right)C_P=0. \tag{12}$$

Then:

$$\text{At } r=R,\ -D_E\frac{\partial C_P}{\partial r}+\left(v_{pT}\cos\theta\right)C_P=0 \text{ for } 0\leq \theta\leq 2\pi. \tag{13}$$

The convective/dispersive model is a linear second order parabolic partial differential equation and cannot be solved analytically by the method of separation of variables. Since in the convective term of Eq. 6 there is a variable coefficient (the term v_r is a function of the angle θ), it is not possible to divide the variables into two ordinary differential equations. However, the equation can be solved by a large variety of numerical methods: finite differences, finite element, finite volume, characteristics methods, discontinuous Garlekin methods, etc.

As a complimentary study for this research, the convective/dispersive model was solved at steady state. Two different numerical solutions for the model were presented and compared: one based in finite differences equations and the other in the finite element method. The solution based on the finite element method presented better convergence and stability properties due to the regular grid structure and its flexibility with respect to the adaptation to the geometry domain. Based in these results, the finite element method was used to solve the convective/dispersive model. Computational fluid dynamics software was used in order to apply the finite element method. The solution of the convective/dispersive model is presented in the Results section of this work.

EXPERIMENTS

The objective of the experiments is to observe the behavior of dispersed particles immersed in a fluid media. Alumina particles with an average size of 75 μm were used. The fluid medium was a 50:50 (v/v) mixture of water and glycerol. The alumina particles were painted with a dye in order to enhance the visibility in the fluid medium.

The experiments were performed in a custom made plexiglass cylinder (that was designed for this work) with an internal diameter of 4.44 cm and 6.45 cm in length. These dimensions represent 100 ml of volume inside the cylinder. The cylinder was closed at both ends with two small square plexiglass slabs glued to the open ends of the cylinder. For these experiments the relation of particles-fluid used was 1 g of particles in 100 ml of fluid. The molar concentration of the particles in the cylinder is 97.8313 mol/m^3 if the density of the alumina is taken as 3970 kg/m^3 [13].

The cylinder was filled with the solution of water and glycerol. Then the solid particles were introduced from a small orifice in the top of the cylinder. The orifice was covered with a plug and the cylinder was agitated. The cylinder was then left in repose and the phenomenon was photographed and recorded with a digital camera. A series of photographs that show the complete process of the particles deposition is presented in Fig. 1.

One important parameter measured during the experiments was the time that the particles need to sediment completely. This is the time that was necessary to reach the steady state, that is, when there was no change in the particle concentration. In this set of experiments the time needed to reach the steady state was about 74 s.

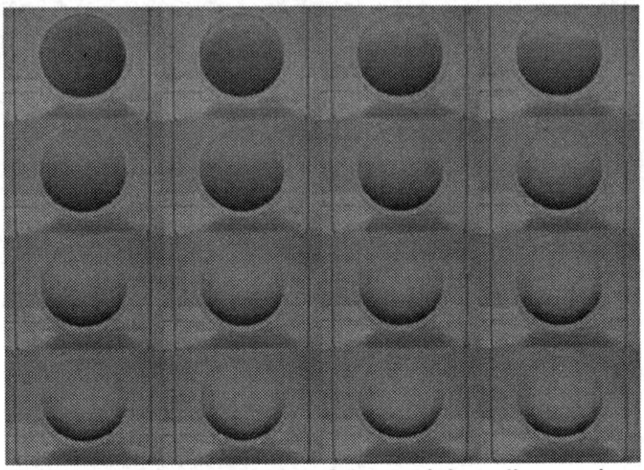

FIGURE 1. Series of photographs of the particle sedimentation process.

RESULTS AND DISCUSSION

Parameters of the model

In Eq. 6 (which represents the convective/dispersive mathematical model) it can be seen that there are two important parameters: the radial velocity, v_r, and the effective dispersion coefficient, D_E.

The radial velocity is related to the vertical velocity by an angular projection of the latter over the radius of the cylinder ($v_r = v_{pT} \cos\theta$). The vertical velocity is considered to be the particle sedimentation velocity, v_p. If v_p stays constant during the sedimentation of the particle, then it is equal to the terminal particle sedimentation velocity, v_{pT}, which depends on parameters such as the density of the particles, ρ_p, and of the fluid, ρ_L, the viscosity of the fluid, μ_L, the diameter of the particles, d_p, and the acceleration due to gravity, g.

The data obtained from experiments in our lab was used as the input parameters to obtain a numerical solution of the convective/dispersive model. The density of the alumina particles (ρ_p) was 3970 kg/m^3 [13], and the diameter of the particles (d_p) was 75 μm. The fluid medium is a 50:50 (v/v) mixture of water and glycerol. The density of this mixture (ρ_L) at 25°C is 1261 kg/m^3, and the viscosity (μ_L) is 6.05×10^{-3} Pa s [14].

With these values it is now possible to calculate the terminal velocity of the particles. Using the Stokes law (Eq. 1) the obtained value is $v_{pT} = 1.37 \times 10^{-3}$ m/s. The Stokes law is valid for solutions with small values of solid volume fraction [11] and Reynolds numbers below 2 [10]. The solid volume fraction in this case was 0.002513 and the Reynolds number associated to the terminal velocity is calculated to be 0.0214, which confirms the validity of the use of the Stokes law for the terminal velocity.

Equation 2 can be used to know how long it takes to reach the terminal sedimentation velocity. The time to reach the 99.99% of the terminal velocity was calculated to be 1.015×10^{-3} s. Since the time to reach the terminal sedimentation was very short, it can be considered that the particle sedimentation velocity on Eq. 6 was practically constant.

The effective dispersion coefficient of the particles, D_E, depends on the characteristics of the turbulent flow and the size of the channel. It can be calculated using various empirical equations for various Reynolds numbers [15-17]. However, these empirical relations are only valid for certain materials and ranges of particle size. The best way to obtain this value is to perform an adjustment of this parameter using concentration values from experimental data and the results of the mathematical model at different points of the cross-section of the cylinder and then apply nonlinear parameter estimation.

Since there are no experimental data or empirical correlations reported in literature for the materials used in this work, the following procedure was performed in order to obtain the value of the effective dispersion coefficient of the particles.

As a first approximation, a particular value of the dispersion coefficient was used to solve the convective/dispersive model. Once a solution was obtained, the time at which the simulation did not give any more changes in the concentration (the time at which the steady state is reached) was recorded and compared with the one obtained from the experiments. Then, the dispersion coefficient was changed in the simulation, until the time necessary to reach the steady state in the simulation became equal to the one obtained from the experiments.

The value of dispersion coefficient used in the simulation, as an initial guess, was the one reported by Snyder and Lumley [18]. It was calculated for solid particles with similar density and size than the ones used in this work for Reynolds numbers below 5 in an air flow. The value of the reported effective dispersion coefficient of the particles, D_E, was 1.38×10^{-4} m^2/s.

Solution of the convective/dispersive model using the finite element method implemented with computational fluid dynamics

A mass transfer time-dependent two dimensional convection-dispersion simulation was carried out using computational fluid dynamics software (COMSOL MULTIPHYSICS™). The results were obtained as follows.

The initial concentration, C_{P0}, considered for the solution was the same one that was used for the experiments (97.8313 mol/m^3). As mentioned above a first guess of the dispersion coefficient, D_E, was used to initially solve the model and compare the results with the experiments. After several comparisons it was found that the dispersion coefficient that produced the closest solution to the experiments was 7.88×10^{-6} m^2/s.

The modeled cross-section of the cylinder has the same diameter (4.44 cm) as the one used in the experiments. The selected boundary condition from the computational fluid dynamics software was the insulation/symmetry boundary condition for all the boundaries present in the model, that is, the complete circumference of the channel. For the solution of the model 2968 grid points were used. The solution of the time dependent convective/dispersive model at different times is presented in Fig. 2.

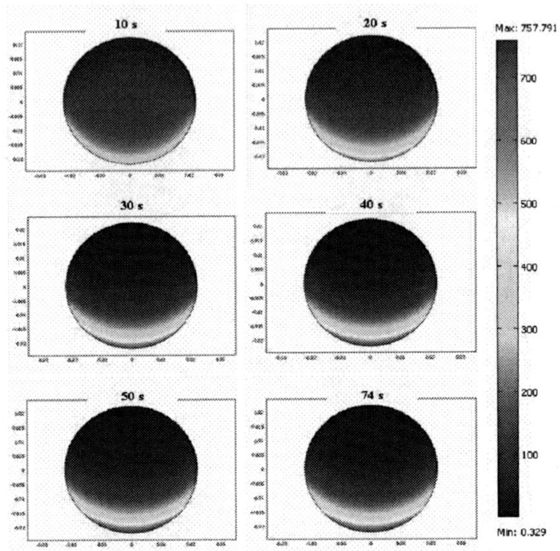

FIGURE 2. Surface response of the results of the sedimentation mathematical model at different times. Concentration in mol/m^3.

Discussion of the results

The solution of the model shows that the particles gathered at the bottom of the cylinder. Therefore, the concentration in this part increases to values higher than the initial concentration. On the other hand, the particles rapidly leave the top of the cylindrical channel, resulting in very low values of concentration in this zone.

According to the solution of the model, the maximum concentration at the bottom of the cylinder at steady state was 757.791 mol/m^3. This concentration decreases as the angle increases from 0 to π, at the top of the cylinder the minimum concentration was 0.329 mol/m^3.

The photographs of the experiments (Fig. 1) show that the shape in which the particles are distributed in the cross-sectional part of the cylinder is the same as predicted by the convective/dispersive model (Fig. 2).

Effect of the dispersion coefficient and particle diameter in the deposition of particles inside a horizontal cylinder

One of the targets of this research is to unveil the conditions that are necessary to maintain particle suspension inside a medium with cylindrical geometry. In this part of the work the convective/dispersive simulation model is applied to a system of interest for the petroleum upgrading and refining industry: MoO_3 hydroprocessing catalytic particles (density = 4500 kg/m^3) flowing in Athabasca bitumen. The physical properties of this bitumen can be calculated with thermodynamic models that were developed by us in a parallel research [19].

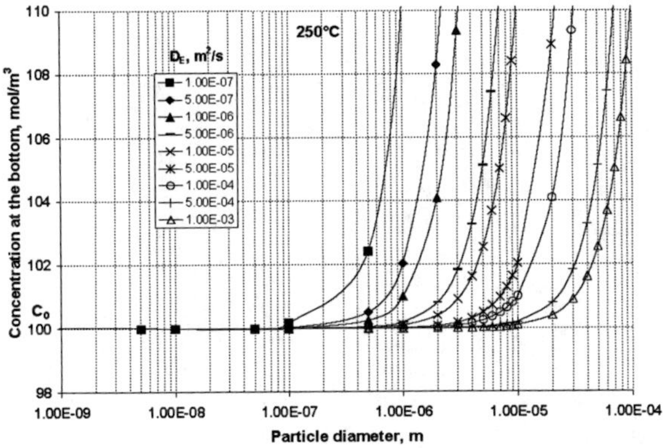

FIGURE 3. Effect of the dispersion coefficient and particle diameter. Athabasca bitumen properties at 250° C: Density = 938.08 kg/m^3, Viscosity = 4.27 mPa s.

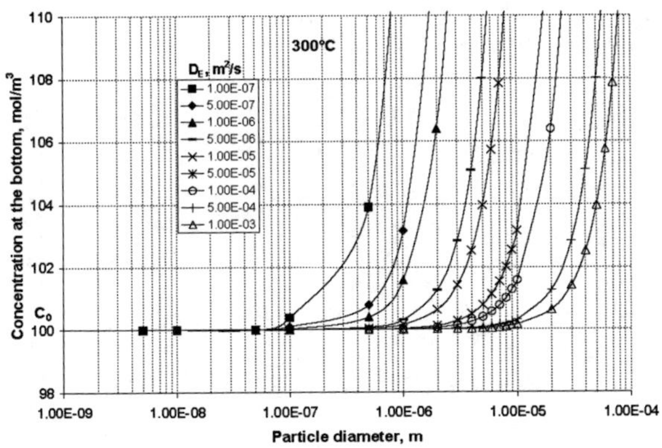

FIGURE 4. Effect of the dispersion coefficient and particle diameter. Athabasca bitumen properties at 300° C: Density = 911.1 kg/m^3, Viscosity = 2.78 mPa s.

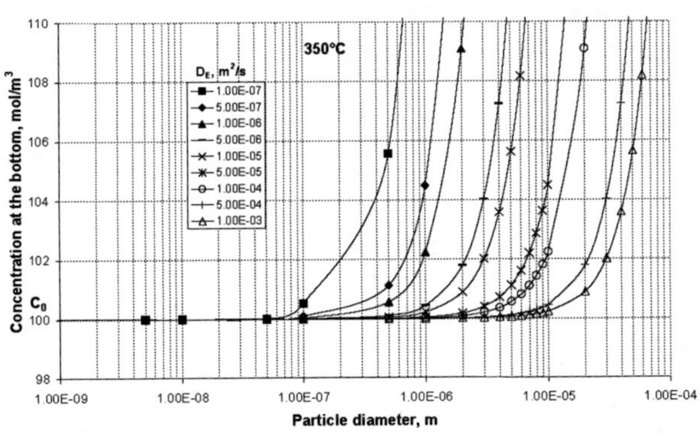

FIGURE 5. Effect of the dispersion coefficient and particle diameter. Athabasca bitumen properties at 350° C: Density = 877.21 kg/m^3, Viscosity = 1.98 mPa s.

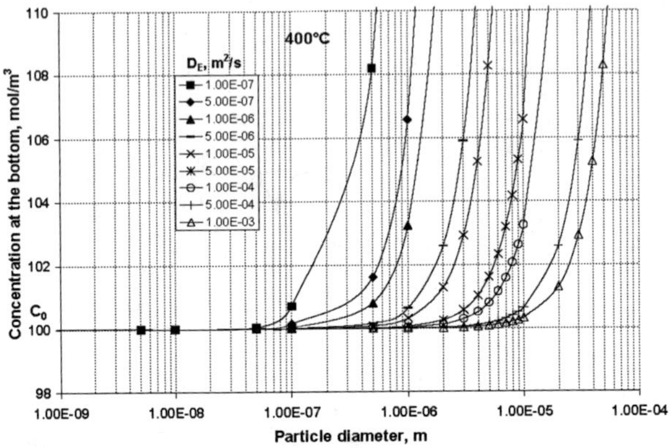

FIGURE 6. Effect of the dispersion coefficient and particle diameter. Athabasca bitumen properties at 400° C: Density = 826.28 kg/m^3, Viscosity = 1.38 mPa s.

The objective of these simulations is to find the critical values of particle diameter and dispersion coefficient that permits that the MoO_3 particles flowing in Athabasca bitumen remain suspended in the cross-section part of a horizontal cylinder (which can represent a pipeline).

Figures 3, 4, 5 and 6 show the results of several simulations carried out with different particle diameters and dispersion coefficients. The maximum particle concentration inside a horizontal cylinder, that is, the concentration at the bottom of the cylinder for a very long time (steady state) is represented in the vertical axis. Each figure represents the results of the simulations at different temperatures (and therefore different densities and viscosities of the bitumen) with an initial concentration of 100 mol/m^3, thus, the fact that the maximum concentration tends to this value means that there is not particle deposition at the bottom of the pipe.

The first effect that can be seen in the simulations is that as the dispersion coefficient increases the concentration of the particles tends to the initial one, that is, for high dispersion coefficients settling of particles will not occur. In the case of the particle diameter there is a zone for values lower than 1×10^{-7} m (100 nm) where the maximum concentration is the initial one independently of the dispersion coefficient value, this means that particles smaller than 100 nm will remain suspended in the fluid for any temperature, viscosity or dispersion coefficient in the medium.

CONCLUSIONS

The proposed convective/dispersive model is a useful tool for determining the concentration distribution of ultradispersed particles at different times in a given zone of a cylindrical channel.

As confirmed by the experiments and the model for particles sizes in the micrometer range, the concentration of particles in horizontal cylindrical channels has a nonsymmetrical distribution due to the effect of the gravity force.

The experiments conducted in this work are a new alternative for obtaining values of the dispersion coefficient. It was demonstrated that the dispersion coefficient can be determined by simply measuring the time required by the particles to reach the steady state. This is an easier method than the typical parameter adjustment, which consists on the use of the concentration values from experimental data and the results from the mathematical model at different points of the cross-section of the cylinder to apply nonlinear parameter estimation. This procedure is complex in the case of models that deal with partial differential equations.

The convective/dispersive simulation model can predict the ranges of particle sizes and dispersion coefficients at which sedimentation of the particles will not occur. In the system analyzed in this work (MoO_3 particles flowing in Athabasca bitumen) it was found that particles smaller than 100 nm will remain suspended in the fluid for any temperature, viscosity or dispersion coefficient present in the medium.

ACKNOWLEDGMENTS

This work was supported in part by the National Council for Science and Technology of Mexico, Alberta Ingenuity Centre for In Situ Energy and Schulich School of Engineering at the University of Calgary.

REFERENCES

1. Z. Adamczyk and T.G.M. Van de Ven, *Journal of Colloid and Interface Science* **80**, (2) 340-356 (1981).
2. E. Michaelides, *Particles, Bubbles & Drops: Their Motion, Heat and Mass Transfer*, New Jersey: World Scientific, 2006.
3. R. Clift, J.R. Grace and M.E. Weber, *Bubbles, Drops and Particles*, New York: Academic Press, 1978.
4. L.J. Forney and L.A. Spielman, *Journal of Aerosol Science* **5**, 257-271 (1974).
5. G.A. Sehmel, *J. Geophys. Res.* **75**, (9) 1766-1781 (1970).
6. N. Yoshioka, C. Karaoka and H. Emi, *Kagaku Kogaku* **36**, (9) 1010-1016 (1972).
7. J.E. Laurinat and T.J. Hanratty, Phys. Chem. Hydrodyn. **6**, 179-195 (1985).
8. B. Molls and R.V.A. Oliemans, *Int. J. Multiphase Flow* **24**, (1) 55-75 (1998).
9. A. Sarimeseli and G. Kelbaliyev, *Powder Technology* **140**, 79-85 (2004).
10. R.S. Ramalho, *Introduction to Wastewater Treatment Processes* (2nd ed.), New York: Academic Press, 1983.
11. A. Zeidan, S. Rohani, A. Bassi, and P. Whiting, *Reviews in Chemical Engineering* **19**, (5) 473-530 (2003).
12. R.B. Bird, W.E. Stewart and E.N. Lightfood, *Transport Phenomena* (2nd ed.), New York: John Wiley & Sons Inc., 2002.
13. R.H. Perry and D.W. Green, *Perry's Chemical Engineering Handbook* (7th ed.), New York: McGraw Hill, 1997.
14. Rudolph, S. "Dynamic Viscosities and Densities of Newton Liquids", (2006), http://www.a_m.de/englisch/lexikon/viskositaet-tafel1.htm, accessed May 2006.
15. C.S. Lin, R.W. Moulton and G.L. Putnam, *Ind. Eng. Chem.* **45**, (3) 636-640 (1953).
16. A. Altunbaş, G. Kelbaliyev and K. Ceylan, *Journal of Aerosol Science* **33**, 1075-1086 (2002).
17. P. Tandon and M.A. Adewumi, *Journal of Aerosol Science* **29**, (1) 141-156 (1998).
18. W.H. Snyder and J.L. Lumley, *Journal of Fluid Mechanics* **48**, (1) 41-47 (1971).
19. H. Loría, P. Pereira and M. Satyro, *Journal of Petroleum Science and Engineering*, Submitted for publication (2007).

Fixed-Bed Adsorption Study of Metal Ions on Bagasse Fly Ash (BFA)

Chandra Wahyu Purnomo and Agus Prasetya

Chemical Engineering Department, Engineering Faculty, Gadjah Mada University
Grafika Street no.2 Yogyakarta Indonesia 55281.

Abstract. Bagasse fly ash (BFA) has become a prospective low cost adsorbent preference for remediating wastewater containing many types of contaminant from organic compounds to toxic metal ions. The abundant availability and its unique characteristics such as large surface area and mesoporous pore size become the major reasons for utilizing BFA as adsorbents. In this paper, the continuous adsorption of Cr(VI), Cu(II) and Ni(II) into fixed bed column of bagasse fly ash (BFA) at room temperature were conducted. The experimental data are represented by breakthrough curves. Fundamental constants which govern the rate of adsorption, such as effective diffusivity of metal ions, have estimated by fitting the data with a breakthrough curve model. The effective diffusivity can be used to predict breakthrough curves in any other adsorption conditions. Meanwhile, the intensive material characterizations have been conducted before the adsorption experiments which successfully reveal the material uniqueness.

Keywords: Bagasse fly ash, Breakthrough curve, Effective diffusivity, Fixed-bed adsorption.
PACS: 68.43.Mn

INTRODUCTION

Bagasse Fly Ash (BFA) drives many attentions recently and has been intensively studied for various types of application. Mostly the research was conducted in several countries such as India, Thailand and Indonesia due to abundant availability of this material which very much depends on the existing sugar cane industries. BFA is a waste material obtained from the flue gas of the bagasse-fired boilers of sugar mills. Previously, many researchers have been fascinated by Coal Fly Ash (CFA) for its composition, properties and plentiful availability in almost every industry which utilize coal as one of their energy source. CFA has very wide applications from construction material component to advanced materials raw material such as synthetic zeolite and mesoporous silica materials.

BFA has several advantages over CFA in terms of larger surface area and less metal containments [1,2]. The metal substituent in fly ash might be released by simple leaching processes or ion exchange reactions. This is the reason why several regulations have defined coal fly ash into dangerous waste categories. The advantages of BFA drew many researcher attentions for developing this material into several applications which are quite similar to CFA. Mostly the utilization of BFA is for adsorbents several contaminants for example synthetic and natural dyes [3-8], metal ions and anions [9-15], organic and carcinogenic compounds [16-18] and also several

CP1007, *Current Themes in Engineering Technologies*, edited by S.-I. Ao, M. A. Amouzegar, and S.-S. Chen

pesticides [19-21]. Another utilization of BFA is as constituent in concrete mixture and cements.

In developing countries, the heavy metal contained wastewaters have been produced by electroplating and leather tanning industries, which are mostly categorized as small and medium scale industries. Unfortunately, these small scale industries in the region have a lack of capital for installing an appropriate wastewater treatment unit. This condition led the industries to release their waste directly without any pretreatment to the environment. Thus a simple and low cost adsorbent must be found to overcome this problem.

Although a number of adsorption studies as mentioned above have been performed, fundamental studies on adsorption kinetics were rarely undertaken thoroughly especially for continuous adsorption system. Understanding the characteristic and capacity of adsorption through breakthrough curves will enable us to better design continuous wastewater treatment units. These curves are the most common basis for assessing the performance of adsorbers. In this paper, breakthrough curves of hexavalent chrome ions (Cr(VI)), Ni(II) and Cu(II) into BFA are studied. Several breakthrough curves have also been simulated using fixed bed adsorption model.

FIXED BED ADSORPTION MODEL

Mass Balance on an Element of Column

The model is developed from mass balance on the solute for the flow of solution through a differential adsorption-bed length, dZ over a differential time interval, dt. Thus, the generated equation gives the concentration of solute in the bulk fluid as a function of time and location in the bed.

$$-D_L\frac{\partial^2 c}{\partial Z^2}+\frac{\partial(uc)}{\partial Z}+\frac{\partial c}{\partial t}+\frac{(1-\varepsilon_b)}{\varepsilon_b}\frac{\partial q}{\partial t}=0 \tag{1}$$

In equation (1) the first term represents the axial dispersion with eddy diffusivity D_L, the second term accounts an axial variation in fluid velocity, and the fourth term that is based on q, the mass-average adsorbate loading per unit mass accounts for the variation of q throughout the adsorbent particle.

The analytical solution of a simplified form of equation (1), in which constant fluid velocity u, negligible axial dispersion, and the linear driving force mass-transfer model were assumed, was provided by Klinkenberg [22]. An adopted approximate solution is given by equation (2):

$$\frac{c}{c_F}\approx\frac{1}{2}\left[1+erf\left(\sqrt{\tau}-\sqrt{\xi}+\frac{1}{8\sqrt{\tau}}+\frac{1}{8\sqrt{\xi}}\right)\right] \tag{2}$$

where $\xi = (kKZ/u)((1-\varepsilon_b)/\varepsilon_b)$ is dimensionless distance coordinate, $\tau = k(t-(Z/u))$ the dimensionless time coordinate and *erf(x)* is the error function.

Relationship for the Coefficients

For estimating several variables, the value can be obtained by employing following equations [23].

$$\frac{1}{kK} = \frac{R_p}{3k_c} + \frac{R_p^2}{15D_e} \tag{3}$$

where k_c is the external mass-transfer coefficient in cm/s, *D*e is effective diffusivity in cm^2/s and R_p is adsorbent particle radius in cm. Meanwhile, *k* is overall mass-transfer coefficient (s^{-1}) and *K* is adsorption equilibrium constant for metal ions in a linear adsorption isotherm.

The first term in Eq. (3) is the overall mass-transfer resistances, the second and third term is external one and internal one, respectively. The external transport coefficient of particles in a fixed-bed column can be correlated by [23]:

$$Sh = 2 + 1.1Re^{0.6} Sc^{1/3} \tag{4}$$

where Sh = Sherwood number = $k_c D_p / D_i$, Re =Reynolds number =$Dp.G/\mu$, and Sc = Schmidt number =$\mu/\rho Di$. By Eq. (4), external mass-transfer coefficient k_c can be estimated from *Sh*.

The fitting method between experiment and calculation data was performed using numerical method of error minimization by sum of square error (SSE) principle. The chosen numerical method was multi variables minimization. Firstly the value of *k* and ξ were guessed and then iterated to get lowest SSE. The best obtained value of the both variables can be used to determine D_e by equation (3) and (4).

EXPERIMENTAL

Material

Fly ash used in the present study was bagasse fly ash which is generated from a furnace unit of PT. Madukismo, a local sugar cane industry located in the city of Yogyakarta, Indonesia. The bulk fly ash was washed by distilled water to remove surface dust and dirt and then had been oven dried at 120°C for 3 hours. Dry fly ash was then sieved. The fly ash of the size range from 20 mesh to 30 mesh was collected and chemically activated by putting it into 10 ml solution of 3 N HCl per gram dry-fly ash under thorough mixing in a flask for 2 hrs. The fly ash was then filtered, washed by distilled water and oven dried at 120°C for 3 hours.

Cr(VI) solution were synthesized by diluting $K_2Cr_2O_7$, while Cu(II) and Ni(II) was from their nitrate compound. All the chemicals including HCl 37% were from Merck and used as purchased. The composition of Cr(VI) in solutions has been measured by UVmini 1240 (UV spectrometer), while for Cu(II) and Ni(II) being done by Unicam 929 (Atomic Absorption Spectrophotometer). For BFA characterization, the equipment for characterization were Quantachrome NOVA 2000 (surface area analyzer), Rigaku RI3000 (X-ray Fluorescence), and JEOL JSM5400LV (Scanning Electron Microscopy).

Column Adsorption Experiment

Breakthrough curve has been obtained by the continuous adsorption experiment using two different fixed bed glass columns filled by fly ash as the adsorbent. The bed porosity and density were determined before a column test was begun. The bed porosity of the both column was comparable at 0.44. The first column has a 42 cm of height of the bed and contain of 65 g of fly ash. The second column height is 21.5cm and packed with 10 grams of fly ash. Metal ion solutions flow continuously upward the column at superficial velocity of 0.1 cm/s. The equipment layout is shown in the following figure.

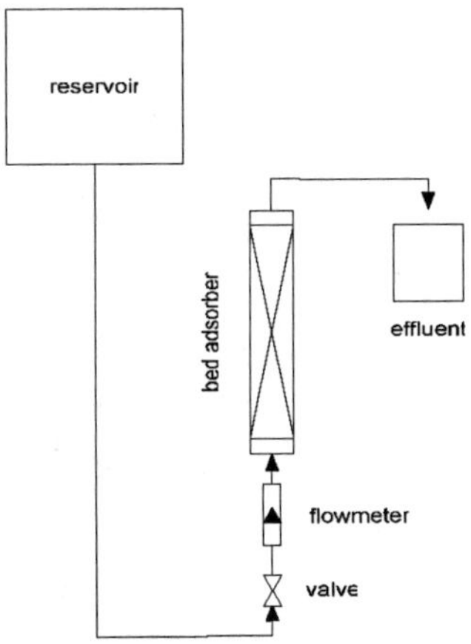

FIGURE 1. Experimental apparatus for measuring breakthrough curves.

In the experiment, the solution flowed by gravity. The flow rate was adjusted by the valve and was maintained in the desired value indicated by the flowmeter. The effluent was collected and sampled periodically for measuring concentration of each ion. The pH for all experiments was keeping at neutral and the temperature at ambient temperature. Fly ash was supported by glass wool at the bottom and upper level of the column to avoid washed out.

RESULTS AND DISCUSSION

Material Characterization

The composition of each material has been measured by X-ray Fluorescence (XRF), for non activated BFA. The table below summarized all the composition data obtained for BFA before activation.

TABLE 1. Composition of BFA from XRF analysis

Element	% weight
SiO_2	49.98
Al_2O_3	2.2
CaO	2.78
MgO	1.645
FeO	1.218
K_2O	3.97
Carbon	36.5

From the above composition table, BFA has a significant amount of total carbon. The carbon content will provide large effective sites for adsorption. The silica and alumina content of BFA also give a unique contribution to adsorption process.

SEM coupled with EDX analysis of the materials has been conducted as well. The results have revealed a unique characteristic of BFA particles. There are two types of grains, the first is porous particles which contain large carbon content and the other is dense particles which are rich with silica but low of carbon content. The two types are described in the following figures and table.

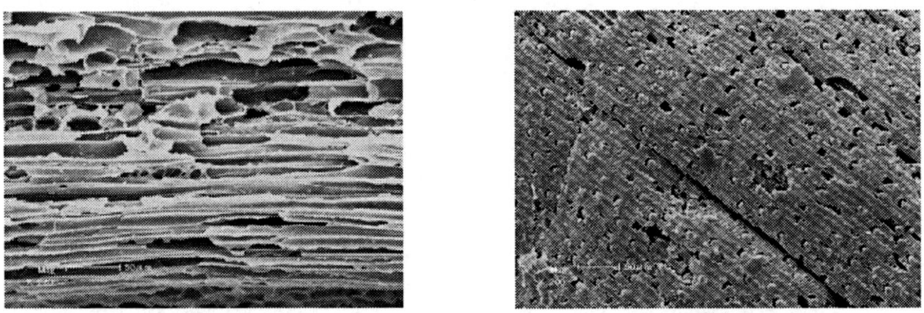

(a) (b)

FIGURE 2. SEM images of BFA: (a). carbon rich grain; (b). silica rich grain of BFA

It can be said that BFA is a mixture of two different grains which are porous carbon rich grain and dense silica rich grain. The surface concentration of each grain type has been measured by Energy Dispersive X-ray (EDX) and the results are shown in the table below. The result is quite different from XRF measurement above because EDX only measure the average local surface composition. The dense grains are mostly smaller than porous grain. Therefore, BFA of different size fractions will have different composition due to different proportion of the two type of the grains. Usually, the smaller particle size of BFA, the higher the silica content is.

TABLE 2. EDX results of the two type of BFA grains

Element	**% weight in C rich grain**	**% weight in Si rich grain**
C	92.69	29.08
O	5.84	24.62
Si	1.56	42.11
K	0.00	1.80
Ca	0.00	0.31
Mg	0.20	0.52
Al	0.00	0.46

For ensuring the structure and distribution of the main elements i.e. Silica and Carbon, elemental mapping were conducted. The images of each atom mapping were shown in the figure bellow. It can be seen that carbon atom mostly located in the center of the materials grain structure; meanwhile silica constructed the frame of the structure. The surface of material having adsorbed Cr(VI) was also mapped and the result shows that chrome ions are distributed uniformly in the material surface.

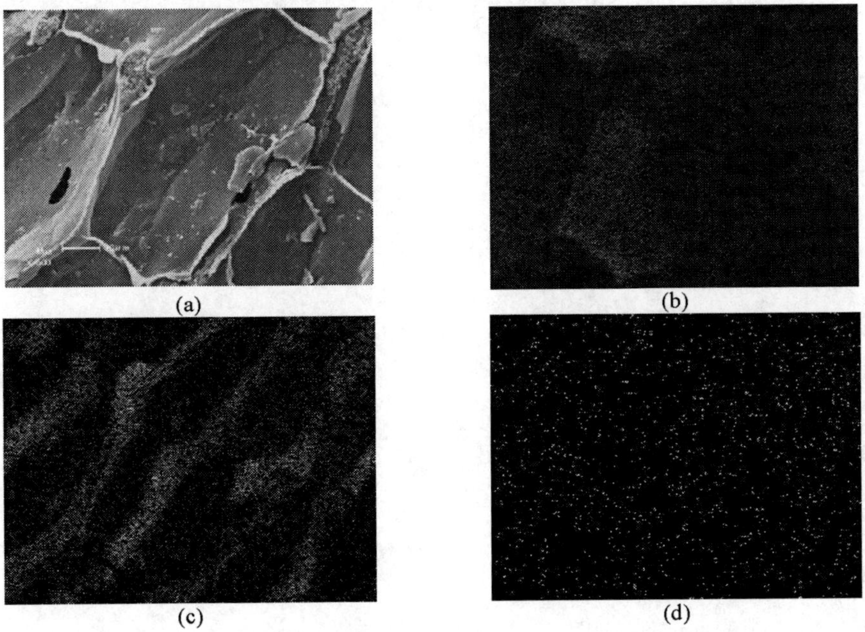

FIGURE 3. BFA Atomic mapping : (a) SEM Image, (b) C mapping, (c) Si mapping, and (d) Cr mapping,.

Breakthrough Curves

Effect of Bed Height

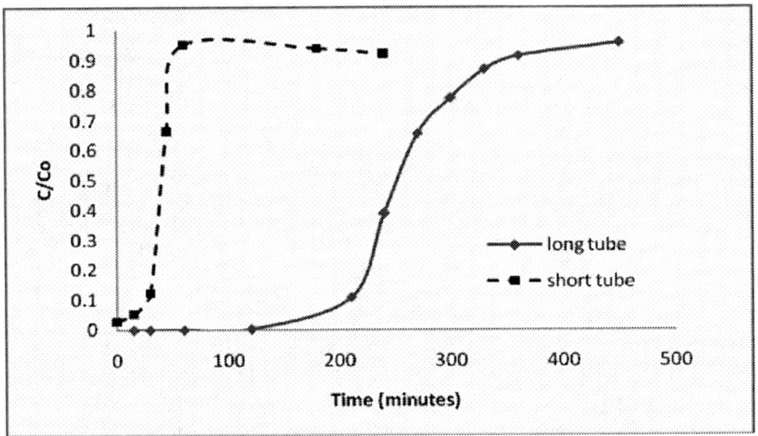

FIGURE 4. Breakthrough curve of Cr(VI) in different tube length

From the figure 4, it shows that longer bed column exhibit longer breakthrough time. The curve are shifted to the right for longer tube length, which represents a larger capacity.

Effect of Metal Ion Type

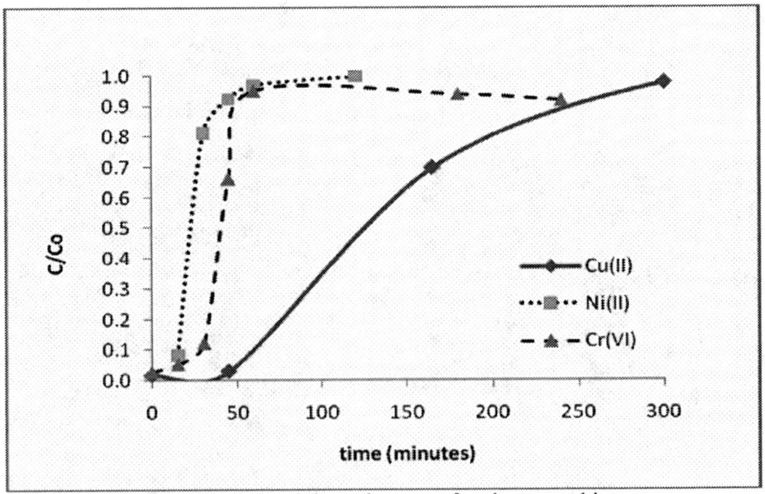

FIGURE 5. Breakthrough curve of various metal ions

Figure 5 has shown the breakthrough data otbtained from 21.5 cm column length for different ion metals. The material exhibits the lowest capacity in adsorbing Ni (II) compared to the other ions. The better breakthrough curve of BFA for Cr(VI) compared to Ni(II) as showed above has also been documented previously [11]. On the other hand, the BFA shows higher capacity in Cu(II) adsorption. However, from shape of the breakthrough curve of Cu(II), it can be concluded that the less step curve may represent low interaction between adsorbent and adsorbate molecules.

Breakthrough Curve Simulation

Figure 6 shows the adsorption breakthrough curve of Cr(VI) ions from the first column. The dimensionless breakthrough time τ ranges between 20 and 50. These values are significantly higher than those obtained from the second column as depicted in Fig. 7. This is mainly due to larger adsorbent capacity in a larger column, leading to higher amount of Cr(VI) ions adsorbed prior to a breakthrough.

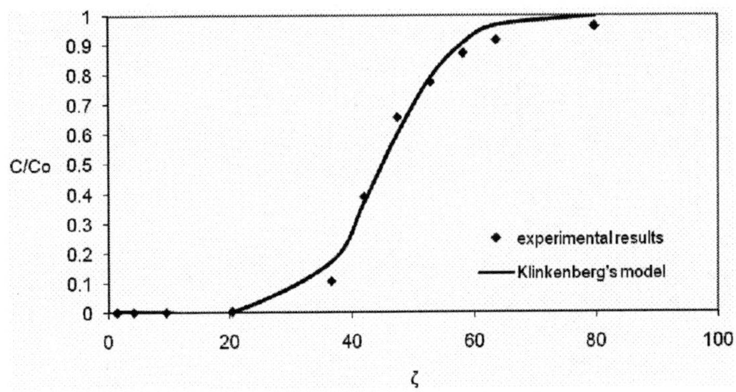

FIGURE 6. Breakthrough curve of Cr(VI) in 42 cm height column

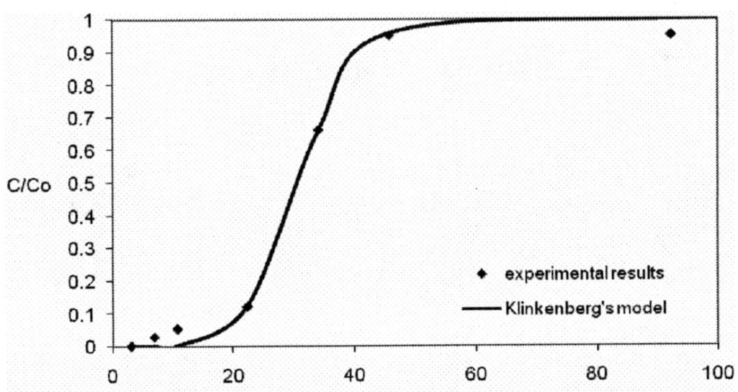

FIGURE 7. Breakthrough curve of Cr(VI) in 21.5 cm height column

Figures 6 and 7 show different saturation time of the two columns. The first column was almost completely saturated at dimensionless time of about 70, while the shorter column is already saturated at dimensionless time of about 50.

Fitting the data with the model, it can be concluded that the model represents the data satisfactorily. The fitted values of k, ξ and then calculated D_e are listed in Table 3.

TABLE 3. Fitted and calculated values of parameters.

Column length (cm)	k (s^{-1})	ξ	*De* ($cm^2/s \times 10^6$)
42	0.0032	47.8227	4.0283
21.5	0.0031	27.4919	4.0195

All of the experiments were conducted in the same temperature, using the same effective velocity u. Therefore, the value of D_e and k must remain unchanged, as has been shown in Table 3. If the value of D_e and k can be obtained, it will allow an easily prediction of the behavior of breakthrough curves in a specific adsorption operating conditions and column dimensions without doing any adsorption equilibrium experiments as adsorption isotherm is assumed linear.

APPENDIX

c : concentration of Cr(VI) ($mol.cm^{-3}$)
c_F : concentration of Cr(VI) in feed solution ($mol.cm^{-3}$)
Di : molecular diffusivity ($cm^2\ s^{-1}$)
Dp : diameter of adsorbent (cm)
G : mass velocity of liquid ($g.m^{-2}.s^{-1}$)
k : overall mass-transfer coefficient (s^{-1})
kc : external mass-transfer coefficient (cm/s)
K : adsorption equilibrium constant for Cr(VI) in a linear adsorption isotherm
u : effective velocity of Cr(VI) (cm/s)

ACKNOWLEDGMENTS

The authors thank to Hi-Link Project JICA for funding the research. Special acknowledgement is addressed to Prof. Tsuyoshi Hirajima and his research group in Kyushu University, Japan for valuable discussion, suggestion and support for analysis.

REFERENCES

1. C.W. Purnomo, A. Prasetya and K. I. Sugianto, "The performance comparison of Bagasse Fly Ash (BFA) and Coal Fly Ash (CFA) utilization for removal Cr(VI) ions from aqueous solution" Pertemuan Nasional Insinyur Kimia Indonesia Conference Proceedings, BKKPII, Yogyakarta, Indonesia, 2006.

2. C.W. Purnomo and A. Prasetya, "Study of Zeolit and Fly Ash as Heavy Metal Ions Adsorbent in Wastewater", Seminar Nasional Teknik Kimia Universitas Parahyangan Conference Proceedings, Universitas Parahyangan, Bandung, Indonesia, 2007.
3. I. D. Mall, V. C. Srivastava, N. K. Agarwal and I. M. Mishra, *Chemosphere* **61,** 492–501 (2005) .
4. I. D. Mall, V. C. Srivastava and N. K. Agarwal, *Dyes and Pigments* **69**, 210e223 (2006)
5. I. D. Mall and V. C. Srivastava, *Colloids and Surfaces A: Physicochem. Eng. Aspects* **264**, 17–28 (2005).
6. V. S. Mane, I. D. Mall and V. C. Srivastava, *Dyes and Pigments* **73,** 269e278 (2007).
7. M. Rachakornkij, S Ruangchuay and S Teachakulwiroj, *Songklanakarin J. Sci. Technol.* **26**(Suppl. 1), 13-24 (2004).
8. I. D. Mall, V. C. Srivastava and N. K. Agarwal, *Journal of Hazardous Materials,* xxx (2006) xxx–xxx.
9. V. C. Srivastava, I. D. Mall and I. M. Mishra, *Chemical Engineering Journal* **117**, 79–91 (2006).
10. V. K. Gupta and I. Ali, *Separation and Purification Technology* **18**, 131–140 (2000).
11. M. Rao, A.V. Parwate and A.G. Bhole, *Waste Management* **22**, 821–830 (2002).

12. G.M. Taha, *Ground Water Monitoring & Remediation* **26**, 137–141(2006).
13. V. C. Srivastava, I. D. Mall and I. M. Mishra, *Ind. Eng. Chem. Res.* **46**, 5697-5706 (2007).
14. V. K. Gupta and S. Sharma, *Ind. Eng. Chem. Res.* **42**, 6619-6624 (2003).
15. V. K. Gupta, C.K. Jain, I. Ali, M. Sharma, V.K. Saini, *Water Research* **37**, 4038–4044 (2003)
16. D. H. Lataye, I. M. Mishra and I. D. Mall, *Ind. Eng. Chem. Res.* **45**, 3934-3943 (2006)
17. V. K. Gupta, S Sharma, I. S. Yadav and D. Mohan, *J. Chem. Technol. Biotechnol.* **71**, 180È186 (1998).
18. V. C. Srivastava, M. M. Swamy, I. D. Mall, B. Prasad, I. M. Mishra, *Colloids and Surfaces A: Physicochem. Eng. Aspects* **272**, 89–104 (2006).
19. M. Akhtar, S. M. Hasany, M.I. Bhanger, S. Iqbal, *Chemosphere* **66**, 1829–1838 (2007)
20. Vinod K. Gupta, C.K. Jain, I. Ali, S. Chandra and S. Agarwal, *Water Research* **36**, 2483–2490 (2002)
21. V. K. Gupta and I. Ali, *Water Research* **35**, 33-40 (2001)
22. K. Adriaan, *Ind. Eng. Chem.* **40**, 1992–1994 (1948)
23. J. D. Seader and E. J. Henley, *Separation Process Principles,* John Wiley & Sons, Inc., 2002, pp. 814.

Computational Flow Modeling of Hydrodynamics in Multiphase Trickle-Bed Reactors

Rodrigo J. G. Lopes and Rosa M. Quinta-Ferreira*

GERSE – Group on Environmental, Reaction and Separation Engineering, Department of Chemical Engineering, Faculty of Sciences and Technology, University of Coimbra Pólo II, Rua Sílvio Lima, 3030-790 Coimbra, Portugal
**Corresponding author: Tel.:+351 239798723; fax: +351 239798703*
E-mail addresses: rosaqf@eq.uc.pt (R. M. Quinta-Ferreira), rodrigo@eq.uc.pt (R. J. G. Lopes)

Abstract. This study aims to incorporate most recent multiphase models in order to investigate the hydrodynamic behavior of a TBR in terms of pressure drop and liquid holdup. Taking into account transport phenomena such as mass and heat transfer, an Eulerian *k*-fluid model was developed resulting from the volume averaging of the continuity and momentum equations and solved for a 3D representation of the catalytic bed. Computational fluid dynamics (CFD) model predicts hydrodynamic parameters quite well if good closures for fluid/fluid and fluid/particle interactions are incorporated in the multiphase model. Moreover, catalytic performance is investigated with the catalytic wet oxidation of a phenolic pollutant.

Keywords: CFD, Euler-Euler model, Hydrodynamics, Multiphase Flow.

INTRODUCTION

Multiphase flow systems are described by the cocurrently downward flow of gas and liquid through a packed bed and its commercial applications arises in processing of fuels and chemicals such as desulfurization, hydrotreating, hydrocracking, distillation and filtration. Over the last three decades, the research conducted for the design of such multiphase systems still relies on simplified empirical models rather than on a theoretical basis. The lack of knowledge about the detailed flow picture in packed beds relies in the complex mechanisms governing the fluid flow so that pilot scale experiments is often carried out to perform scale-up studies. Therefore, in reactor design, the coupling between different flow regimes as well as mass and heat transfer rates are directly linked with the hydrodynamics of multiphase reactors such as trickle-bed reactors (TBR).

The success of the modeling of multiphase flow processes is virtually related with recent advances achieved in computational fluid dynamics (CFD) given that nowadays computers offer unprecedented numerical power to address complex chemical process operational and design issues. Our case study outlines an alternative CFD modeling method to investigate the hydrodynamic behavior of a TBR in terms of

CP1007, *Current Themes in Engineering Technologies*, edited by S.-I. Ao, M. A. Amouzegar, and S.-S. Chen

pressure drop, liquid holdup and catalyst wetting efficiency. After a brief review of modeling approaches, further details of an Eulerian two-fluid model is provided discussing velocity 3D maps and catalyst surface temperature profiles evaluated in unsteady state after performing the CFD validation.

STATE OF THE ART

The large number of studies that have been reported in the literature on various hydrodynamic aspects of trickle-bed reactors rely on several correlations and models of pressure gradient and liquid holdup. The existing hydrodynamic models can be broadly classified into two different categories [1]-[9]. The first category uses an empirical approach based on dimensional analysis to produce explicit correlations for pressure drop and holdup. These correlations have several parameters for fitting the experimental results in which the predicted values of pressure gradient and liquid saturation vary considerably. The second category involves the development of models resulting from equations of motion and considers determination of drag forces of gas and liquid phases at various operating regimes. In this category it has been used three distinct approaches: the relative permeability model, the fundamental force balance model and the slit model [4]. More recently, the effort has been routed for advanced CFD models based on macroscopic mass and momentum conservation laws, in which the drag force has a contribution to both particle-liquid and gas-liquid interactions [5].

The present trend is to develop models based on the fundamental approach to exploit their wider range of applicability unlike the correlative models, which are system specific. Therefore, our model is based on the fundamental physics based approach containing the mechanistic details of the system coupled with reaction kinetics for the catalytic degradation of liquid pollutants performed elsewhere [10] and attempts to model pressure drop and liquid holdup at high-pressure operation for the trickle-bed reactor. The CFD model was then developed focusing a unified approach in modeling of the hydrodynamics with incidence for the downflow mode to present a more realistic picture of the complex hydrodynamics prevailing in the reactor.

CFD MODEL

In the present work, the flow in the trickle-bed reactor was modeled using a multiphase CFD approach incorporated in the FLUENT 6.1 (FLUENT INC. USA) [11] software that is the Eulerian multiphase model. In the Eulerian two-fluid approach, the different phases are treated mathematically as interpenetrating continua. The derivation of the conservation equations for mass, momentum and energy for each of the individual phases is done by ensemble averaging the local instantaneous balances for each of the phases. The current model formulation specifies that the probability of occurrence of any one phase in multiple realizations of the flow is given by the instantaneous volume fraction of that phase at that point where the total sum of all volume fractions at a point is identically unity. Fluids, gas and liquid, are treated as incompressible, and a single pressure field is shared by all phases. In multiphase flows, the continuity (1), momentum (2) and species continuity equations (3) are solved for each phase and the momentum transfer between the phases is modeled through a drag term [12].

$$\frac{\partial U_x}{\partial x}+\frac{\partial U_y}{\partial y}+\frac{\partial U_z}{\partial z}=0 \tag{1}$$

$$\frac{\partial}{\partial t}(\alpha_i\rho_i\vec{U}_i)+\nabla\cdot(\alpha_i\rho_i\vec{U}_i\vec{U}_i)=-\alpha_i\nabla p+\nabla\cdot\hat{\tau}_i+\alpha_i\rho_i\vec{g}+\sum_{p=1}^{n}(\vec{R}_{ij}+\dot{m}_{ij}\vec{U}_{ij}-\dot{m}_{ji}\vec{U}_{ji})+\vec{F}_{ij} \tag{2}$$

$$\frac{\partial \rho_i}{\partial t}+\nabla\cdot\rho_i\times\vec{U}+\nabla\cdot\vec{j}_i-w_i=0 \tag{3}$$

$$\vec{j}_i=-\rho\, D_i\,\nabla\cdot\alpha_i$$

ρ is density, U the mass average velocity, h is specific enthalpy and g is gravity; subscripts i and k represents different species. The arrow overbar signifies a vector, and ^ is a second-order tensor. μ and b are the first and second coefficients of viscosity, respectively. κ is the thermal conductivity, and D_i is the multi-component diffusion coefficient and α_i is the mass fraction. In turbulent flows, the incompressible transport equations are given by (4)-(6).

$$\frac{\partial \rho k}{\partial t}+\sum_i\frac{\partial \rho U_i k}{\partial x_i}=\sum_j\frac{\partial}{\partial x_j}\left[(\mu+\mu_t)\frac{\partial k}{\partial x_j}\right]+\sum_i\sum_j\tau_t\frac{\partial U_i}{\partial x_j}-\rho\varepsilon \tag{4}$$

$$\frac{\partial \rho\varepsilon}{\partial t}+\sum_i\frac{\partial \rho U_i\varepsilon}{\partial x_i}=\sum_i\frac{\partial}{\partial x_j}\left[(\mu+0.77\mu_t)\frac{\partial \varepsilon}{\partial x_j}\right]+C_{\varepsilon 1}\frac{\varepsilon}{k}+\sum_i\sum_j\tau_t\frac{\partial U_i}{\partial x_j}-\frac{C_{\varepsilon 2}\rho\varepsilon^2}{k} \tag{5}$$

$$\tau_t=\mu_t\left(\frac{\partial U_i}{\partial x_j}+\frac{\partial U_j}{\partial x_i}\right)-\frac{2}{3}\delta_{ij}\left(\rho k+\mu_t\frac{\partial U_k}{\partial x_k}\right) \tag{6}$$

δ_{ij} is Kronecker delta and k subscript indicates a summation over the x_k Cartesian coordinates. The thermal energy balance and the heat flux are expressed in (7) and (8), respectively.

$$\frac{\partial \rho h}{\partial t}+\sum_i\frac{\partial \rho U_i h}{\partial x_i}=\frac{\partial P}{\partial t}+\sum_i U_i\frac{\partial P}{\partial x_i}+\sum_i\left\langle U_i''\frac{\partial P'}{\partial x_i}\right\rangle_m+\sum_i\frac{\partial q_i}{\partial x_i}-\sum_i\frac{\partial}{\partial x_i}\left\langle \rho U_i i''\right\rangle_m$$
$$-\sum_i\sum_j\left[\tau_{ij}\frac{\partial U_i}{\partial x_j}+\left\langle\tau_{ij}\frac{\partial U_i''}{\partial x_j}\right\rangle_m\right] \tag{7}$$

$$q_j = -\kappa \frac{\partial T}{\partial x_j} - \sum_{j=1}^{n} \frac{\mu}{Sc}\left(h_i \frac{\partial \alpha_i}{\partial x_j}\right) = -C_p \frac{\mu}{\Pr} \frac{\partial T}{\partial x_j} - \sum_{j=1}^{n} \frac{\mu}{Sc}\left(h_i \frac{\partial \alpha_i}{\partial x_j}\right) \quad (8)$$

NUMERICAL METHOD

The numerical simulation was performed on a cylindrical grid (L – 1m, ID – 0.05m) and the mesh adopted in the TBR reactor is tetrahedral around and over the catalyst particles and hexahedral elsewhere with 800,000 cells with the first ten catalytic layers shown in Fig. 1. Catalytic bed grid generation for the trickle-bed reactor was created using the integrated solid modeling and meshing program Gambit (Fluent Inc., USA) [14]. The approach consists in dividing the domain in subdomains, each of which is represented by a boundary-fitted coordinate mapping to a specific region in which a uniform grid is generated. In order to manage with the geometric complexity of the catalytic bed, the subdomain decomposition was unstructured, leading to multiblock block-structured grids. The local grids for a sample catalyst particle are structured and the flexibility of this kind of grid was covered for trickle flow in which boundary parts move relatively to each other. Furthermore, this is a way to include adaptivity in the structured grid context. The discretization takes place on boundary-fitted structured grid and the flow is governed by the incompressible *Navier-Stokes* equations. The discretized continuity equation serves as an algebraic constraint and often a semi-heuristic turbulence model is used to predict time-averaged flow variables based on a compromise between accuracy, memory requirements and computing time. The local refinement and coarsening of unstructured tetrahedral meshes in our case study require local grid modifications to efficiently resolve solution features for computing three-dimensional problems that arises in TBR. However, repeated anisotropic subdivision can significantly deteriorate the quality of a tetrahedral mesh demonstrating that isotropic subdivision is mandatory if mesh quality is to be controlled effectively for arbitrary refinement levels in tetrahedral meshes, without resorting to local mesh regeneration. Experimentally, it is observed that trickle-bed reactors present random directional flow fields imposing serious limitations to tetrahedral meshes that could lead to an inefficient distribution of grid points in the final mesh. The momentum equations are solved with the coupling SIMPLE algorithm and the second upwind discretization scheme. The pressure is computed by means of the PRESTO scheme. Model equations were solved in a transient fashion with a time step of 1 s for the Eulerian simulations and a number of sub-iterations were performed within each time step to ensure continuity. The residuals convergence was accelerated by under-relaxation parameters, 0.4 for pressure and 0.8 for velocity vector field. Inlet boundary conditions are assigned at the top distributor and outlet conditions at the free surface.

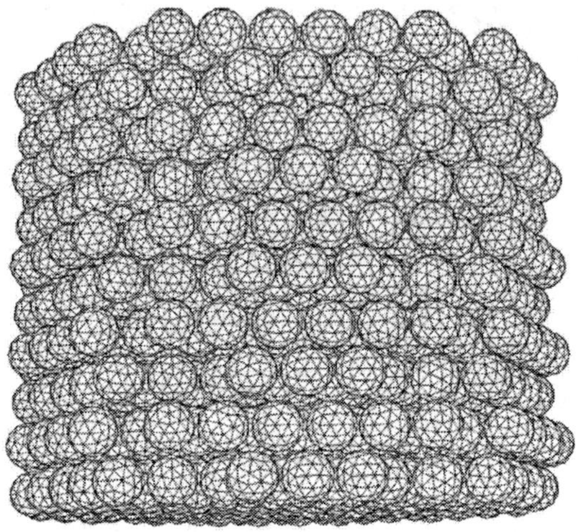

FIGURE 1. Trickle-bed reactor computational mesh

The turbulent flow is modeled through a set of modified k-ε equations with terms that include interphase turbulent momentum derived from the instantaneous equation of the continuous phase and involves the velocity covariance. The equations discussed above are solved using an extension of the SIMPLE algorithm. The momentum equations are decoupled using the full elimination algorithm available in FLUENT in which the variables for each phase are eliminated from the momentum equations for all other phases. The pressure correction equation is obtained by summing the continuity equations for each of the phases. The equations are then solved in a segregated, iterative fashion and are advanced in time. At each time step, with an initial guess for the pressure field, the primary- and secondary-phase velocities are computed. These are used in the pressure correction equation and based on the discrepancy between the guessed pressure field and the computed field, the velocities, L/G holdups and fluxes are suitably modified to obtain convergence in an iterative manner. In the first stage, several runs were computed with sufficiently fine meshes to evaluate this dependency. At this point, it was possible to check the near-wall mesh in the post-processing treatment. The solution independency was then established after several assays with the definition of turbulence boundary conditions available in k-ε model. The boundary conditions at the walls are internally taken care by FLUENT, which obviates the need for boundary condition inputs for k and ε supplied by inlet boundaries, specifically velocity inlet. Pressure inlet was also tested but the results seem to be well described by the first which specify more realistic boundary conditions at the inlet. It should be pointed that inlet turbulence can significantly affect the downstream flow as observed in high pressure trickle-bed reactor [15]. In the trickle-bed simulations performed, the fidelity of the results for turbulent flows is largely determined by the turbulence model being used and in order to enhance the quality of turbulent flow simulations, the mesh generation accounts for wall-bounded flow, at least on catalyst particle, since the wall is expected to significantly affect the flow.

HYDRODYNAMIC VALIDATION

Since the CFD methodology is not specifically designed for application in constrained geometries, such as particle packed beds, it is necessary to verify if the simulated results are valid. Although the CFD code is based on fundamental principles of flow and heat transfer, some of the boundary issues are modeled using empirical data not necessarily appropriate for fixed bed applications. Validation of CFD flow field calculations has generally taken one of the two forms. In the first, noninvasive velocity measurements inside the packed bed have been made, and compared to velocities computed from a model of either the entire experimental bed or a representative part of it. In the second form, computed pressure drops have been compared to either measured values or established correlations for pressure drop in fixed beds, such as the Ergun equation. The present case study employed the last method to assess the Eulerian model. Therefore, the numerical methodology is validated against experimental data available from literature related to the hydrodynamic information for TBR operation. Indeed, the actual strategy is to compare CFD results in terms of well known parameters such as liquid holdup and pressure drop that are the two most employed characteristics in TBR development study. The experimental conditions and the parameters commonly measured in high pressure TBRs are evaluated extensively in topics such as: pressure effect on physicochemical properties, phenomenological analysis of two-phase flow, flow regime transition, single-phase pressure drop, two-phase pressure drop, liquid holdup, gas-liquid interfacial area and mass transfer, catalyst wetting efficiency as well as catalyst dilution with inert fines in laboratory scale TBRs. In this context, the mesh was validated by checking the mesh sensitivity and by comparing the numerical results against the single-phase and two-phase experimental data. The validation of CFD codes using pressure drop is most reliable when actual experimental data are taken in equipment identical to the situation that is being simulated. Existing literature correlations such as the Ergun equation are known to have shortcomings with respect to wall effects, particle shape effects, application to ordered beds and validity at high Reynolds numbers (Re). The applicability of literature correlations to typical CFD simulation geometries needs to be examined critically before fruitful comparisons can be made because pressure drop measurements can provide an indirect means of checking on the computations at higher flows, although most comparisons have been made at relatively low flow rates. Therefore, pressure gradient and liquid holdup was the two fundamental hydrodynamic parameters evaluated for the design, scale-up, and performance studies of TBR. Pressure drop is an important parameter in the design of two-phase concurrent reactors because if affects the energy supply and it has been use to correlate the gas-liquid and solid-liquid mass transfer whereas liquid holdup is the liquid volume contained in a unit column volume. It should be pointed that in the model validation it was taken into consideration that almost all the holdup data available in open literature refers to laboratory columns [15].

Several runs simulating operating pressures in the range 10 to 30 bar runs were carried out for the vector field of liquid and gas velocity and for liquid hold up and pressure drop using spherical catalysts with 2 mm diameter. Simulated CFD liquid holdup and pressure drop are represented in Figs. 2 and 3 by lines as a function of liquid mass flux water when the reactor operates with air as the gas phase at different gas flow rates. The experimental data plotted were obtained with one reactor with dimensions described by Nemec and Levec [15]. In the high interaction regime modeled for the gas and liquid phases, the predictions are in good agreement with experimental values which enables the validation of our CFD model. In fact, the computational fluid dynamic model validation was carried out first in single-phase pressure drop simulations with only the gas phase flowing downward the bed; afterwards, two-phase flow is simulated to perform the final comparison between predicted hydrodynamic parameters and experimental data. In the whole range of Re numbers for gas phase, pressure drop predictions are within 10% error when comparing with the literature measurements. The resulting pressure drop is given by the addition of laminar flow local losses with frictional losses. At very low velocities, exclusively laminar or viscous contributions to pressure drop are observed but at higher velocities the laminar term from *Blake-Kozeny-Carman* equation and the inertial term from *Burke-Plummer* equation are additive. This mutual contribution that represents the ratio between the static pressure and the hydrostatic pressure is plotted in Fig. 2. The operational region of flow rates ($10<Re_G<400$) is that of particular interest to TBR and in this ambit Eulerian model fits the pressure drop data as well as liquid holdup quite well within acceptable limits of 10%. Furthermore, taking into account that the detailed knowledge of liquid holdup is essential for safe processing to prevent hot-spot formations and possible runaways that could have significant influence on the liquid residence time distribution, mass- and heat-transfer processes as well as wetting efficiency, our results in terms of liquid holdup were also successfully validated in Fig. 3 avoiding many correlations that have been published in multiphase reactors.

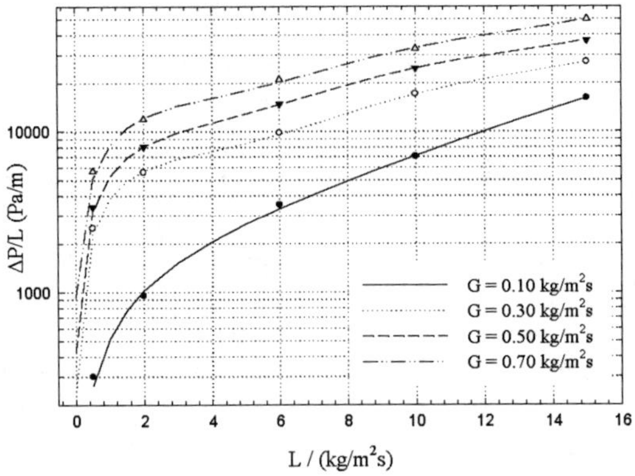

FIGURE 2. Comparison of simulated pressure drop as a function of liquid mass flux at constant gas flow rates with experimental data [15]

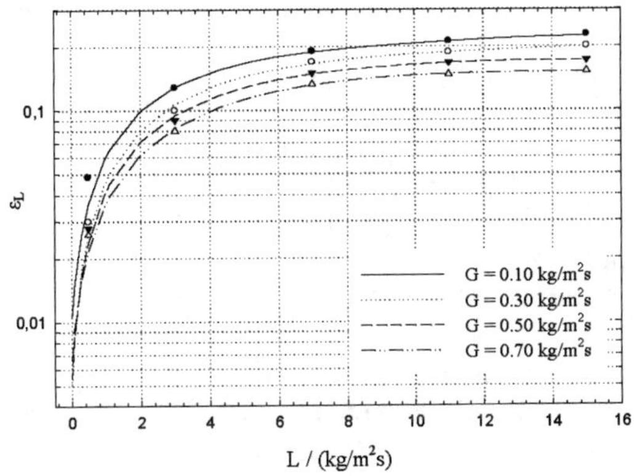

FIGURE 3. Comparison of simulated liquid holdup as a function of liquid mass flux at constant gas flow rates with experimental data [15]

According to Fig. 3, liquid holdup decrease when the pressure was increased for given gas and liquid superficial velocities. This decrease is interpreted as due to a shift in the reactor fluid dynamics from a state predominantly controlled by gravity to a state controlled by gas-liquid shear stress or pressure drop. Comparing Figs. 2 and 3, where the pressure gradient per unit reactor length has been plotted as a function of liquid mass flux, we see that for very low values of pressure drop the liquid holdup are equally small. With the increase of pressure drop due to higher reactor pressures, the total driving force enlarges noticeably and, hence, the liquid holdup growth rate reduces when the liquid mass flux increases. On the other hand, the investigation of liquid distribution in the cocurrently gas/liquid system at elevated pressures could also be related to the results plotted in Figs. 2 and 3, in which it can be seen that liquid holdup values at elevated gas flow rates are much lower in comparison with those accounted in lower gas flow rates conditions. This effect can be explained by means of the ratio between the driving forces, shear and gravitational forces, and the retarding viscous force. As the viscosity of liquid increases exponentially with increasing pressure, the ratio of driving forces and viscous force increases also. This causes a gradual reduction of the liquid saturation in the packed bed. Moreover, the comparison between the hydrodynamic parameters determined at 10 and 40 bar shows that the effect of the reactor pressure has greater influence on the pressure drop than it has on the liquid holdup, as expected. Furthermore, CFD model results state that liquid holdup is slight insensitive to low gas flow rates. This fact could be interpreted by the evolution of liquid holdup as a function of gas density by plotting liquid holdup as a function of superficial mass liquid velocity demonstrating that the theoretical model is able to predict quite well the significant influence of the gas flow rate on the hydrodynamic parameters when comparing the theoretical results with the experimental data sets. It should be also emphasized that the fixed-bed modeled in this work had the tube to particle diameter ratio higher than 10 so the available geometry and data taken from literature should not be affected by the reactor column wall. In accordance to Fig. 3, when the liquid mass flux increases, the liquid holdup also increases for L higher than 8 kg/m^{2}s being the growth rate smaller for the same total pressure value whereas an increase of the total pressure results in a considerable decrease of liquid holdup. The influence of the gas flow determined by a different operating pressure on the liquid holdup is less pronounced at low values of liquid mass fluxes. For example, in case the reactor operates with a gas flow rate at 0.7 kg/m^2s, the liquid holdup is substantially lower when compared with the case it operates at 0.1 kg/m^2s. These higher differences at higher liquid flow rates result from the fact that a further increase of the reactor pressure at a constant gas velocity corresponds to a higher driving force. The theoretical predictions from the model correctly account for the strong influence of the gas flow on the hydrodynamic behavior of the trickle-bed reactors, as stated by several authors [1]-[4]. The important influence of the gas flow is attributed to the interactions phenomena exerted by the gas phase on the liquid phase. These interactions clearly appear to be significant at high superficial gas mass velocities.

Finally, in order to address CFD flow streamlines, the 3D map taken with a vertical catalyst layer illustrated in Fig. 4 indicate that the velocity is higher at points where the flow is processed downward in axial direction. In accordance to these results, the maximum gas velocity is about 0.5 cm/s (whereas the liquid velocity is about 0.005 cm/s) which is in the range of well accepted trickle flow maps reviewed in the literature [1]. In fact, the TBR hydrodynamics are affected differently in each flow regime and the operating conditions that are of particular interest in the industry is the extensively used trickle flow encountered at low gas and liquid superficial velocities. In the 3D map, it is shown the uniformity in packing structure with spherical particles, but the gas/liquid distribution depends not only on the superficial gas and liquid velocities but also in particle shape and particle equivalent diameters in order to study the effect of particle geometry. Spheres were used by virtue of their unique shape and are incapable of influencing the structure of the bed by their orientation. Some additional differences between the porosities of beds, despite the same packing procedures were due to wall effect, which as mentioned before did not affect the overall pressure drop. With regards to the porosity dependence within the inertial regime, it should be reported in the basis of theoretical simulations of flow through random arrays of spheres that the porosity function is also well taken into account as long as the porosity is around 0.4 as is indeed the case for packed bed reactors when made up of spheres. The values of porosity distribution function for the present CFD Eulerian model were applied in the range from 0.38 to 0.40. It should be also pointed that when the superficial velocity of gas is sufficient to interact comparatively with that of liquid, liquid distribution improves significantly and the pressure drop arises as described elsewhere [13]. In our simulation activities, it was assumed that the trickle-bed reactor has a uniform distributor at the top and we can state that liquid distribution do not depends on the design of the distributor.

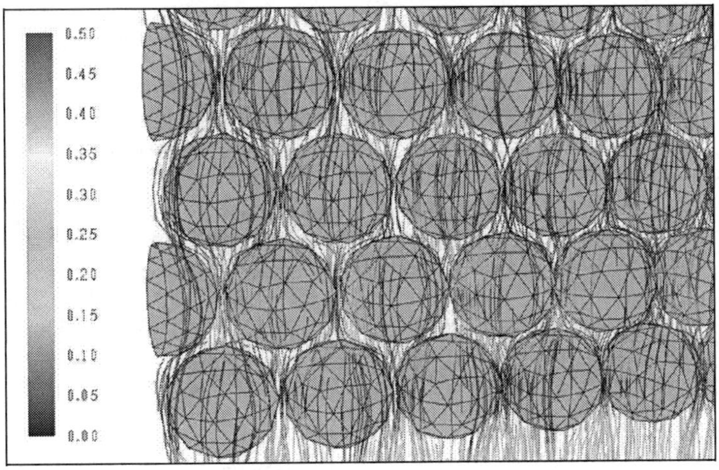

FIGURE 4. Gas streamlines colored by axial velocity (*cm/s*)

REACTION STUDIES

Aiming to assess the TBR reaction behavior, the catalytic wet oxidation of a model phenolic acid solution was simulated in continuous mode by means of CFD codes. The kinetic expressions of a mixture of six phenolic acids previously calculated [10] were then integrated in the TBR computational model where it was assumed that chemical reaction occurs namely on the catalyst surface. This assumption is expected to be mostly reasonable because of the hydrodynamic interaction regime achieved by the Eulerian model. The CFD model has also taken into account external mass transfer limitations which is the most suitable when operating at large scale pilot plant units. According to Fig. 5, a temperature color map was taken with a flow time of two hours. As the operation is modeled in unsteady state, after evaluating successive temporal temperature color maps it is possible to conclude that steady state of TBR unit is achieved in this time.

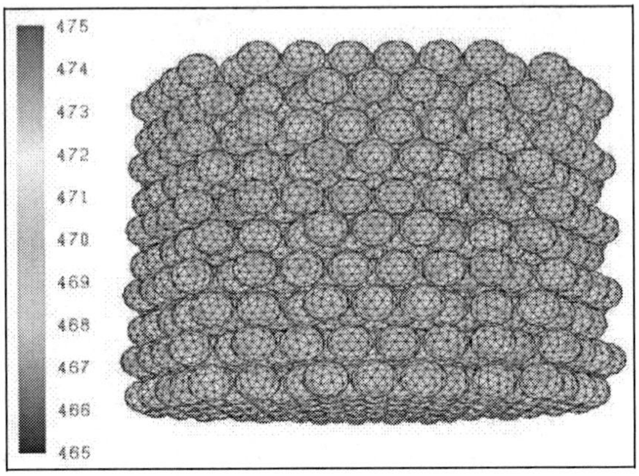

FIGURE 5. Catalyst surface temperature (K)

Furthermore, as the liquid holdup is directly related to the catalyst wetting efficiency that also might affect the reaction yield, in accordance to Fig. 5, the different temperatures ranging from 470 to 474 K attained in different locations of the catalyst particles indicate different reaction rates. These results in terms of catalyst surface temperature for the exothermic oxidation process of the pollutants reflect different wetting levels of the solid by the liquid effluent. Therefore, in the TBR design and scale-up studies external catalyst wetting efficiency is also a hydrodynamic parameter that indicates the utilization degree of catalyst surface area. However, the coupling nature of transport phenomena and kinetics in TBR is far from being completely understood so that general scale-up and scale-down rules for the quantitative description of multiphase flows depends on how phenomenological analysis is correlated with available numerical power to address complex chemical process operational and design issues.

CONCLUSION

A unique physics-based model has been proposed for modeling trickle-bed reactors at elevated pressures aiming to predict the hydrodynamic parameters. The unified approach includes fundamental point force balance and takes into account the influence of gravity in the force balance. The model consists in an Euler-Euler treatment for the fluid phases coupled with the energy equation. The numerical simulations are compared against experimental data to validate the predicted pressure drop and liquid holdup. Operating conditions were simulated with 10-30 bar of reactor pressure while gas and liquid mass flow rate were in the range 0.10 – 0.70 and 0.05 – 15 kg/m^{2}s, respectively.

The novel hydrodynamic model has been found to predict with a reasonable accuracy the experimental data, pointing out that the liquid holdup increases as the liquid mass flux increases and decreases for higher operating pressure values. At low values of pressure drop the liquid holdup is small but with an increasing value of pressure drop due to an increase of the reactor pressure, the liquid holdup growth rate reduces when the liquid mass flux increases. The influence of operating pressure on liquid holdup is less pronounced than in pressure drop. Finally, CFD runs performed in unsteady state for the catalytic wet air oxidation of one phenolic solution demonstrated the effect of temperature illustrated by catalyst surface temperature 3D map.

ACKNOWLEDGMENTS

The authors gratefully acknowledged the financial support of REMOVALS – 6th Framework Program for Research and Technological Development – FP06 Project no. 018525 and *Fundação para a Ciência e Tecnologia*, Portugal.

REFERENCES

1. M. H. Al-Dahhan, F. Larachi, M. P. Dudukovic, and A. Laurent, "High pressure trickle-bed reactors: A Review" Ind. Eng. Chem. Res. 36 (8), 1997, 3292-3314.
2. R.G. Carbonell, "Multiphase flow models in packed beds" Oil & Gas Science and Technology – Revue de l'IFP 55, 2000, 417-425.
3. M. P. Dudukovic, F. Larachi, and P. L. Mills, "Multiphase catalytic reactors: A perspective on current knowledge and future trends" Catal. Rev. 44 (1), 2002, 123.
4. A. Lakota, J. Levec, R. G. Carbonell, "Hydrodynamics of trickling flow in packed beds: relative permeability concept" A.I.Ch.E. J., 48, 2002, 731.
5. P. R. Gunjal, M. N. Kashid, V. V. Ranade, and R. V. Chaudhari, "Hydrodynamics of trickle-bed reactors: experiments and CFD modeling" Ind. Eng. Chem. Res. 2005, 44, 6278.
6. S. T. Sie and R. Krishna, "Process development and scale up: III. Scale-up and scale-down of trickle bed processes" Rev. Chem. Eng. 14, 1998, 203-252
7. S. Goto and J. M. Smith, "Trickle bed reactors performance: I hold-up and mass transfer effects" A.I.Ch.E. Journal 21, 1975, 706.
8. R. A Holub, M. P. Dudukovic, and P. A. Ramachandran, "Pressure drop, liquid hold-up and flow regime transition in trickle flow" A.I.Ch.E. Journal 39, 1993, 302.
9. A. E. Saez and R. G. Carbonell, "Hydrodynamic parameters for gas liquid cocurrent flow in packed beds" A.I.Ch.E. J. 31, 1985, 52.
10. R. J. G. Lopes, A. M. T. Silva, and R. M. Quinta-Ferreira, "Screening of catalysts and effect of temperature for kinetic degradation studies of aromatic compounds during wet oxidation" Appl. Catal B: Environmental, 73 (1), 2007, 193-202.
11. FLUENT 6.1., 2005. User's Manual to FLUENT 6.1. Fluent Inc. Centrera Resource Park, 10 Cavendish Court, Lebanon, USA.
12. A. Attou and G. A. Ferschneider, "Two-fluid model for flow regime transition in gas–liquid trickle-bed reactors" Chem. Eng. Sci. 54 (21), 1999, 5031-5037.
13. R. J. G. Lopes, A. M. T. Silva, and R. M. Quinta-Ferreira, "Kinetic Modelling and Trickle-Bed CFD Studies in the Catalytic Wet Oxidation of Vanillic Acid" Ind. Eng. Chem. Res., 46 (25), 2007, 8380-8387.
14. GAMBIT 2, 2005. User's Manual to GAMBIT 2. Fluent Inc. Centrera Resource Park, 10 Cavendish Court, Lebanon, USA.
15. Nemec and J. Levec, "Flow through packed bed reactors: 2. Two phase concurrent downflow" Chem. Eng. Sci. 60 (24), 2005, 6958-6970.

Single and Catalytic Ozonation for Phenolic Wastewaters Remediation

Rui C. Martins, Hélder M. Leal and Rosa M. Quinta-Ferreira*

GERSE – Group on Environment, Reaction and Separation Engineering, Department of Chemical Engineering, Faculty of Sciences and Technology, University of Coimbra, Pólo II, Rua Sílvio Lima, 3030-790 Coimbra, Portugal
**Corresponding author: Tel.:+351 239798723; fax: +351 239798703*
E-mail addresses: rosaqf@eq.uc.pt (R. M. Quinta-Ferreira), martins@eq.uc.pt (R. C. Martins)

Abstract. Aiming the improvement of single ozonation of a mixture of six phenolic acids normally present in Olive Mill Wastewaters, Manganese and Cerium oxides based catalysts were used in the catalytic ozonation process. Regarding the total phenolic content degradation no significant differences were found in the activity of MnO_2, CeO_2 and the bimetallic catalyst with a molar ratio Mn/Ce of 70/30. However, concerning the total organic carbon removal, the best results were obtained for the Mn/Ce catalyst, explained by.the synergetic effect between Mn and Ce. During the preparation of the catalysts, the increase of the calcination temperature leads to a decrease in the Brunauer-Emmet-Teller area resulting in a slight inhibitory effect over the total organic carbon removal. A higher importance of the catalyst in the organic intermediates mineralization than in the parent phenolic compounds degradation was observed. The comparison between single ozonation and catalytic ozonation in the presence of the Mn-Ce-O (70/30) catalyst revealed a higher enhancement in both total phenolic content and total organic carbon oxidation. After 120 min of reaction low carbon adsorption was detected in the recovered catalyst meaning that the pollutants removal occurred by oxidation and not by adsorption to the solid phase. Moreover, a low Mn quantity was detected in the liquid phase confirming the stability of the catalyst in what concerns leaching behaviour of the active phase to the liquid. Therefore, Mn-Ce-O showed to be an interesting catalyst to improve the efficiency of ozonation in the treatment of phenolic wastewaters.

Keywords: Catalytic ozonation, Environment protection, Phenolic acids, Environmental reaction engineering.

INTRODUCTION

Public concern for environmental problems arising from industrial effluents is increasing and as consequence environmental legislation is nowadays imposing more restrictive regulations for effluents discharge. In Mediterranean sea area countries, Olive Mill Wastewaters (OMW) constitute an important source of pollution due to the increasing volume of such effluents that are yearly produced (estimated at 7-30 million m^3 each year) [1]. These wastewaters have a high pollutant character, mainly due to the large organic fraction which includes sugars, tannins, acids, pectins, lipids, and especially phenols and polyphenols. In some cases, these effluents are directly dumped into the environment untreated. When treated, there are various common

CP1007, *Current Themes in Engineering Technologies*, edited by S.-I. Ao, M. A. Amouzegar, and S.-S. Chen

processes: a) evaporation ponds where the effluents are retained for water removal by solar energy, which cause unpleasant odours and the possibility of ground waters contamination; b) direct application in soil that is an inexpensive way of disposing and recycling nutrients and organic matter; nevertheless, this procedure can have an unfavourable environmental impact, since the toxic bio-refractory pollutants present may reach the water beds [2]; c) a possible alternative method due the high organic load of the olive mill wastewaters is incineration, though inherent disadvantages have to be balanced like fuel costs, gas emissions and the fact that some by-products still remain; d) other applications deal with the use of these effluents as a source of fermentation products, such as fat and oils preservatives. Most of the studies related with olive mill wastewater treatment were focused on biological depuration systems. However, many problems were identified in these systems due the high toxicity and low biodegradability mainly related to the phenolic content of the OMW, since these compounds are known bio-refractory agents; thus these treatments generally lead to unsatisfactory results. Within this scenario, other techniques have to be implemented in order to achieve the depuration degree needed to accomplish environmental regulations. In the context of OMW, ozone is very efficient to remove phenolic compounds with high reaction rates due to the strong electrophilic nature of molecular ozone which is able to react with the nucleophilic sites of the aromatic rings. Therefore, ozonation shows to be an interesting pre-treatment for biological depuration. However, when the objective is the wastewater mineralization, single ozonation is not a suitable process due to the formation of refractory saturated intermediates during the process; as a consequence, Total Organic Carbon (TOC) is kept practically unchanged leading to a low mineralization level of the effluent. Within this scenario, ozonation enhanced by an heterogeneous catalyst reveals to be a feasible treatment technique due to its potentially effectiveness in the degradation of refractory organic pollutants. In what concerns phenolic wastewaters, gallic acid catalytic ozonation was studied using perovskite [3] and activated carbon [4] as catalyst. In the former case, the presence of the catalyst reduced the time required to gallic acid elimination, enhancing TOC reduction even when used at low concentrations, whereas in the later, the contribution of activated carbon is mostly noticed in the intermediate products evolution and low advantage was gained with the use of the catalyst during ozonation to remove gallic acid itself. Perovskite showed also to be an active and stable catalyst when used in ozonation of complex simulated phenolic acids mixture and real wastewaters as well [5].

The aim of this research work addresses the study of the efficiency of ozonation enhanced by a catalyst prepared in our laboratory (Mn-Ce-O) in a model effluent degradation. To the best of our knowledge this is the first study regarding the catalytic ozonation treatment using the bimetallic Mn-Ce-O catalyst, which showed to be very active in the catalytic wet oxidation of phenolic compounds [6] as well as of low weight carboxylic species [7], [8].

EXPERIMENTAL

Materials and Catalysts

Syringic acid, vanillic acid, 3,4,5-trimethoxybenzoic acid, veratric acid, protocatechuic acid and 4-hydroxybenzoic acid were obtained from Sigma-Aldrich and used as received without further purification. These six phenolic acids typically found in OMW were used to prepare the synthetic effluent with a concentration of 100 mg/L of each compound. Mn-Ce-O, MnO_2 and CeO_2 catalysts were prepared by precipitation of a mixed aqueous solution of the respective metal nitrates as precursors (Riedel-de-Häen and Labsolve) by the procedure described elsewhere [9]. The metals salts solution was poured into 200 mL of a 3 M NaOH solution. The precipitate was filtrated and washed five times with 500 mL of ultra pure water and dried at 100°C over night, followed by calcination during 3h, at 300°C or 450°C. Before calcination the catalysts were crushed into fine powder in order to provide maximum specific surface area for reaction.

Oxidation Procedure

Ozonation and catalytic ozonation were carried out in a semi-batch mode being the liquid pollutant charged at the beginning of the experiment and the gas continuously fed. The reactor consisted in a glass vessel (1 L of capacity) provided with magnetic stirring and four top inlets for bubbling the gas feed, by means of two gas diffusers placed in the liquid bulk, sampling and venting. In every experiment, the reactor was filled with 500 mL of the model solution. Ozone was produced from a pure oxygen stream (99.999% Praxair, Portugal) in an ozone generator (MICROLAB 1579, SPO_3, Portugal) with a flow rate of 2 gO_3/h and the remaining ozone was trapped into a gas absorption bottle containing 10% (w/v) KI solution. The catalytic experiments were carried out in slurry conditions with 1.5 g of powder catalyst (3 g/L) introduced into the reactor just before turn on the ozone generator. The stirring speed was maintained at 750 rpm, in order to maintain the external mass transfer resistances negligible and during the experiments, samples were withdrawn periodically and immediately filtered with 0.45 μm disposable filters for further analysis.

Analytic Techniques

Total Organic Carbon (TOC) was determined with a *Shimadzu 5000A Analyser*, which operates based on the combustion/non-dispersive infrared gas analysis method. Total carbon (TC) was first measured followed by the measurement of the inorganic carbon (IC) and TOC was then determined by subtracting IC from TC. Total Phenolic Content (TPh) was measured colourimetrically using a Folin-Ciocalteau reactant obtained from Fluka. The procedure used is described elsewhere [10]: 20 μL of sample were introduced in a 2 mL cuvette and 100 μL of the Folin-Ciocalteau reagent as well as 1.58 mL of water were added. After 3-6 min, 300 μL of a saturated sodium carbonate solution was introduced and the resulting solution was left in the dark for 2 h. The absorbance was determined in a spectrophotometer (T60 U, PG instruments) at 765 nm against a blank solution containing 20 μL of water instead of 20 μL sample. The calibration curve was prepared using gallic acid at different concentrations and therefore the phenolic content values are reported as gallic acid equivalent. pH was monitored with a HANNA instrument-HI8711E. *Elemental analysis* was used to detect carbon adsorption in catalysts with a *Fisons Instruments EA 1108 1108* CHNS-O equipped with a pre-packed ox/red quartz reactor, operating with a flash combustion and using a thermal conductivity detector (TCD). *Atomic absorption* was used to measure the leaching of manganese to the liquid phase and the analysis was performed in a spectrometer *Perkin-Elmer 3300*, with hollow cathode lamps (Cathodeon). The catalyst *Brunauer-Emmett-Teller (BET) surface area analysis* (S_{BET}) was determined with an accelerated surface area and porosimetry analyzer (ASAP 2000) from Micrometrics using nitrogen at a constant temperature (-196°C).

RESULTS AND DISCUSSION

Manganese and Cerium Oxides Catalysts

The use of Mn as an active homogenous or heterogeneous catalyst in the catalytic ozonation of organic pollutants is well spread in literature. Andreozzi *et al.* (1996) [11] studied the applicability of MnO_2 as heterogeneous catalyst in catalytic ozonation of oxalic acid, a known refractory compound to ozone oxidation; Andreozzi *et al.* (1998) [12] showed that the reactivity of pyruvic acid with ozone was strongly enhanced by both suspended MnO_2 and dissolved Mn ions. Villaseñor *et al.* (2002) [13] tested different catalysts of MnO_2 supported in TiO_2 for phenol catalytic and photocatalytic ozonation of phenol. The enhancement in the mineralization achieved in the ozonation treatment of precursors of trihalomethanes using MnO_2 was studied by Alsheyab and Muñoz (2007) [14]. Also cerium was used with satisfactory results as homogeneous catalyst in the ozonation of phenol solutions [15]. Moreover, the use of CeO_2 as support to a Ru catalyst in catalytic ozonation of succinic ozonation was studied by Delanöe and co-workers (2001) [16] with two different catalyst preparation methods: impregnation and acid exchange. However a bimetallic catalyst using both Manganese and Cerium oxides was never tested in the catalytic ozonation field. As the Mn-Ce-O catalyst with molar relative proportion between Mn and Ce of 70/30 has, recently, showed good results when used in catalytic wet oxidation of several organic pollutants [6]-[9], the efficiency of this catalyst in the catalytic ozonation of a complex simulated mixture of phenolic acids present in OMW was compared with MnO_2 and CeO_2 as it can be observed in Fig. 1 for both TPh and TOC reduction. The difference of using different catalysts for the catalytic ozonation of the pollutants is more evident in the case of TOC reduction. In fact, concerning to TPh removal (Fig.1a), all the catalysts led to total degradation and though the initial times of reaction were similar, the instants where total phenolic content degradation is achieved were different, since TPh is totally removed after 60 min of reaction using Mn-Ce-O 70/30, whereas 75 min with MnO_2 and almost 120 min when ozonation is enhanced by CeO_2 were required. The difference between the activity of the three catalysts in the ozonation of the phenolic wastewater is more evident in what concerns TOC degradation (Fig.1b), meaning that the presence of the catalyst is more important in the ozonation of the intermediates than for the degradation of the parent phenolic compounds. Ozone by himself is already highly reactive concerning the degradation of benzenic derivates, as the case of the phenolic compounds [17-18], but the problem is that the degradation of these compounds leads to the formation of saturated intermediates, which are no further reactive with ozone. Therefore, the differences between the activities of the various catalysts will be more evident in what respects the oxidation of intermediates and consequently in TOC depletion than in the initial acids decompositions. Fig. 1b shows that for the initial times of the process, both Mn-Ce-O (70/30) and CeO_2 allow a pronounced TOC reduction while MnO_2 has a lower degradation rate. However, at higher reactional times, TOC removal by cerium oxide tends to lower leading to the lowest mineralization degree achieved for the three catalysts tested after 120 min (50 %). This is due to the formation in this oxidation system of refractory intermediates along time which resulted in a lower final efficiency of the treatment. The reactional system with ozone and manganese oxide reaches up to 56 % of mineralization after 2h of catalytic ozonation. Indeed, a synergetic effect between Mn and Ce is highlighted by the fact that the best result for TOC, 66 % of mineralization at the end of the experiment, was observed for the system where the catalyst was composed by the two metals. In fact, it is reported in literature that in Mn-Ce composites the concentration of Ce^{4+} is higher than Ce^{3+} and the remaining electrons seem to be transferred to Mn due to the high CeO ability as promoter of stored oxygen, which improves the mobility of electrons. In this context, the higher activity of the bimetallic catalyst can be explained by an efficient equilibrium of the electrons transfer which enhances the oxidation reactions leading to higher treatment performances [19].

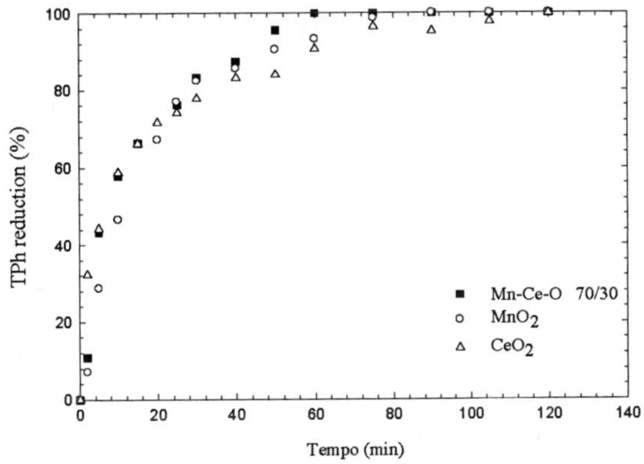

(a)

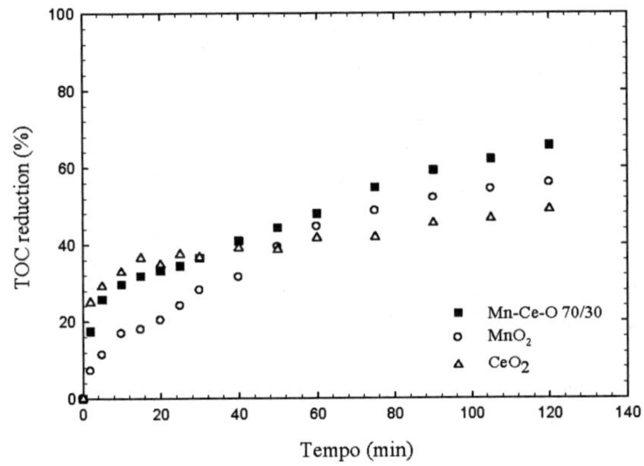

(b)

FIGURE 1. (a) Total Phenolic Content, TPh and (b) Total Organic Carbon, TOC abatement for catalytic ozonation over different catalysts. 3 g/L of catalyst, 2 gO_3/L and pH=3.5.

In order to analyse the influence of the calcination temperature over the Brunauer-Emmet-Teller (S_{BET}) areas and on the catalyst efficiency, Mn-Ce-O 70/30 was prepared by using two different calcination temperatures, namely 300°C and 450°C. As expected, the area of BET decreased, from 102.48 to 98.52 m^2/g, when the calcination temperature was raised. In what regards the activity of the catalyst in the catalytic ozonation of the phenolic mixture, no significant differences were detected for TPh (Fig.2a) while a slight inhibitory effect can be observed in TOC removal when the catalyst calcinated at 450°C is used (Fig. 2b). In fact, after 120 min of reaction the catalyst calcinated at 300°C led to a mineralization degree 6% higher than the catalyst calcinated at 450°C, which may be explained by the decrease in the BET area. The similar results for the catalysts performance in TPh removal and the slight differences for TOC degradation, point out once more that the presence of the catalyst is more important in the oxidation of the reaction intermediates than in the phenolic acids degradation.

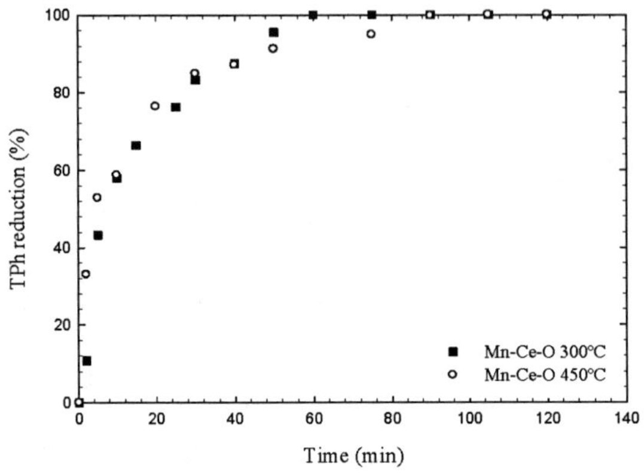

(a)

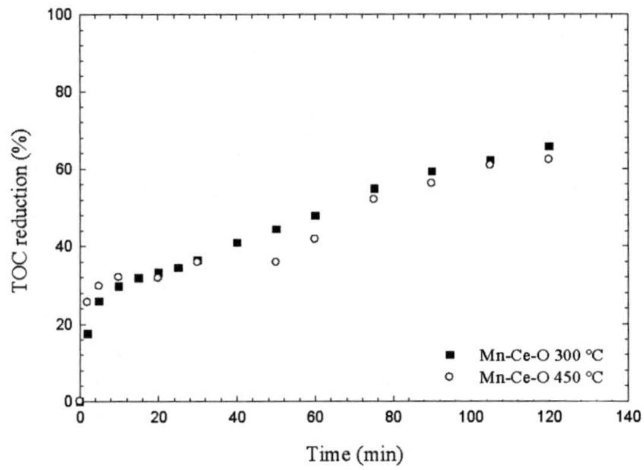

(b)

FIGURE 2. (a) Total Phenolic Content, TPh and (b) Total Organic Carbon, TOC abatement for catalytic ozonation over the Mn-Ce-O (70/30) catalyst calcinated at different temperatures. 3 g/L of catalyst, 2 gO_3/L and pH=3.5.

Catalytic Ozonation *versus* Single Ozonation

Within these results, a comparison between the performance of single ozonation and catalytic ozonation enhanced by Mn-Ce-O (70/30) calcinated at a temperature of 300ºC is shown in fig. 3 for both TPh (fig.3a) and TOC (fig. 3b). As observed, single ozonation can be an efficient technology to eliminate the phenolic content of the effluent with a reduction up to 88% of the TPh after 120 min. However whit these conditions, only a slight mineralization degree is achieved with less than 24% of TOC removed after the same period of time. It is well known that ozone is very efficient in benzenic compounds degradation, especially when the aromatic ring is activated with electron-donating groups, as is the example of hydroxyl (OH) and methoxy (CH_3O) groups in the specific case of the phenolic acids used in this work. This is because the ozone molecule preferentially attacks following electrophilic pathways, being then reaction enhanced when the benzenic ring is provided with a high electronic density. In fact, due to the strong electrophilic nature of the ozone molecule, this oxidant specie reacts quickly with the nucleophilic positions of the activated rings [4]. Nevertheless, the accumulation of different ozone refractory intermediates (mostly aliphatic saturated carboxylic acids), during the ozonation process, keeps TOC concentration practically unchanged leading, as a consequence, to a low level of mineralization. The addition of Mn-Ce-O truly enhanced the performance of ozone in the degradation of the pollutants, specially concerning to the mineralization degree achieved, confirming that the importance of the presence of the catalyst is higher for the oxidation of the intermediates than for the degradation of the parent phenolic compounds. Even at 2 min of reaction, the degradation of TOC was already 17% while only 2% was achieved for the same experimental time by single ozonation; at the end of 120 min more than 66% of removal was detected, corresponding to an improvement higher than 42% when comparing with single ozonation at the same experimental time. Therefore, although non-catalytic ozonation showed to be appropriated for TPh abatement, the use of the catalyst enhanced the process leading to total removal after 60 min while for single ozonation total degradation is not achieved even after 120 min of reaction.

Our results also revealed that the total organic carbon content was removed mainly by oxidation and not by adsorption in the solid Mn-Ce-O catalyst, since low carbon adsorption was detected by elemental analysis after the experiment (less than 8% of the total carbon load into the reactor). This is important since catalyst deactivation may occur due to the formation of carbonaceous deposits in the catalyst surface irreversibly adsorbed on the active sites. Another important issue which has to be taken into account concerning to the chemical stability of ceramic catalysts involves the analysis of the leaching behaviour of the active metals to the liquid phase during reaction. The occurrence of this phenomenon may also lead to the deactivation of the catalyst due the irreversible loss of the catalytic species. Besides, diffusion of metals from the solid matrix may contaminate the liquid wastewater being then required a post-treatment to separate the metals from the final effluent. However in our studies, less than 1 mg Mn/L (being the typical threshold of 2 mg Mn/L) was detected by atomic absorption after 120 min of oxidation, which corresponds to a leaching lower than 0.05% of the total Mn charged at the reactor. Catalytic ozonation using a Mn-Ce-O catalyst seems then to be an interesting pre-treatment technique for biological processes in order to enhance the ozone degradation of phenolic compounds which are known to cause severe problems in OMW treatment.

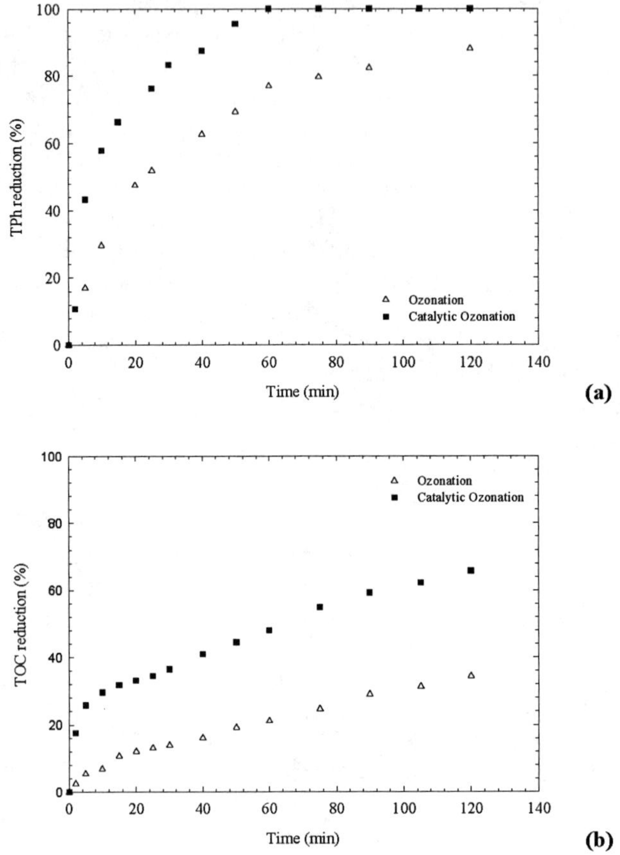

FIGURE 3. Comparison between Ozonation and Catalytic Ozonation in the (a) reduction of the Total Phenolic Content, TPh and (b) Total Organic Carbon, TOC. 3 g/L of Mn-Ce-O (70/30) catalyst with a calcination temperature of 300ºC, 2 gO_3/L and pH=3.5.

CONCLUSIONS

The aim of this work addressed the study of the activity of a laboratorial catalyst (Mn-Ce-O) in the catalytic ozonation of a complex synthetic wastewater composed by a mixture of six phenolic acids present in Olive Mill Wastewaters. The comparison between the activity of a bimetallic catalyst (Mn-Ce-O 70/30) and the respective metal oxides (MnO_2 and CeO_2) was performed. No significant differences were observed between the three catalysts concerning to TPh degradation; nevertheless, Mn-Ce-O led to total removal after 60 min, whereas MnO_2 and CeO_2

needed 75 min and around 120 min, respectively. The synergetic effect between Mn and Ce is more evident when TOC removal is analyzed. Thus, the following order of mineralization was achieved after 120 min of reaction: Mn-Ce-O (66 %) > MnO_2 (56 %) > CeO_2 (50 %). Therefore, the presence of the catalyst is more important in intermediates total mineralization than in the phenolic content depletion. Moreover, the calcination temperature is an important parameter in the catalyst conception; the increase of the calcination temperature from 300°C to 450°C showed no influence on TPh degradation and a slight inhibitory effect over TOC removal. Comparing ozonation with catalytic ozonation enhanced by Mn-Ce-O (70/30) calcinated at a temperature of 300°C it was possible to notice that whereas ozonation achieved 88% of TPh degradation after 120 min, catalytic ozonation allowed total degradation after 60 min of reaction. Concerning TOC, an improvement of more than 42% in the removal of this parameter is obtained when the ozonation of the pollutants is performed in the presence of the Mn-Ce-O catalyst. Less than 8% of the total carbon loaded at the reactor was found adsorbed at the catalyst after 120 min of reaction, meaning that TOC was indeed removed by oxidation and not by adsorption into the solid phase. Moreover less that 1 mg/L of Mn was detected at the liquid after the treatment process, pointing out that the catalyst is stable in what concerns diffusion of the active metal from the solid matrix to the liquid effluent. Therefore, Mn-Ce-O showed to be an interesting catalyst for the enhancement of ozonation of organic pollutants.

ACKNOWLEDGMENTS

The *Fundação para a Ciência e Tecnologia,* Portugal is grateffully acknowledged by the financial support.

REFERENCES

1. A. Amat, A. Arques, H. Beneyto, A. García, M. Miranda "Ozonisation coupled with biological degradation for treatment of phenolic pollutants: a mechanistically study" *Chemosphere* vol. 53, 2003, pp. 79–86.
2. A. Nastri, N. Ramieri, R. Abdayem, R. Piccaglia, C. Marzadori, C. Ciavatta "Olive pulp and its effluents suitability for amendmenent" *J.Hazard. Mater.* vol. A138, 2006, pp. 211–217.
3. M. Carbajo, F. Beltrán, F. Medina, O. Gimeno, F. Rivas "Catalytic ozonation of phenolic compounds: the case of gallic acid" *Appl. Catal. B-Environ.* vol. 67, 2006, pp. 177–186.
4. F. Beltrán, J. García-Araya, I. Giráldez "Gallic acid water ozonation using activated carbon" *Appl. Catal. B-Environ.* vol. 63, 2006, pp.249–259.
5. M. Carbajo, F. Beltrán, O. Gimeno, B. Acedo, F. Rivas "Ozonation of phenolic wastewaters in the presence of a perovskite type catalyst" *Appl. Catal. B-Environ.* vol. 74 (3-4), 2007, 203-210.
6. R. Lopes, A. Silva, R. Quinta-Ferreira "Screening of catalysts and effect of temperature for kinetic degradation studies of aromatic compounds during wet oxidation" *Appl. Catal. B-Environ.* vol. 73, 2007, pp.193–202.

7. A. Silva, R. Marques, R. Quinta-Ferreira "Catalysts based in cerium oxide for wet oxidation of acrylic acid in the prevention of environmental risks" *Appl. Catal. B-Environ.* vol. 47, 2004, pp. 269–279.
8. A. Silva, A. Oliveira, R. Quinta-Ferreira "Catalytic wet oxidation of ethylene glycol: kinetics of a reaction on a Mn-Ce-O catalyst" *Chem. Eng.Sci.* vol.59, 2004, pp. 5291–5299.
9. A. Silva, I. Castelo-Branco, R. Quinta-Ferreira, J. Levec "catalytic studies in wet oxidation of effluents from formaldehyde industry" *Chem. Eng. Sci.* vol. 58, 2003, pp. 177–186.
10. A. Silva, E. Nouli, N. Xekoukoulakis, D. Mantzavinos "Effect of key operating parameters on phenols degradation during H_2O_2 – assisted TiO_2 photocatalytic treatment of simulated and actual olive mill wastewaters" *Appl. Catal. B-Environ.* vol. 73, 2007, pp. 11–22.
11. R. Andreozzi, A. Insola, V. Caprio, R. Marotta, V. Tufano "The use of manganese dioxide as a heterogeneous catalyst for oxalic acid ozonation in aqueous solution" *Appl. Catal. A* vol.138, 1996, pp. 75-81.
12. R. Andreozzi, V. Capri, A. Insola, R. Marotta, V. Tufano "The ozonation of pyruvic acid in aqueous solutions catalyzed by suspended and dissolved manganese" *Water Res.*, vol. 32(5), 1998, pp. 1492–1496.
13. J. Villaseñor, P. Reyes, G. Pecchi "Catalytic and photocatalytic ozonation of phenol on MnO_2 supported catalysts" *Catal. Today*, vol. 76, 2002, pp.121–131.
14. M. Alsheyab, A. Muñoz "Comparative study of ozone and MnO_2/O_3 effects on the elimination of TOC and COD or raw water at the Valmayor station" *Desal.* vol. 207, 2007, pp. 179-183.
15. M. Matheswaran, S. Balaji, S. Chung, I. Moon "Studies on cerium oxidation in catalytic ozonation process: A novel approach for organic mineralization" *Catal.Communications*, vol. 8, 2007, 1497-1501.
16. F. Delanöe, B. Acedo, N. Karpel Vel Leitner, B. Legube "Relationship between the structure of Ru/CeO_2 catalysts and their activity in the catalytic ozonation of succinic acid aqueous solutions" *Appl. Catal. B.*, vol. 29, 2001, pp. 315-325.
17. F. Benítez, J. Beltran-Heredia, J. Acero, M. Pinilla "Ozonation kinetics of phenolic acids present in wastewaters from olive oil mills" *Ind. Eng. Chem. Res.*, vol. 36, 1997, 638-644.
18. J. Beltran-Heredia, J. Torregrosa, J. Dominguez, J. Peres "Kinetics of the reaction between ozone and phenolic acids present in agro-industrial wastewaters" *Water Res.*, vol. 35(4), 2001, 1077-1085.
19. H. Chen, A. Sayaru, A. Adnot, F. Larachi "Composition-activity effects of Mn-Ce-O composites on phenol catalytic wet oxidation" *Appl. Catal. B.*, vol. 32(3), 2001, pp. 195-204.

Two-stage Atomic Layer Deposition: Deposition of Aluminum Oxide Films on Hydrophilic and Hydrophobic Surfaces

Nobuhiko P. Kobayashi[a] and R. Stanley Williams[b]

[a]*Baskin School of Engineering, University of California Santa Cruz, 1156 High Street, Santa Cruz, California 95064 USA, email: nobby@soe.ucsc.edu*
[b]*Quantum Science Research, Hewlett-Packard Laboratories, 1501 Page Mill Road, Palo Alto, California 94034 USA, email: stan.williams@hp.com*

Abstract. We describe the growth of aluminum oxide (AlO_x) on strong hydrophobic surfaces that consist of CH_3-terminated self-assembled monolayers (CH_3-SAMs) by utilizing atomic layer deposition (ALD) with H_2O as the oxygen source. The evolution of AlO_x on the CH_3-SAMs was studied by comparing with that on hydrophilic OH-terminated silicon dioxide (OH-SiO_2) surfaces. The AlO_x grown on the CH_3-SAM surfaces underwent growth instability and developed significantly rough surface morphologies while the AlO_x on the OH-SiO_2 maintained atomically smooth surface morphologies. The structural integrity of the CH_3-SAMs was also found to be disturbed substantially at the onset of the ALD process with H_2O. To improve the surface morphology of AlO_x on CH_3-SAM surfaces, a two-stage ALD process was developed. In the two-stage ALD process for AlO_x, the first stage utilized *n*-propanol as the oxygen source and the second stage proceeded with H_2O. The optimized two-stage ALD process significantly improved the surface morphology of AlO_x films and effectively protected the structural integrity of underlying CH_3-SAMs.

Keywords: atomic layer deposition, aluminum oxide, self-assembled monolayer, surface morphology
PACS: 82.33.Ya, 68.55.-a, 61.30.Hn

INTRODCUTION

In optimizing functional devices in which organic materials are used as active components, employing encapsulation films that shield organic materials during processing and effectively prevent water and oxygen from penetrating into organic materials during device operation is often desirable. Although the formation of organic films on inorganic substrates (i.e. organic on inorganic) has been studied extensively in the development of functional organic devices [1-3], detailed studies on the deposition of inorganic films onto organic layers (i.e. inorganic on organic) are still limited to a few reports on metal oxides / nitrides [4-7] and elementary metals [8-10] deposited on self-assembled monolayers (SAMs). In this paper, we describe the growth of aluminum oxide (AlO_x) on CH_3-terminated SAMs (hydrophobic surfaces), i.e. inorganic on organic, at low temperatures. Our objective is to study the growth of an inorganic film on organic layers in the context of developing an encapsulation layer that can cover organic layers uniformly at low temperatures. We utilize a deposition process such as atomic layer deposition (ALD) at low temperatures so that organic layers are not chemically and physically disturbed. This paper consists of two parts.

CP1007, *Current Themes in Engineering Technologies,* edited by S.-I. Ao, M. A. Amouzegar, and S.-S. Chen

First, a comparative study of a conventional ALD process using H_2O as the oxygen source on hydrophobic (CH_3-SAMs) and hydrophilic (OH-SiO_2) surfaces is described. In the conventional ALD process, as the nominal thickness of the AlO_x films increases, the films on the CH_3-SAM exhibited a growth instability accompanied with a rough surface morphology, while the films on the OH-SiO_2 maintained an atomically smooth surface. To improve the surface morphology of AlO_x on the CH_3-SAM, we attempted to control surface wettability in the early stage of ALD processes. In the attempt, AlO_x was deposited onto hydrophobic CH_3-SAM surfaces by a newly developed two-stage ALD process in which the first stage utilized *n*-propanol as the oxygen source and the second stage proceeded with water.

SUBSTRATE PREPARATION

Shown schematically in Fig. 1(a) and (b) are two types of substrates, representing hydrophilic and hydrophobic surfaces, respectively, used in this study. On these two types of substrates having different characteristics of wettability, AlO_x was deposited by ALD to study how the growth of AlO_x progressed on different surfaces. Panel (a) represents a hydrophilic surface provided by a hydroxyl terminated silicon dioxide surface (OH-SiO_2). The silicon dioxide was thermally grown on a silicon (100) surface by a standard high temperature oxidation process used in complementary metal-oxide-semiconductor (CMOS) processes.

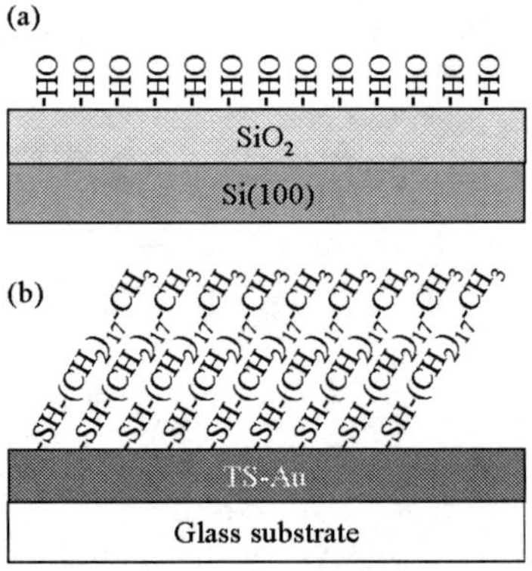

Figure 1. Schematics show two types of substrates on which AlO_x was deposited. (a) OH-SiO_2 and (b) CH_3-SAMs.

Panel (b) represents a hydrophobic surface provided by CH_3-terminated self-assembled monolayers (CH_3-SAMs). In preparing the CH_3-SAM surfaces, first a gold film with an atomically smooth surface was prepared on glass substrates via the template-stripping process [11]. Subsequently, SAMs of the alkanethiolate CH_3-$(CH_2)_{17}$SH were formed on the template-stripped gold (TS-gold) film by immersion into an ethanol solution containing the alkanethiolate at a molar concentration of 0.01 M/l for 24 hours at room temperature. The TS-gold surface having atomically smooth surface profiles ensures that CH_3-SAMs are formed with minimum number of structural defects, resulting in atomically smooth surface morphologies of CH_3-SAMs. The CH_3-SAM surface is expected to be strongly hydrophobic that is manifested in the early stage of the deposition of AlO_x for which H_2O is used as a source of oxygen.

Conventional Atomic Layer Deposition Process

AlO_x films were deposited by conventional atomic layer deposition (ALD), using H_2O and trimethylaluminum (TMAl) as sources for oxygen and aluminum respectively. The ALD process was performed by alternatively supplying pulses of nitrogen gas containing either H_2O or TMAl vapor. A set of key deposition parameters are listed in Table 1. The substrate temperature was set to 45 °C for all samples, which was lower than those temperatures reported to cause SAMs to degrade structurally [12]. Dissociative reaction, including ligand exchange, of TMAl on SAMs with CH_3 functional group has been found to be nearly quenched due to the small thermodynamic driving force and large kinetic barrier [6], however as mentioned earlier, our primary objective was to study the deposition of AlO_x on SAMs at low temperature to form a protective barrier that does not disrupt the SAM during deposition, and thus the relative inertness of TMAl on CH_3-SAMs was, in fact, a potential advantage. In addition, H_2O used as the oxygen source is not reactive with alkane chains in the SAMs, thus the CH_3-SAMs were expected to favor the formation of water droplets on the surface of the densely packed SAMs that enabled pyrophoric TMAl to dissociate and form AlO_x nuclei on the surface of the CH_3-SAMs.

TABLE 1. Deposition parameters of a conventional ALD process using H_2O.

Deposition conditions	Conditions used in the experiment
Temperature (°C)	45
Sources	$Al(CH_3)_3$ and H_2O
Chamber base pressure (torr)	~$2x10^{-1}$
H_2O pulse time (ms)	140
H_2O purge time (s)	160
$Al(CH_3)_3$ pulse time (ms)	140
$Al(CH_3)_3$ purge time (s)	16
Deposition rate (nm / cycle)	0.12

One cycle of the ALD process consists of a 140 ms water pulse followed by a 160 s nitrogen purge period, after that, a 140 ms TMAl pulse was given and followed by another 16 s nitrogen purge period. This specific deposition conditions were previously optimized for AlO_x at 45 °C, and confirmed to maintain a deposition rate of self-limited 0.12 nm / cycle on an OH-terminated silicon surfaces, which ensured a precise and reproducible amount of source materials to be delivered to the SAM surface during the conventional ALD process.

Aluminum Oxide Deposited by Conventional ALD

Fig. 2 show representative non-contact mode AFM images collected on the AlO_x deposited on the OH-SiO_2 in (a)(b) and on the CH_3-terminated SAM in (c)(d) for the

0, and 400 ALD cycle samples, respectively. As shown in (a) as-prepared OH- SiO_2 surface exhibits an atomically smooth surface morphology and, as seen in (b), the surface morphology of the AlO_x on the OH-SiO_2 surface essentially maintained the atomically smooth surface morphology with slight increases in root-mean-square roughness (R_{rms}) from (a) 0.11 nm to (b) 0.37 nm. As seen in Fig. 2(c), the surface of the freshly prepared CH_3-SAM exhibited atomically smooth morphology with R_{rms} of (c) 0.24 nm. However, Rrms increased significantly to 8.09 nm after 400 ALD cycles was completed as seen in (d). The rough surface developed during the 400 cycles of ALD appears to be covered with granular-like surface features separated by voids.

Figure 2. AFM images collected on (a) as-formed OH-SiO_2, (b) 400-ALD cycle AlO_x on the OH-SiO_2, (c) as-formed CH_3-SAMs, and (d) 400-ALD cycle AlO_x on the CH_3-SAMs.

Shown in Fig. 3 are water contact angle (θ_{cont}) measurements on the surfaces of AlO_x on the CH_3-SAM (solid circles) and the OH-SiO_2 (solid triangles). On the CH_3-SAM, the water contact angle on the surface of freshly prepared alkanethiol SAMs was approximately 106 degree, indicating high-quality SAM. Once AlO_x was deposited on the CH_3-SAM, the water contact angle decreased steadily as the thickness increased until the thickness reached 25-30 nm. Subsequently, the water contact angle appeared to increase slowly. In contrast, the water contact angle on the OH-SiO_2 remained nearly unchanged from the initial number for the 0 ALD cycle sample. The gradual decrease in the water contact angle on the CH_3-SAM suggests that the surface of the CH_3-SAM is not uniformly covered with AlO_x, as indicated by the AFM image in Fig. 2(d).

Reflection absorption infrared spectroscopy spectra (RAIRS) collected from the AlO_x sample after 200 ALD cycles is shown in Fig. 4 (spectrum (a)) with a reference RAIRS spectrum collected from freshly-prepared CH_3-SAM (spectrum (b)). In the reference spectrum, the four peaks that are well-resolved and numbered 1-4 are associated with asymmetric CH_3 (a-CH_3), asymmetric CH_2 (a-CH_2), symmetric CH_3

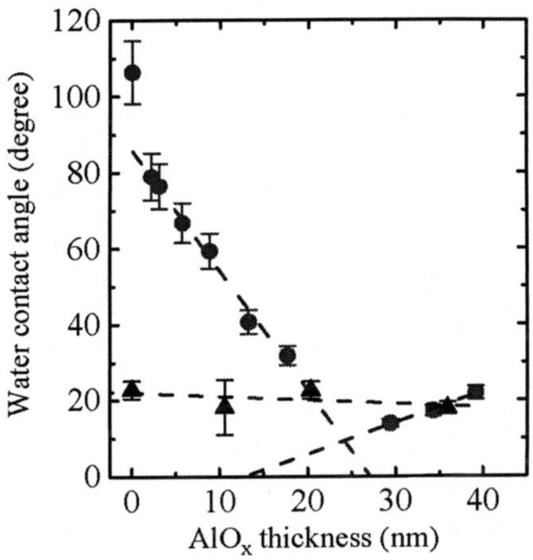

Figure 3. Plot of water contact angle on the surfaces of the AlO_x on the hydrophobic CH_3-SAM surfaces (solid circles) and the hydrophilic OH-SiO_2 surfaces (solid triangles).

(s-CH_3), and symmetric CH_2 (s-CH_2) stretching modes, respectively, suggesting that the CH_3-SAM was well-ordered on the TS-Au surface. The RAIRS spectrum collected on the 200-cycle ALD sample showed significant contrast to the reference spectrum. The CH_3 peaks (peaks 1 and 3) disappeared completely, implying that the CH_3 functional group of the SAMs experienced a substantial perturbation as a result of the

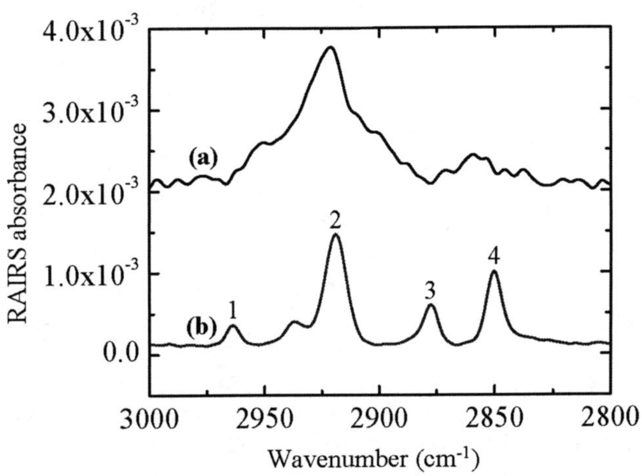

Figure 4. (a) RAIRS spectra collected from AlO_x samples after 200-ALD cycles on the CH_3-SAMs. The spectrum (b) is a reference collected from as-formed CH_3-SAMs without AlO_x deposition.

AlO_x deposition. As mentioned earlier, the reaction of TMA with CH_3-SAMs was found to form no adsorbed complex and the ligand exchange reaction to form methane needs to go through a large kinetic energy barrier [6]. In contrast to the instantaneous disappearance of the CH_3 peaks, the CH_2 peaks were found to be present even on the 200-ALD cycle sample. The CH_2 peaks (peaks 2 and 4) were still present, however they were substantially perturbed. Both peaks 2 and 4 showed significant broadening and peak shifts. Both a-CH_2 and s-CH_2 mode peaks exhibited significant broadening and peak shift even after only 25-ALD cycles (not shown), suggesting that the ALD process in the early stages of the deposition (up to 25 ALD cycles) resulted in the formation of structural disorder within the alkane chains of the SAMs. The AFM and the RAIRS results, by comparing the way AlO_x evolves on the CH_3-SAMs (hydrophobic) and OH-SiO_2 (hydrophilic) surfaces, clearly suggests that it is the wetting characteristics of H_2O pulses in the early stage of the deposition that dictates the surface morphology of the AlO_x in the later stage. On CH_3-SAMs, numerous small water droplets would form during H_2O pulses, and TMA is expected to react spontaneously with these water droplets pre-existing on the CH_3-SAMs surface. Clearly, atomic-scale studies on the early stage of the AlO_x deposition on CH_3-terminated SAMs need to be done to address several questions raised in our experiment such as the physical and/or chemical origin of the results that the CH_3 vibration modes disappeared and the CH_2 vibration modes broadened significantly at the onset of the ALD process.

TWO-STAGE ATOMIC LAYER DEPOSITION

n-propanol as an Oxygen Source

The growth instability observed in the evolution of AlO_x films on the CH_3-SAMs, in contrast to those on the OH-terminated SiO_2, eventually resulted in rough surface morphologies. The comparative studies on the CH_3-SAMs (hydrophobic) and the OH-SiO_2, (hydrophilic) surfaces suggested that the instability be associated with the characteristics of surface wetting by H_2O in a conventional ALD process, presumably in the early stages. In other words, deposition kinetics on hydrophobic surfaces in the early stage of ALD processes could be actively modified by adding chemical species that promote surface wetting by water on hydrophobic surfaces. Therefore, we examined the wetting properties of a mixture of water and *n*-propanol, instead of pure water, on the CH_3-SAMs.

Fig. 5 shows the contact angles measured at room temperature with a mixture of water and *n*-propanol on the CH_3-SAMs. The contact angle are plotted as a function of the volume percentage of *n*-propanol in the mixture (*R*). The contact angle sharply dropped within the range of R = 0-30 % and slowly saturated to approximately 43 degree afterward. The wetting characteristics progressively improved until R reached roughly 40 %, which suggests that utilizing *n*-propanol would improve surface morphology of AlO_x films on hydrophobic surfaces such as CH_3-SAMs during ALD processes. Clearly this new concept in ALD process at low temperatures would benefit other materials deposited on hydrophobic surfaces. Given the contact angle data in

Fig. 5, we attempted to explicitly control the surface wettability on the CH_3-SAM by utilizing *n*-propanol as the oxygen source in the early stage of a deposition. The vapor pressure and the heat of vaporization of *n*-propanol at 25 °C are comparable to those of water.

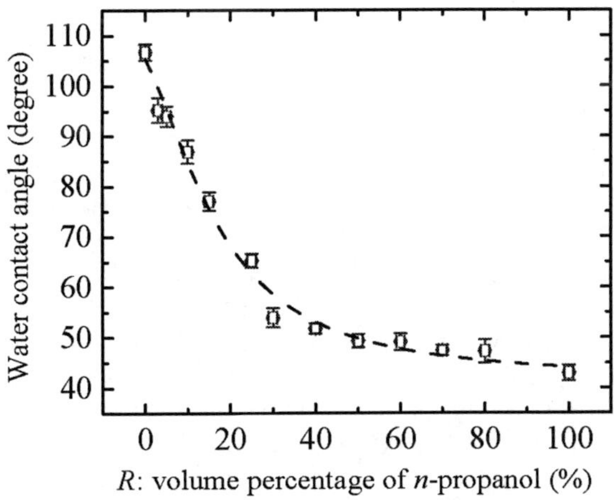

Figure 5. Contact angles measured on the CH_3-SAMs using a mixture of water and *n*-propanol with various concentration of *n*-propanol are plotted.

Two-stage Atomic Layer Deposition Process

In the two-stage atomic layer deposition process proposed here, we utilized pure *n*-propanol as the oxygen source to nucleate AlO_x on the CH_3-SAM in the first stage. Subsequently, in the second stage, H_2O was used as the oxygen source to further deposit AlO_x. In both stages, trimethylaluminum (TMAl) was used as the aluminum source. One ALD cycle in the initial stage was performed by supplying a 140 ms *n*-propanol pulse, a 160 s nitrogen purge period, a 140 ms TMAl pulse, and a 16 s nitrogen purge period. In the second stage, one cycle was performed in the same sequence using water instead of *n*-propanol. These specific deposition conditions were calibrated for AlO_x on OH-terminated silicon surfaces at 45 °C in order to maintain deposition rates self-limited at 0.02 and 0.12 nm / cycle in an *n*-propanol and a water ALD stage, respectively. All samples described below were identified by the cycle fraction $R_c = N_{n\text{-}propanol} / (N_{n\text{-}propanol} + N_{water})$, where $N_{n\text{-}propanol}$ and N_{water} are the total number of ALD cycles in an *n*-propanol (i.e. the first stage) and a water stage (i.e. the second stage), respectively. The substrate temperature was set to 45 °C for all samples to minimize structural damage on the CH_3-SAMs. The total number of pulses utilized was adjusted for each R_c to produce the same total film thickness from each deposition. Nominal (average) thicknesses of the AlO_X films measured by spectroscopic ellipsometry on all samples were within 31 ± 1 nm. In Table 2, a set of key deposition parameters are listed.

TABLE 2. Deposition parameters of the two-stage ALD process

Deposition conditions	First stage / Second stage
Temperature (°C)	45
Sources	$Al(CH_3)_3$ and *n*-propanol / $Al(CH_3)_3$ and H_2O
Chamber base pressure (torr)	~$2x10^{-1}$
H_2O pulse time (ms)	140
H_2O purge time (s)	160
$Al(CH_3)_3$ pulse time (ms)	140
$Al(CH_3)_3$ purge time (s)	16
Deposition rate (nm / cycle)	0.02 / 0.12

Aluminum Oxide Deposited by the Two-stage ALD

Show in Fig. 6 are representative AFM images (scan area of 500 x 500 nm) collected from AlO_X samples with (a) $R_c = 0.002$ (11 nm r.m.s. roughness) and (b) $R_c = 0.301$ (2.4 nm r.m.s. roughness). The r.m.s. roughness of the AlO_X films was found to be a strong function of R_c, which revealed a clear minimum in R_c within the range of $0.2 < R_c < 0.4$. Panel (a) reveals the similar morphological characteristics to those seen in Fig. 2(d), which reflects the fact that the sample with $R_c = 0.002$ was deposited with only one *n*-propanol ALD cycle and effectively identical to the sample represented in Fig. 2(d). The rough surface seen in (a) consists of multiple granules with a diameter and a height in the range of 43 nm and 12 nm, respectively. Several voids are also clearly seen. In contrast, the surface morphology shown in (b) is characterized by much smaller granules densely covering the surface and leaving a smaller number of voids, accounting for the smaller r. m. s. roughness on the surface of the sample with $R_c = 0.301$.

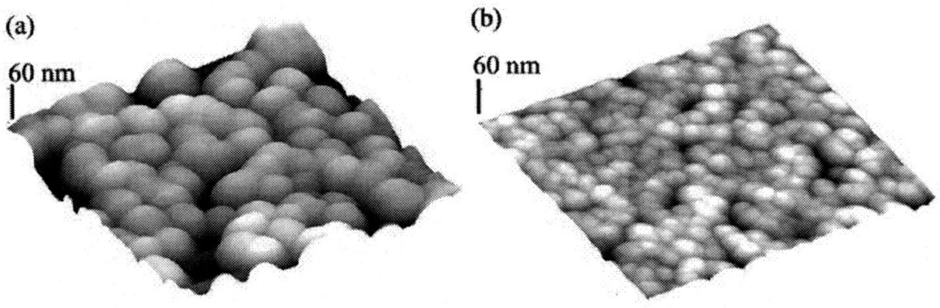

Figure 6. AFM images collected on R_c = (a) 0.002 and (b) 0.301 samples are shown. Each image is 500 x 500 nm scan.

RAIRS measurement further revealed striking differences between the two samples discussed in Fig. 6. Fig. 7 are RAIRS spectra collected from the AlO_X samples with R_c = 0.002 (bottom spectrum) and 0.301 (top spectrum), respectively. Since the absolute absorbance of RAIRS spectrum is highly sensitive to total structural characteristics of each sample, the absorption peaks of each spectrum were normalized with respect to the absorbance of the asymmetric CH_2 peak, i.e. peak 2 of each spectrum. As in Fig. 4, the four peaks 1-4 seen in the spectra represent asymmetric CH_3, asymmetric CH_2, symmetric CH_3, and symmetric CH_2 stretching modes, respectively. For the $R_c = 0.301$ sample (top spectrum), the well-resolved four vibrational modes indicate that the characteristics of the as-formed CH_3-SAM were well preserved. The peak positions of peaks 2 and 4 revealed no discernible peak shift with respect to the corresponding peak positions for the as-formed SAM, which further indicates that the alkyl chains were still well-ordered. A broadened shoulder on the lower wavenumber side of peak 2 is an indication of slight mode-softening within the alkyl chains [13]. The position of peaks 1 (2954.4 cm^{-1}) and 3 (2849.3 cm^{-1}) were shifted toward lower wavenumbers compared to those for the as-formed SAM (2964.1 and 2850.3 cm^{-1}), which indicates mode-softening of the terminal methyl groups [14]. In contrast, the spectrum of the $R_c = 0.002$ sample (bottom spectrum) reveals that two vibrational modes (i.e. peaks 1 and 4) were strongly perturbed by the AlO_x deposition. Although peaks 2 and 3 were still well-resolved, peak 4 was completely absent, which could be attributed to symmetry-breaking in the alkyl chains [15]. In addition, peak 1 appears to consist of two small peaks, suggesting the presence of both in-plane and out-of-plane asymmetric vibrational modes of CH_3 [16]. For the $R_c = 0.002$ sample, the RAIRS spectrum was consistent with our previous observations for AlO_x deposition with water only [17].

The two stage ALD with $0.2 < R_c < 0.4$ process produced dramatically superior results compared to the growth using either water or *n*-propanol as the sole oxygen source for 31 nm AlO_x films. A major unanticipated bonus was that the CH_3-SAM was

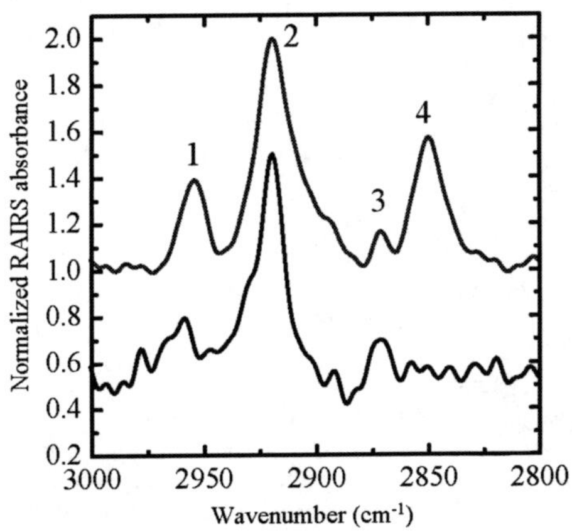

Figure 7. RAIRS spectra collected from $R_c = 0.002$ (bottom spectrum) and 0.301 (top spectrum) samples are shown. Four peaks shown are associated with C-H stretching modes for the CH_3-SAMs underneath the AlO_x films.

much less perturbed by the two stage AlO_x deposition, as shown by the RAIRS spectrum in Fig. 7 for $R_c = 0.301$ for which the r.m.s. roughness was minimum coincidently. One possibility for the improvement in the film roughness within the window of $0.2 < R_c < 0.4$ could be that the wettability of a thin AlO_x film deposited with *n*-propanol is better for water than for *n*-propanol. However, we performed contact angle measurements on a 14 nm thick AlO_x film deposited only by *n*-propanol (corresponding to the $R_c = 0.826$ sample), and found that the contact angles for *n*-propanol and water were approximately 15° and 29.4°, respectively. Thus, there is some other driving forces for the film roughening for $R_c > 0.4$, which may be the increasing incorporation of carbon into the growing film with thickness.

SUMMARY

The deposition of AlO_x on CH_3-SAMs (hydrophobic) and OH-SiO_2 (hydrophilic) surfaces by ALD were studied. During a conventional ALD using H_2O as an oxygen source, the AlO_x on CH_3-SAMs was found to undergo growth instability and develop substantially rough surface morphologies, which were attributed to the inefficient surface wetting on the hydrophobic CH_3-SAMs by H_2O in the early stage of the ALD. In the optimized two-stage ALD, using *n*-propanol in the first stage improved the surface morphology of AlO_x significantly and maintained the structural integrity of underlying CH_3-SAMs, suggesting the importance of explicitly controlling surface wetting in the early stage of a ALD process, in particular, on hydrophobic surfaces.

REFERENCE

1. C. D. Dimitrakopoulos and D. J. Mascaro, *IBM J. Res. & Dev.* **45**, 11-27 (2001).
2. R. Toniolo and I. A. Hümmelgen, *Mater. Eng.* **289**, 311-314 (2004).
3. K. Morii, M. Ishida, T. Takashima, T. Shimoda, Q. Wang, Md. K. Nazeeruddlin, and M. Grätzel, *Appl. Phys. Lett.* **89**, 183510-3 (2006).
4. H. Shin, R. J. Collins, M. R. De Guire, A. H. Heuer, C. N. Sukenik, *J. Mater. Res.* **10**, 692-698 (1995).
5. P. Wohlfart, J. Weiβ, J. Käshammer, M. Kreiter, C. Winter, R. Fisher, S. Mittler-Neher, *Chem. Vap. Deposition* **5**, 165-170 (1999).
6. Y. Xu and C. B. Musgrave, *Chem. Mater.* **16**, 646-653 (2004).
7. J. P. Lee, Y. J. Jang, M. M. Sung, *Adv. Funct. Mater.* **13**, 873-876 (2003).
8. A. Hooper, G. L. Fisher, K. Konstadinidis, D. Jung, H. Nguyen, R. Opila, R. W. Collins, N. Winograd, D. L. Allara, *J. Am. Chem. Soc.* **121**, 8052-8064 (1999).
9. J. J. Senkevich, F. Tang, D. Rogers, J. T. Drotar, C. Jezewski, W. A. Lanford, G.-C. Wang, T.-M. Lu, *Chem. Vap. Deposition* **9**, 258-264 (2003).
10. G. L. Fisher, A. E. Hooper, R. L. Opila, D. L. Allara, N. Winograd, *Phys. Chem. B* **104**, 3267-3273 (2000).
11. J. J. Blackstock, Z. Li, M. R. Freeman, D. R. Stewart, *Surf. Sci.* **546**, 87-96 (2003).
12. N. Prathima, M. Harini, N. Rai, R. H. Chandrashekara, K. G. Ayappa, S. Sampath, S. K. Biswas, *Langmuir* **21**, 2364-2374 (2005).
13. M. J. Hostetler, W. L. Manner, R. G. Nuzzo, and G. S. Girolami, *J. Phys. Chem.* **99**, 15269-15278 (1995).
14. A. Michaelides and P. Hu, *J. Chem. Phys.* **114**, 2523-2526 (2001).
15. M. Yamamoto, Y. Sakurai, Y. Hosoi, H. Ishii, E. Ito, K. Kajikawa, Y. Ouchi, and K. Seki, *Surf. Sci.* **427-428**, 388-392 (1999).
16. R. A. MacPhail, H. L. Strauss, R. G. Snyder, and C. A. Elliger, *J. Phys. Chem.* **88**, 334-341 (1984).

17. N. P. Kobayashi, C. L. Donley, S.-Y. Wang, and R. S. Williams, *J. Crystal Growth* **299**, 218-222 (2007).

Experimental Investigation and Simulation of Ferrous Biooxidation and Flow Hydrodynamics in a Bubble Column Bioreactor Using CFD Approach

S.M. Mousavi [a,b], A. Jafari [c], S. Yaghmaei [a], M. Vossoughi [a], and I. Turunen [b]

[a]*Biotechnology Group, Chemical Engineering Department, Faculty of Engineering, Tarbiat Modares University, P.O. Box 14115-143, Tehran, Iran*
[b]*Department of Chemical Technology, Lappeenranta University of Technology, Lappeenranta, Finland*
[c]*Laboratory of Engineering Thermodynamics, Department of Energy and Environmental Technology, Lappeenranta University of Technology, Lappeenranta, Finland*

Abstract. In the present attempt a set of experiments and a 3D simulation using a commercially available computational fluid dynamics package (FLUENT) were adopted to investigate complex behavior involving hydrodynamics and ferrous biological oxidation in a gas–liquid bubble column reactor. By combining the hydrodynamics and chemical species transport equations, the velocity field, air volume fraction and ferrous biooxidation rate in the column were simulated. The kinetic model proposed by Nemati and Webb (1998) was used to simulate the biooxidation rate in the column. Gas-liquid interactions were modeled using an Eulerian model in three dimensions. The effects of inlet air velocity and initial substrate (Fe^{+2}) concentration on the velocity field, air volume fraction and biooxidation rate of ferrous iron in the column were investigated. To validate the model, simulation was compared with the experimental data in the presence of *Acidithiobacillus ferrooxidans* in an aerated column where the superficial gas velocity was adjusted between 0 and 0.5 m/s. It was found that the initial ferrous concentration and the inlet air velocity had a pronounced effect on the ferrous biooxidation rate. The results indicated that the maximum biooxidation rate can be obtained at superficial air velocity of 0.1 m/s and initial ferrous concentration of 6.7 g/L.

Keywords: CFD simulation; velocity field; air volume fraction; ferrous biooxidation rate; bubble column.

PACS: 07.05.Tp; 47.11.–j; 47.63.-b

CP1007, *Current Themes in Engineering Technologies*, edited by S.-I. Ao, M. A. Amouzegar, and S.-S. Chen

INTRODUCTION

Bubble column reactors are attracting increasing academic and industrial research interest in view of their many potential applications. Bubble columns are widely used as gas–liquid contactors and as reactors in the chemical, petrochemical and biochemical industries. Effective mixing as well as high interfacial area between the phases, leading to improved heat and mass transfer characteristics, relatively cheap to install and the lack of moving parts, are the factors that render under bubble columns an attractive choice as reactors for the described processes.

In chemical and biochemical process engineering, much works have been devoted to either kinetic modeling or fluid dynamics. Most works have been performed under the assumption of idealized conditions. Either the reactor was assumed to be an ideal model case or the fluid dynamics was considered in a realistic way, but mass transfer and chemical reaction were omitted. Despite the widespread application of bubble columns and substantial research efforts devoted to understand their behavior, detailed knowledge on the fluid flow, mass transfer and chemical reactions as well as their interactions are still lacking. However, the scale-up of bubble column is still poorly understood because of the complexity of the flow patterns and their unknown behavior under different sets of design parameters.

Understanding the complexity of the fluid dynamics in bubble column and airlift reactors is important due to their application in the chemical and bioprocess industries. Knowledge of the hydrodynamics of such reactors helps to determine the efficiency of biochemical production rates through transport processes such as inter-phase oxygen transfer, mixing of nutrients and reactants plus the effects that pH has on micro-organisms growth, metabolic pathways and cell lyses. Also of importance is knowledge of the influence of the biomass on the gas phase through inter-phase interactions and the impact the biomass has on the liquid phase viscosity [1].

Study of the flow hydrodynamics in bubble columns involves the use of experimental techniques and computational fluid dynamics (CFD). The potential of CFD for describing the hydrodynamics of bubble columns has been established by several publications, for example [2]-[6]. An important advantage of the CFD approach is that column geometry and scale effects are automatically accounted. The success of the CFD simulation strategy is however crucially dependent on the proper modeling of the momentum exchange, or drag coefficient between the gas and liquid phases. CFD modeling of gas-liquid two-phase flows has shown remarkable progress over the last decade. The methods developed include the volume of fluid methods, dispersed phase models, Eulerian two-fluid models, and the algebraic slip mixture models. The two most common approaches to modeling bubble columns are the Euler-Euler or two-fluid approach and the Euler-Lagrange or discrete bubble approach. In the Euler-Euler approach, both phases (the continuous liquid phase and the dispersed gas phase) are modeled as two interpenetrating continua. In the Euler-Lagrange approach on the other hand, the volume averaged Navier-Stokes equations are used to describe the motion of the liquid phase and each bubble is tracked on basis of a balance of forces acting upon.

In view of the fact that the microbial oxidation of ferrous iron to the ferric form is an essential sub-process in many biological processes, the developing of an understanding to obtain optimum conditions for this sub-process is critical. Bacterial leaching is an established biotechnology that uses the ability of *A. ferrooxidans* to oxidize ferrous iron. This microorganism has the capability to carry out processes of biosorption and bioprecipitation of heavy metal ions, which have important environmental applications [7]. Another application of the acidophilic ferrous iron-oxidizing bacteria can be found in a process proposed for removal of hydrogen sulfide from gases [8]. In this process the removal of hydrogen sulfide is achieved via two steps: the chemical oxidation of the sulfide by Fe^{3+} and the bacterial regeneration of Fe^{3+} as leaching agent [9]-[10].

The objective of this research was to develop a fundamental understanding of the hydrodynamics and interactions with the chemistries involved in the bubble column reactor to maximize overall efficiency for the ferrous biological oxidation according to the following reaction:

$$2Fe^{2+} + 2H^{+} + \tfrac{1}{2}O_2 \xrightarrow{A.ferrooxidans} 2Fe^{3+} + H_2O \qquad (1)$$

In this work computational fluid dynamics (CFD) was used in the Eulerian framework to describe the hydrodynamics of the bubble column bioreactor. A kinetic model was applied to investigate the influence of some variables on biooxidation rate of ferrous ion in an aerated column reactor.

EXPERIMENTAL PROCEDURES

Materials and Methods

Microorganism and Medium

A. ferrooxidans was used as an inoculum in the experimental part. The medium for growth and maintenance of cells was contained (in g/L): $FeSO_4.7H_2O$: 33.4, $(NH_4)_2SO_4$: 0.4, $MgSO_4.7H_2O$: 0.4, and K_2HPO_4: 0.4 [11]. To prepare the medium the iron solution was filtered using Whatman paper 0.2 μm, and the basal salt solution was autoclaved at 121°C and 15 psig for 20 min and after cooling solutions were combined. More details about microorganism and medium can be found in the previous work [12].

Biooxidation Procedure

Biooxidation experiments were carried out in a bubble column reactor consisting of three parts: top, main column with water jacket and bottom section with an air diffuser (Fig. 1). The diameter and height of column were 7.5 and 70 cm, respectively. Total working volume of bioreactor was about 3 L. Compressed air was sparged at five different levels from the bottom into the column. The top part of the reactor had a 15 cm diameter and 10 cm height and was employed to prevent the overflow of liquid from the reactor. In order to provide a constant temperature, bioreactor was maintained at 32°C using a water jacket. The bioreactor contained 2.7 liters of medium and 10% (v/v) inoculum. The initial cells concentration was about 4×10^8 cells/mL. During the experiments, pH was adjusted to 1.5 by addition of 1N H_2SO_4. To obtain variations of ferrous biooxidation rate against initial ferrous concentration and inlet air velocity a set of experiments were carried out at different conditions. Ferrous, ferric and bacterial concentrations were measured regularly. Water was added to the reactor to compensate for evaporation losses. Dissolved oxygen was measured by a digital D.O. meter (Electronics, India). The analysis procedures for the bacterial concentration, determination of iron ions and pH are similar to described in [12].

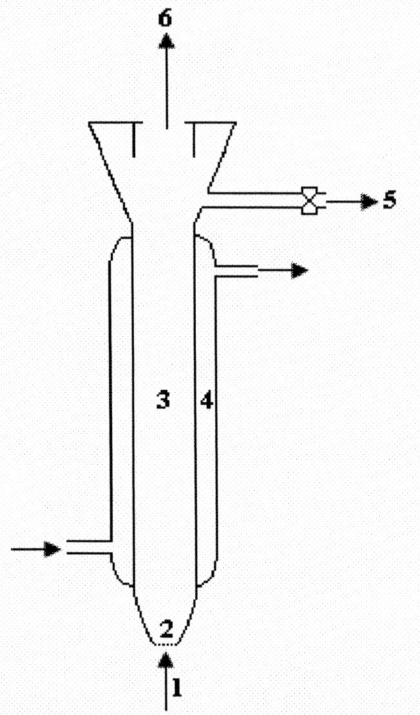

FIGURE 1. Schematic diagram of the bubble column bioreactor used in the experiments (1: input air; 2: air sparger; 3: main column; 4: water jacket; 5: sampling port; 6: output air).

MATHEMATICAL MODELING AND NUMERICAL METHOD

Kinetic Model

The kinetic model proposed by Nemati and Webb [13] (Eq. 2) was used in the simulation. The experimental data obtained from the bioreactor were compared with the kinetic model to ensure the suitability of the model for our simulations. The model includes the effects of temperature, ferrous iron, and bacterial concentrations, also incorporates terms for substrate (Fe^{2+}) and cell inhibition. Values for the constants in the kinetic model are summarized in Table 1.

TABLE 1. Values for the constants in the kinetic model

Coefficient	Value
K_0	114.96 (kmol/m^3.h)/(cells/mL) or 6438 (kg/m^3.h)/(cells/mL)
E_a	68.4 (kJ/mol)
K'_m	1.2 × 10^{-3} (kmol/m^3) or 0.0672 (kg/m^3)
K'_i	2.68 × 10^7 (cells/mL)
α	0.466 (kmol/m^3) or 26.1 (kg/m^3)
β	7.8 × 10^8 (cells/mL)

$$\frac{d\left[Fe^{2+}\right]}{dt}=K\frac{[X]\left[Fe^{2+}\right]}{K'_m\left(1+\frac{[X]}{K'_i}\right)+\left[Fe^{2+}\right]+\left(1-\frac{[X]}{\beta}\right)\frac{\left[Fe^{2+}\right]^2}{\alpha}} \tag{2}$$

where $[Fe^{2+}]$ = ferrous iron concentration (kg/m^3); $[X]$ = bacterial concentration (cells/ mL); K'_m = apparent saturation constant (kg/m^3); K'_i = cell inhibition coefficient (cells/ mL); α and β are constants with values of 26.1 kg/m^3 and 7.8 × 10^8 cells/mL, respectively and K is the rate constant ((kmol/m^3.h)/(cells/mL)) which is defined as:

$$K=K_0e^{-E_a/RT} \tag{3}$$

where E_a = activation energy (kJ/mol); R = universal gas constant (kJ/K.mol); K_0 = frequency factor ((kmol/m^3.h)/(cells/mL)) and T = absolute temperature (K).

Eulerian Approach

The governing equations in this approach can be derived by ensemble averaging the fundamental conservation equations for each phase to describe the motion of liquid and gas in a bubble column. Both the continuous and the dispersed phases are modeled in the Eulerian frame of reference as interpenetrating continua. The mass and momentum balance equations are written separately for each phase. The momentum equations of the phases interact with each other through inter-phase momentum exchange terms. The continuity equation for each phase is written as

$$\frac{\partial(\alpha_k \rho_k)}{\partial t} + \nabla.(\alpha_k \rho_k \boldsymbol{u}_k) = 0 \tag{4}$$

where α_k, ρ_k and $\boldsymbol{u}_k$ are the volume fraction, the density, and the phase averaged velocity of the kth phase, respectively.

The equation of conservation of momentum for each phase is:

$$\frac{\partial(\alpha_k \rho_k \boldsymbol{u}_k)}{\partial t} + \nabla.(\alpha_k \rho_k \boldsymbol{u}_k \boldsymbol{u}_k) = -\alpha_k \nabla p + \alpha_k \rho_k \boldsymbol{g} + \nabla.(\alpha_k \boldsymbol{\tau}_k) + \nabla.(\alpha_k \rho_k \boldsymbol{u}'_k \boldsymbol{u}'_k) + \boldsymbol{F}_{kl} \tag{5}$$

where p is the pressure, $\boldsymbol{g}$ is the gravity acceleration, and $\boldsymbol{F}_{kl}$ is the interfacial momentum exchange between phase k and phase l. $\boldsymbol{\tau}_k$ is the viscous stress tensor, which can be expressed by

$$\boldsymbol{\tau}_k = \mu_k (\nabla \boldsymbol{u}_k + \nabla \boldsymbol{u}_k^T) - \frac{2}{3} \mu_k \nabla.\boldsymbol{u}_k \boldsymbol{I} \tag{6}$$

where μ_k is the viscosity of the kth phase, and $\alpha_k \rho_k \boldsymbol{u}'_k \boldsymbol{u}'$ is the turbulent stress tensor that must be closed by an appropriate multiphase turbulence model. Because of the loss of information in the averaging process, several correlations (such as the interfacial momentum exchange term $\boldsymbol{F}_{kl}$ in Eq. (5)) appear in the macroscopic equations that must be closed in terms of known variables [14].

The momentum exchange between gas phase (subscript G) and liquid phase (subscript L) is given by

$$\boldsymbol{F}_{L,G} = \frac{3}{4} \rho_L \frac{\varepsilon_G}{d_b} C_D (\boldsymbol{u}_G - \boldsymbol{u}_L) |\boldsymbol{u}_G - \boldsymbol{u}_L| \tag{7}$$

here, the inter-phase drag coefficient is calculated from (Clift, 1978):

$$C_D = \frac{2}{3\sqrt{E\ddot{o}}} \tag{8}$$

where

$$E\ddot{o} = \frac{d(\rho_L - \rho_G)d_b^2}{\sigma} \tag{9}$$

and d_b is the equivalent diameter of the bubbles. Note that the rise velocity of bubbles is practically independent of the bubble size in the 3-8 mm range. In the simulations reported here, a bubble diameter 3 mm has been chosen.

In the present analysis only the drag force contribution to $\boldsymbol{F}_{L,G}$ have been included, following the works of Sanyal et al. [2] and Sokolichin and Eigenberger [3] and the added mass and lift force contributions were both ignored.

For the continuous phase (liquid phase) the turbulent contribution to the stress tensor was evaluated by means of $k-\varepsilon$ model described by Sokolichin and Eigenberger [3] using standard single-phase parameters $C_\mu = 0.09$, $C_{1\varepsilon} = 1.44, C_{2\varepsilon} = 1.92, \sigma_k = 1$ and $\sigma_\varepsilon = 1.3$ and no turbulence model was used for calculating the velocity fields of the dispersed bubble phase.

The conservation equation for transport species takes the following general form:

$$\rho(\frac{\partial y_i}{\partial t}) + \rho\nabla.(\boldsymbol{u}y_i) = -\nabla.\boldsymbol{J}_i + S_i \tag{10}$$

where y_i is the local mass fraction of ith species, S_i is the source term for ith species due to the reaction, and $\boldsymbol{J}_i$ is the diffusion flux of species *i*, which arises due to concentration gradients. Diffusion flux can be written as:

$$\boldsymbol{J}_i = -\rho D_{i,m}\nabla y_i \tag{11}$$

here $D_{i,m}$ is the diffusion coefficient for species *i* in the mixture. An equation of this form will be solved for *N*-1 species where *N* is the total number of fluid phase species present in the system.

Numerical Implementation

In this research a commercial grid-generation tool, GAMBIT 2.2 (of Fluent Inc., USA) was used to create the geometry and generate the grids. It is very important to use adequate number of computational cells while numerically solving the governing equations over the solution domain. To divide the geometry into discrete control volumes, more than 4.5×10^5, 3D tetrahedral computational cells and 9.4×10^4 nodes were used. In addition more than

4.1×10^4 triangular elements were used at the inlet and outlet. Figure 2 shows the schematic and dimensions of the computational domain and the reactor grid.

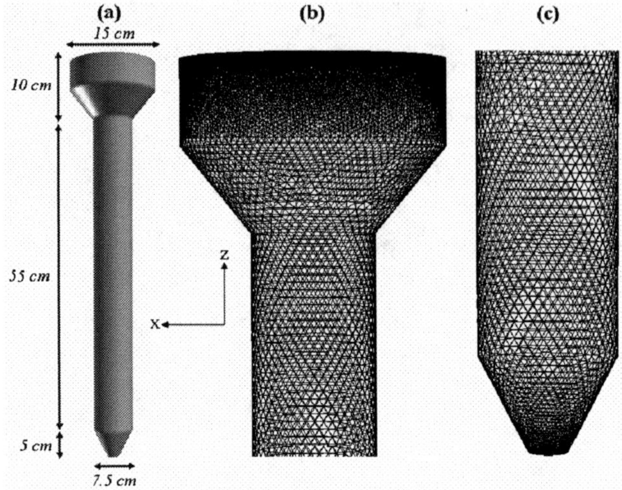

FIGURE 2. (a) Schematic and dimensions of the bubble column used in the simulations. (b) and (c) grids for the top and the bottom sections of the reactor.

The Eulerian approach implemented in the commercial code for computational fluid dynamics, FLUENT, has been used for the simulations. The finite volume method is adopted to solve three dimensional governing equations. The solver specifications for the discretization of the domain involve the first-order upwind for momentum and phase coupled SIMPLE [16] for velocity–pressure coupling, respectively. For the volume fraction and concentration of species in the liquid phase, first-order discretization scheme was used. The under-relaxation factors, which are significant parameters affecting the convergence of the numerical scheme, were set to 0.3 for the pressure, 0.5 for the momentum, and 0.5 for the volume fraction. Using mentioned values for the under-relaxation factors a reasonable rate of convergence was achieved. The equations were solved with the unsteady solver with a time step which was 1×10^{-3} s through the simulations. The relative error between two successive iterations was specified by using a convergence criterion of 1×10^{-5} for each scaled residual component. It is worth noting that refinement of the grids did not produce any significant differences in the results. Air was sparged into the column with different inlet velocities varied from 0 to 0.5 m/s. According to the experimental results the highest biological oxidation rate of ferrous was obtained using 6.7 g/L its initial concentration in the liquid solution, so this initial value was used in the simulation. The bioreactor temperature and the initial bacterial concentration were adjusted equal to the experiments values. The initial ferric concentration was set to zero. Boundary conditions were indicated for solving the governing equations. At the walls, no-slip boundary condition was applied (Dirichlet boundary condition). For inlet flow, constant fluid velocity with given phase concentrations, and for the outlet flow, the pressure were specified as boundary conditions.

RESULTS AND DISCUSSION

The biological oxidation rate of ferrous iron as a function of concentration at a constant temperature of 32°C and inlet air velocity of 0.1 m/s is shown in Fig. 3. The experimental results show an inhibition effect on the activity of bacteria when the ferrous iron concentration was higher than 6.7 g/L. The inhibition effect of ferrous iron on the activity of *A. ferrooxidans* has been reported previously [17]-[19]. Kelly and Jones [17] observed that increases in the ferrous iron concentration up to 5.6 kg m^{-3} increased the rate of biological oxidation, while higher concentrations have the negative effect and inhibited the biooxidation. As can be seen in this figure, the oxidation rate of ferrous iron in the presence of bacteria passed through a maximum which can be explained by a substrate (Fe^{+2}) activation and inhibition effect. In addition an error calculation between experimental data and the chosen kinetic model shows a maximum relative error less than 7.9% which confirms the suitability of this model for simulation of ferrous biooxidation in a column bioreactor.

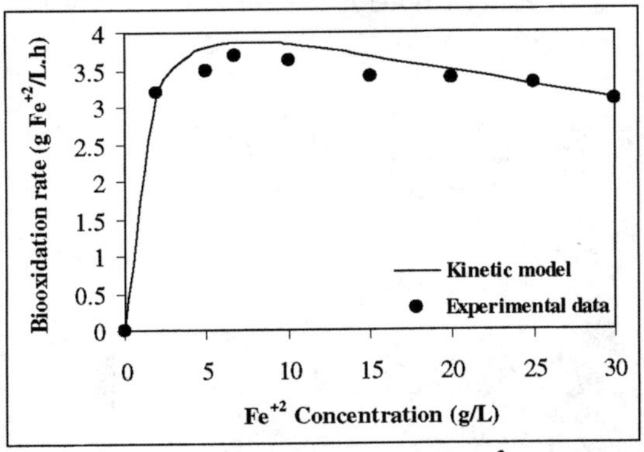

FIGURE 3. Comparison of predicted and experimental Fe^{+2} biooxidation rates at different initial ferrous concentration.

The ferrous biooxidation rate at different inlet air velocities has been presented in Fig. 4. It is clear that inlet air velocity has a significant effect on ferrous iron biooxidation. In this case a comparison between experimental and predicted data shows a maximum relative error about 15%. Both experimental and predicted data indicate a maximum rate for air velocity of 0.1 m/s. In velocities lower than 0.1 m/s a decrease was observed in oxygen concentration in the liquid phase because of low gas–liquid interfacial area and therefore, the reaction rate was decreased. At the gas superficial velocities higher than 0.1 m/s the biological rate was decreased. The reason is that in very high velocities bacteria will damage and therefore the bacterial activity will be decreased.

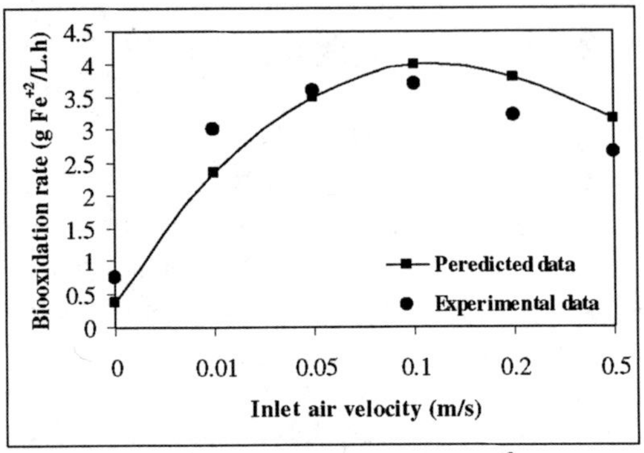

FIGURE 4. Comparison of simulated and experimental Fe^{+2} biooxidation rates at different inlet air velocities.

Figure 5 presents a vector plot of liquid velocity, at the gas superficial velocity of 0.1 m/s. As can be seen the velocity profile gives a maximum value in the center of the column and a negative value close to the wall, which resembles a liquid flow pattern with liquid up-flow in the center region of the column and down flow near the walls.

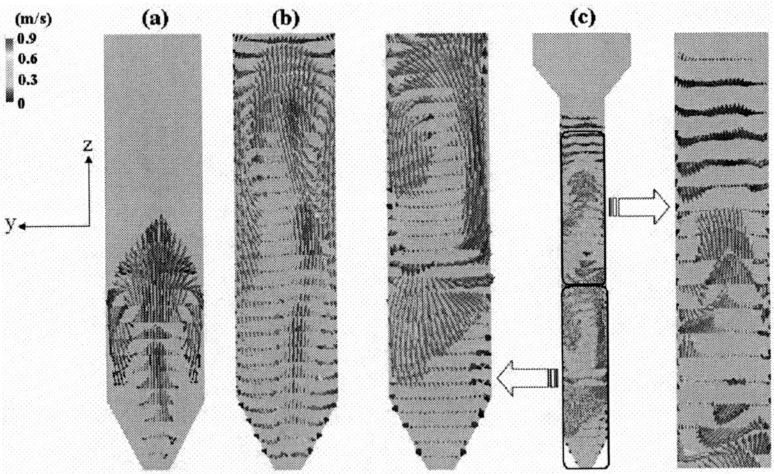

FIGURE 5. (a-c) Snapshots of velocity field in the bioreactor at t=30, 60 and 180 s. Here the inlet air velocity is 0.1 m/s. The velocity field was plotted in the plane (z-y) passing through the center of the column.

Air is injected into an initially quiescent liquid. Shortly after the bubbles are released into the column, air rises through the column and creates a typical mushroom like shape (Fig. 6). Due to the gas–liquid momentum coupling, some motion is induced in the liquid phase, and after a while the entire bulk liquid starts to move. After some time the motion inside the bubble becomes unstable. Strong circulation patterns appear with up flow in the center region and down flow in the corners of the column. Due to the liquid down flow, liquid closes to the wall is dragged downwards to the lower region of the column before being dragged again to the column surface by the upward flow. In many of chemical and biochemical reactions which they occur in column reactors gas holdup is an important factor. It affects on interfacial surface area and also activity of microorganisms and therefore the reaction rate. Fig. 6 indicates snapshots of air volume fraction in the bioreactor at t=10, 30, 60, 120, and 180 s. For the ferrous biological oxidation, air volume fraction in the bioreactor is one of the variables that needs to be maximized to have a high reaction rate. Here the inlet air velocity is 0.1 m/s and the air volume fraction is evaluated in the plane (z-y) passing through the center of the column.

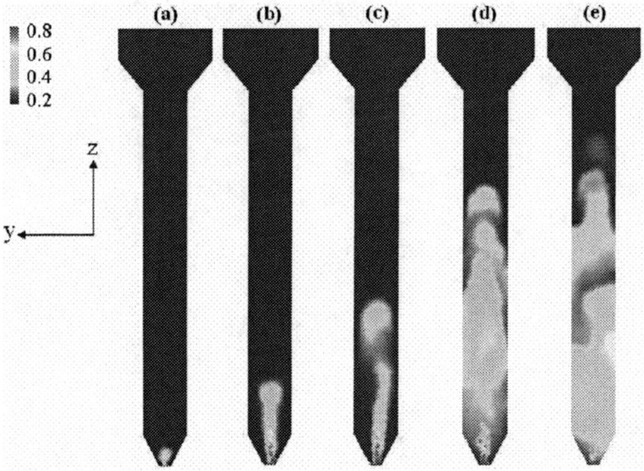

FIGURE 6. (a-e) Temporal evaluation of air volume fraction in the bioreactor at t=10, 30, 60, 120, and 180 s. Here the inlet air velocity is 0.1 m/s. The air volume fraction was evaluated in the plane (z-y) passing through the center of the column.

Fig. 7 represents the velocity and air volume fraction at normalized radial distance in three different cross sections of 15, 30 and 45 cm from the bottom of the column at t=120 s. As it expected the parabolic profiles observed in the simulation results. At each cross section the air volume fraction was reached to the highest value in the center of the column. The higher volume fraction of gas at the center of the column drives the liquid flow upwards at a high velocity, due to gas buoyancy, and the liquid, being in batch mode, returns downwards at the periphery.

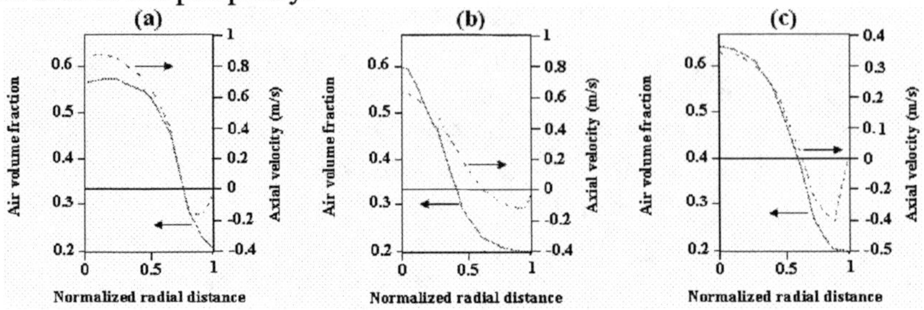

FIGURE 7. 2D plots of velocity and air volume fraction versus normalized radial distance in (a) 15 cm, (b) 30 cm and (c) 45 cm from the bottom of the column at t=120 s.

The measured oxygen concentration in the gas outlet stream at inlet air velocity of 0.1 m/s was higher than 15%, so the biooxidation process was not limited by the availability of oxygen in the liquid medium. Fig. 8 shows that at any time the ferrous biological oxidation rate in the column increases from the bottom to top of the reactor. On the other hand, during the air bobbles moving up through the column because of coalescence, their diameters will increase. Therefore, the ferrous biooxidation depends strongly on the gas–liquid interfacial area, and so, gas–liquid mass transfer dominates.

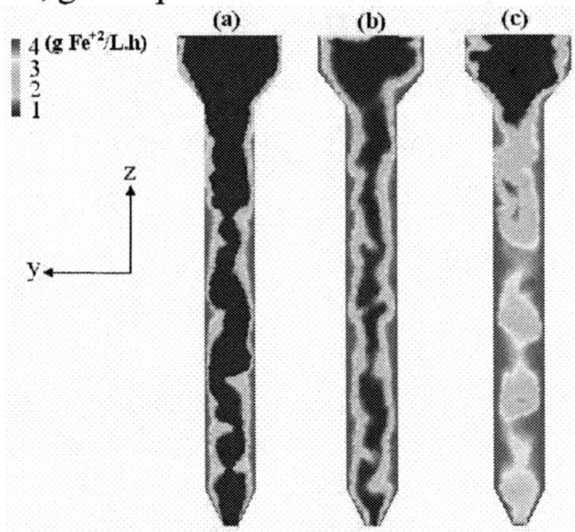

FIGURE 8. (a-c) Snapshots of ferrous bioxidation rate in the bioreactor at t=60, 120 and 180 s. Here the inlet air velocity is 0.1 m/s and initial ferrous concentration set to 6.7 g/L. The figure was plotted in the plane (z-y) passing through the center of the column.

2D plots of biooxidation rate in the bioreactor have been shown in Fig. 9. In this picture the results were taken at t=120 s. Fig 9 (a) shows the computed ferrous biooxidation rate versus distance from the bottom of the bioreactor. According to this part, the maximum predicted biooxidation rate was obtained in the bioreactor outlet which was about 4 g Fe^{+2}/L.h. Fig 9 (b) indicates the rate of biooxidation across several cross sections in the column. As it can be expected at higher sections the reaction rate will increase and on the other hand the maximum value observes in the center of the column because of availability of oxygen. Comparison of Figs. 9 and 4 confirms a good agreement between predicted and experimental data.

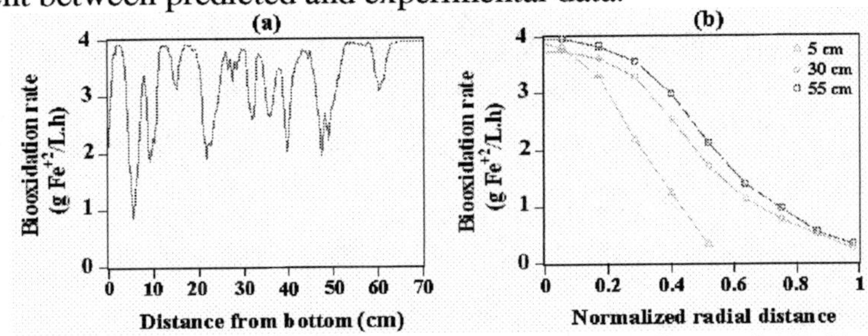

FIGURE 9. 2D plots of ferrous biological oxidation rate versus distance in the bioreactor, (a) at z-y plane and (b) at three different x-y planes at t=120s.

CONCLUSION

In this paper bacterial oxidation of ferrous iron in a bubble column reactor was simulated. The results show that CFD allows gaining new insights in the biological phenomena in the gas-liquid reactors. The simulation results indicate that the Eulerian formulation is a successful approach to predict the hydrodynamics of column bioreactor. It provides good engineering descriptions, and can be used reliably for predicting the flow and holdup patterns in bubble columns. The suggested kinetic model by Nemati and Webb was usable for the prediction of ferrous biooxidation rate in the bioreactor. The simulation results indicate good comparison to experimental data for biological ferrous oxidation rate. It was found that the ferrous biological oxidation depends strongly on the gas–liquid interfacial area, and therefore gas–liquid mass transfer dominates. The initial ferrous concentration and the inlet air velocity had a significant effect on the ferrous biooxidation rate. The superficial air velocity of 0.1 m/s and initial ferrous concentration of 6.7 g/L determined as optimum values. According to the simulation and experiment results maximum ferrous biological oxidation rate was obtained 4 and 3.7 g/L.h at optimum conditions, respectively.

ACKNOWLEDGMENTS

The authors would like to acknowledge Elsevier science LTD for the permission to publish this paper. The original source of this paper has been published in Computers and Chemical Engineering (article in press, 2008).

REFERENCES

1. M. Blažej, G. M. Cartland Glover, S. C. Generalis, and J. Markoš, *Chemical Engineering and Processing* **43**, 137–144 (2004).
2. J. Sanyal, S. Vasquez, S. Roy, and M. P. Dudukovic, *Chemical Engineering Science* **54**, 5071-5083 (1999).
3. A. Sokolichin and G. Eigenberger, *Chemical Engineering Science* **54**, 2273-/2284 (1999).
4. R. Krishna, M. I. Urseanu, J. M. van Baten, and J. Ellenberger, *Chemical Engineering Journal* **78**, 43-51 (2000).
5. R. Krishna and J. M. van Baten, *Chemical Engineering Science* **56**, 6249-6258 (2001).
6. J. M. van Baten and R. Krishna, *Chemical Engineering Science* **56**, 503-512 (2001).
7. G. Cabrera, J. M. Gomez, and D. Cantero, *Process Biochemistry* **40**, 2683–2687 (2005).
8. H. Magota and Y. Shiratori, "Treatment of sour natural gas containing hydrogen sulphide," Japan, 63205124 (1988).
9. M. Boon and J. J. Heijnen, *Hydrometallurgy* **48**, 27–41 (1998).
10. S. Ebrahimi, R. Kleerebezem, M. C. M. van Loosdrecht, and J. J. Heijnen, *Chemical Engineering Science* **58**, 417–427 (2003).
11. M. A. Ronald, *Handbook of Microbiological Media*, Second Edition, Robert Stern publisher, New York, 1997.
12. S. M. Mousavi, S. Yaghmaei, and A. Jafari, *Fuel* **86** (7-8), 993-999 (2007).
13. M. Nemati and C. Webb, *Biotechnology and Bioengineering* **53**, 478-486 (1997).
14. P. Chen, M. P. Dudukovic´, and J. Sanyal, *AICHE* **51** (3), 696-712 (2005).
15. R. Clift, J. R. Grace, and M. E. Weber, *Bubbles, Drops and Particles*, Academic Press, San Diego, 1978.
16. S. A. Vasquez and V. A. Ivanov, A Phase Coupled Method for Solving Multiphase Problems on Unstructured Meshes," In Proceedings of ASME FEDSM'00: ASME 2000 Fluids Engineering Division Summer Meeting, Boston, USA, 2000.
17. D. P. Kelly and C. A. Jones, "Factors affecting metabolism and ferrous iron oxidation in suspension and batch culture of *Thiobacillus ferrooxidans* relevance to ferric leach solution regeneration," in *Metallurgical Applications of Bacterial Leaching and Related Microbiological Phenomena*, edited by L. E. Murr and J. A. Brierley, Academic, New York, 1978, pp. 19-43.
18. J. L. Barron and D. R. Luecking, *Applied Environment and Microbiology* **56** (9), 2801–2806 (1990).
19. L. N. Nikolov and D. G. Karamanev, *Biotechnology Progress* **8**, 252–255 (1992).

A Design of Asynchronous Quasi-Delay Insensitive Computer

Sufian Sudeng and Arthit Thongtak

Digital System Engineering Laboratory (DSEL), Department of Computer Engineering
Faculty of Engineering, Chulalongkorn University
254 Phyathai Road, Pathumwan, Bangkok, Thailand 10330
Tel: +66-2-218-6956, Fax: +66-2-218-6955
E-mail: sovanyy@msn.com, Arthit@cp.eng.chula.ac.th

Abstract. This paper proposes the self-timed circuit design. Quasi-delay insensitive circuit is introduced as asynchronous prototype. The designed circuit focuses on asynchronous processor design. The processor employs asynchronous system bus, asynchronous DMA controller and synchronous interfaces. The system almost completely asynchrony operation except I/O devices and memory interfaces, due to a limitation on the present time devices, the designed has been implemented on spartan-3E FPGA no. 3S500EFG320 by partitioning each module to prevent place and routing conflict, 100-Mhz memory frequency connected, and consumes 141,063 equivalent gate counts. Finally, the timing details of each instruction execution are shown.

Keywords: Asynchronous processor, Asynchronous system bus, Asynchronous DMA controller, QDI-delay model, Dual-rail encoded.

INTRODUCTION

The rapid developments of micro electronics technology, causes the designed circuits are small and high-speed communication. The old design method cannot serve the ability to correct an operation on the high frequency clock. Clock skew is introduced which is cannot distribute high frequency signals to the system. Synchronous circuits also generate worse case delay propagation of global clock, it make much of waiting time to process all parts of circuits, and more energy consumption. Asynchronous design method is restudied. The asynchronous design does not need a global clock to control the combinational circuits to behave on each other. It's depending on event-driven behavior, speed of circuits do not reduced by clock; furthermore, clock skew is avoided [1].

Asynchronous circuits also harder design and verify, no tools supported, made some elements are limited to clock operation. To enhance the design, need to compromise such a hybrid system, both synchronous and asynchronous are working together, Finding synchronization method to synchronize to be well form, and good circuit operation between synchronous and asynchronous one [2].

This work proposes an implementation of asynchronous quasi-delay insensitive circuits on FPGA, and also introduced the interfacing method between synchronous and asynchronous circuits.

CP1007, *Current Themes in Engineering Technologies*, edited by S.-I. Ao, M. A. Amouzegar, and S.-S. Chen

ASYNCHRONOUS CIRCUIT DESIGN

To understanding asynchronous design is familiarity with the assumptions commonly made regarding the delays in the gates and wires within a circuit and the mode in which the circuit operates. The two common delay models, bounded delay model and unbounded-delay model.

Circuit classification

Within the unbounded delay model, there is various different design styles in common use, each with its own problems. In order to increasing number of timing assumptions

Speed – independent (SI) and Delay-insensitive (DI) circuits

If wire delay in a circuits are assumed to be zero and the circuits exhibits correct operation regardless of the delay in any circuit elements, then the circuits is said to be speed-independent. The assumption of zero wire delay is valid for small circuits. A circuit whose operation is independent of the delays in both circuits' elements (gates) and wires is to be delay insensitive.

Quasi delay-insensitive (QDI) circuits

If the difference between signal propagation delays in the branches of a set of interconnecting wires is negligible compared to the delays of the gates connected these branches when the wires are said to from an *Isochronic fork*. Circuits created using the DI design style augmented with the *Isochronic fork* assumption are said to be Quasi-delay assumption.

Signaling protocol

The transfer of information across a channel is negotiated between the sender and receiver using a signaling protocol. Every transfer features a request action where the initiator starts a transfer, and an acknowledge action allowing the target to respond. These may occur on dedicated signaling wires, or may be implicit in the data-encoding used, but in either case, one event indicates data validity, and the other signals its acceptance and the readiness of the receiver to accept further data.

The flow of Information relative to the request event determines whether the channel is classified as a push channel (information flows in the same direction as the request) or pull channel (information flows in the same direction as the acknowledge).

The 4-phase signaling protocol uses the level of the signaling wires to indicate the validity of data and its acceptance by the receiver. When this signaling scheme is used to pass the request and acknowledge timing information on a channel, a return-to-zero phase is necessary so that the channel signaling system ends up in the same state after a transfer as it was in before transfer. This scheme thus uses twice as many signaling edges per transfer than its 2-phase. 4-phase control circuits are often simpler than those of the equivalent 2-phase system because the signaling lines can be used to drive level – controlled latches and the like directly as shown in figure 1.

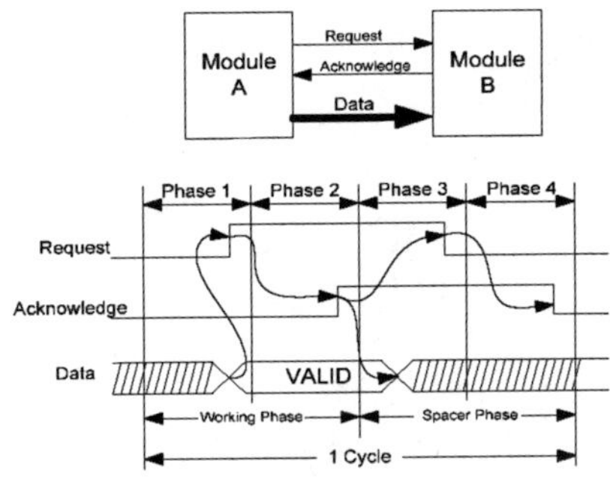

FIGURE 1. 4-phase signaling protocol

Dual-rail encoding

Dual-rail circuits use two wires to represent each bit of information. Each transfer will involve activity on only one of the two wires for each bit, and dual-rail circuits has uses 2n signals to represent n bits of information. 4-phase dual-rail coding is popular for QDI design style.

The Muller C-element

The Muller c-element is commonly used in asynchronous VLSI design where it is used both for synchronizing events and as state holding element. Its represent outputs to 0 when all inputs are 0 and represent outputs 1 when all inputs are 1, otherwise it hold previous state.

ASYNCHRONOUS PROCESSOR DESIGN

The processor architecture[3,5] composes of 8-bit accumulator register A, general purpose register R1 and R2, program counter PC, SP register for CALL instruction , IR register for instruction loading and 2 status flags, carry and zero flag as shown in figure 2.

The processing section composes of shifter used to shift A left or right direction, accumulate based computing (Register A and operand processing, store result to A). The asynchronous processor addressing mode divided into 4 modes, Immediate Mode, Register Mode, Direct Memory Addressing Mode and Indirect Memory Addressing with Register Indexing Mode.

because of accumulate based processing that write the completion result to A. if A value has been changed, it causes to produce a new calculation, need some latching circuit to prevent double writing problem. The processor has 16-bit instruction word with 16K bits location accessing, 1-k location for 8-bit data memory. For dual-rail encoding, need additional circuits to convert the circuit from 1-to-2 signals and from 2-to-1 signal.

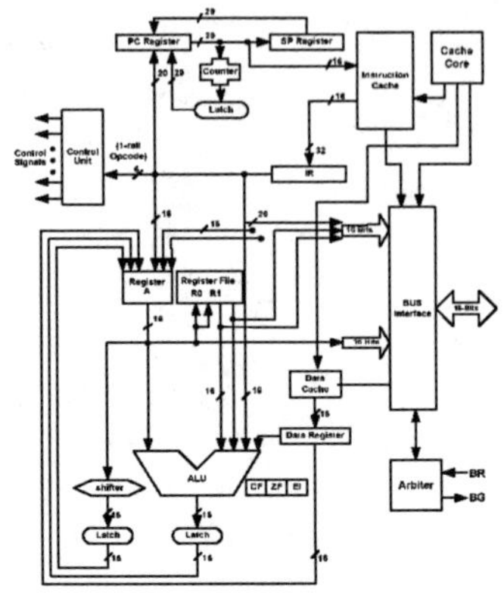

FIGURE 2. Processor Organization

The internal architecture composes of 2 processing logic circuits, arithmetic logic unit and shifting circuit. The main task of arithmetic logic unit is Add, subtract, or, xor. After finished an operation, it passes the result through the latching circuit. The shifting circuit is the circuit that shifts A left or right direction 1-bit precisely.

Control unit processes following 4-phase signal protocol which returns to zero on each computation by using auto sweep module as shown in figure 3 and 6.

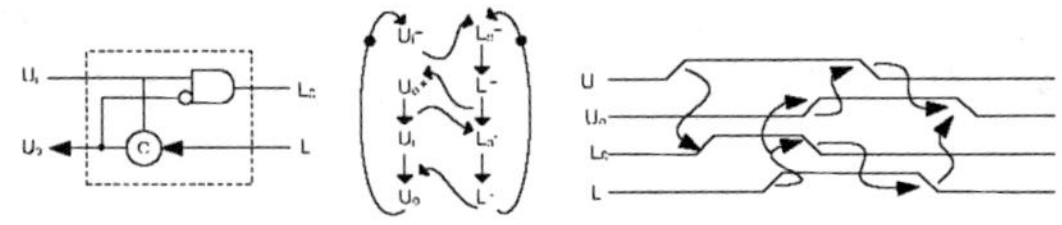

(a) Implementation (b) STG Specification (c) Signal Transition Diagram

FIGURE 3. Auto Sweep Module

From figure 3a, when previous request signal (U_i) changes from 0 to 1, ASM changes current request signal (L_o) from 0 to 1 to enter working phase, after current work has done, the acknowledge signal (L_i) is asserted back to ASM module, provided previous acknowledge signal (U_o) changes from 0 to 1 immediately with current request signal changes from 1 to 0 for idle phase, when working phase has done, current request signal is changes from 1 to 0 automatically, hence ASM was used to design control circuit while it can parallel flip working on previous idle.

The instruction unit has 16-bit wide length, separated to *opcode* and *operand* as shown in figure 4.

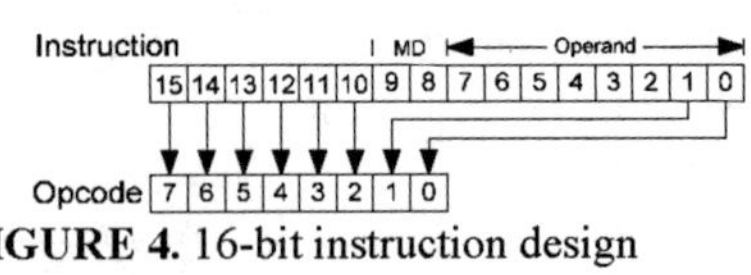

FIGURE 4. 16-bit instruction design

Where MD is memory addressing mode

00 → Immediate modes, 01 → register mode, 10 → direct addressing, 11 → indirect addressing

The processor instruction that used in this system has 16- bit instruction set; the instruction is loaded from cache to IR register. When 16-bit instruction has loaded into IR, IR responses to separate instruction in to two types, *opcode* and *operand*, the *opcode* is loaded to control unit section while operand is loaded to general purpose register or accumulator base on instruction type [5]. The modification benefit to generate a ***virtual instruction*** that occurred during interrupts even.

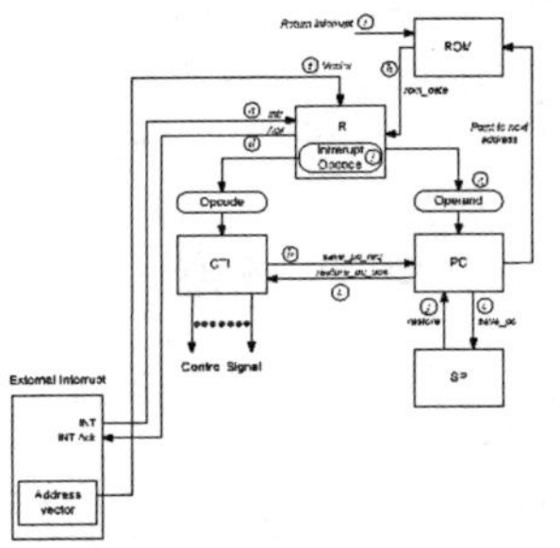

FIGURE 5. Interrupt design

The processor works following sequences, instruction is loaded from IR to PC and processes base on type of instruction. When EI flag is disabled, the interrupt signal does not respond until finished an execution of previous instruction. EI is enabled, if interrupt signal occur by assert *intr* signal by external devices , the IR releases cache connection and produces itself instruction, *virtual instruction* , this instruction is valid for interrupt instruction and separate instruction to two pieces, *opcode* and operand. *opcode* is responsible to save / restore signal to pc register , operand is first address that kept an interrupt service routine, when PC register receives *save* signal from control unit, It save current pointer to SP register and jump to next address that produced by virtual *opcode* as shown in figure 5.

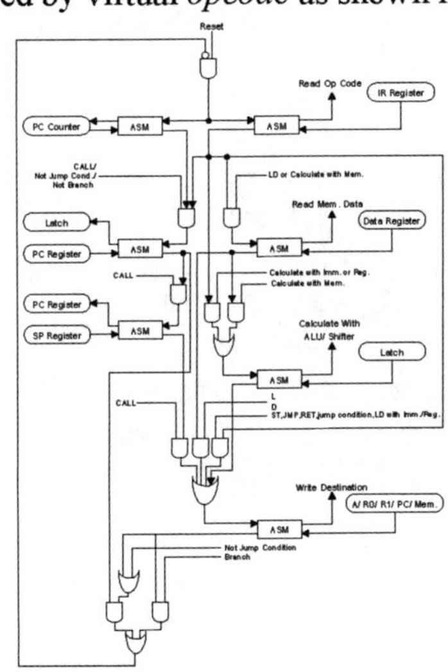

FIGURE 6. Control Unit Design

ASYNCHRONOUS SYSTEM BUS DESIGN

The bus design is multiplex (The data and address transition on the same lines). 8-bit with dual-rail encoded, transition by 4-phase signaling protocol [4]. The designed bus is started with design each component and group them together.

Bus components

The bus is designed as components, each component is separately designed then we group all components together to form a QDI bus as mentioned below.

Bus Interface

Bus interface is designed to interface with asynchronous processor. Due to the multiplex bus design, bus interface also handshakes with bus controller to control the sequence of data, address or spacer transition.

Bus Driver

Bus driver acts like an on/off switch to enable/disable transition on the bus. Designed with tri-state buffer and C-element as shown in figure 7 (the element that produces output 1 when all of inputs are 1, produce output 0 when all of inputs are 0, Otherwise its hold previous state).

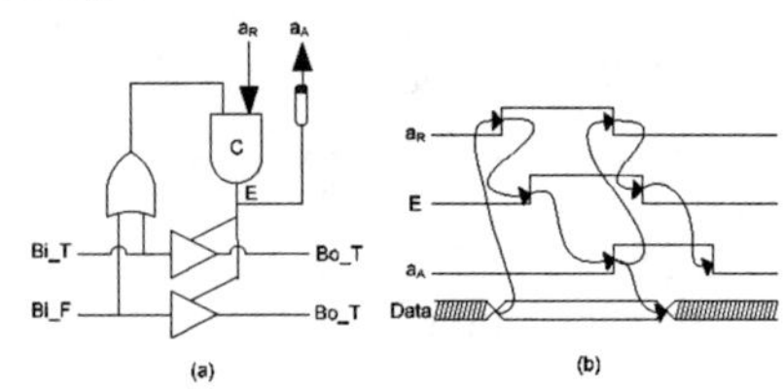

FIGURE 7. Bus Driver

a_R is request line from bus controller, when data arrive at Bi_T and Bi_F , C-element is enabling tri-state buffer to transfer signal to Bo_T and Bo_F , a_A is the acknowledge signal to the bus controller.

Bus lines

Due to high impedance state of Tri-state buffer, there were making some problem for asynchronous circuit. Need to improve to be zero value when it's high impedance by added weak inverter to each line on the bus [5] as shown in figure 8.

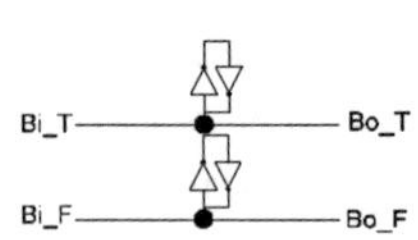

FIGURE 8. Bus lines

Bus Receiver

This component is used to receive data from bus signal and catches all signals before transfer to the destination, designed with C-element as shown in figure 9.

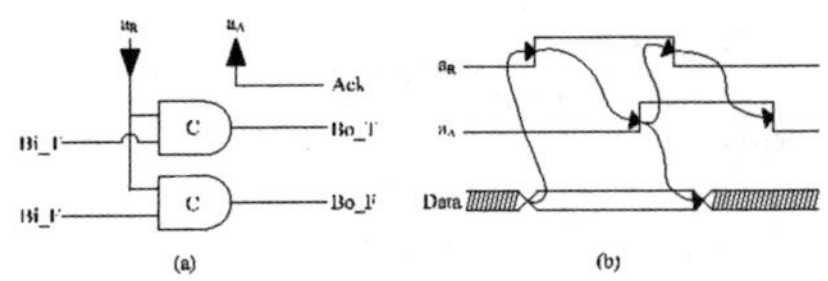

FIGURE 9. Bus Receiver

a_R is request line from bus controller, when data arrive at Bi_T and Bi_F , a_R is transition from 0 → 1 C-element enables signal transition to Bo_T and Bo_F, a_A is the acknowledge signal from synchronous interface to bus controller.

Bus Controller

Designed with STG (Signal Transition Graph, to control the synchronization on the bus, it's responsible to control the sequence of data, address, and spacer transition.

Transaction

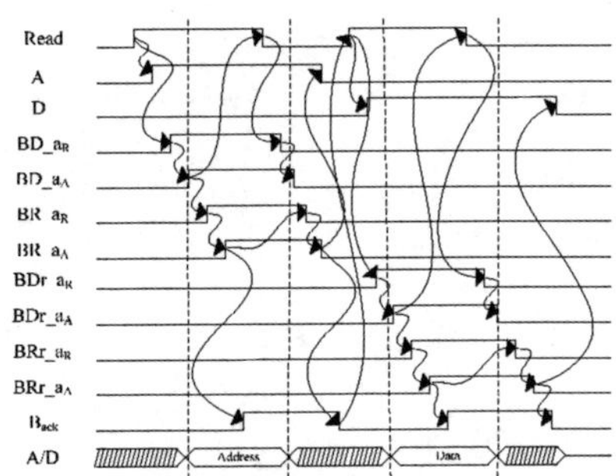

FIGURE 10. Read Transaction

For read transaction, read signal changes from 0→1 then signal A is the address signal changes from 0→1 indicate the address phase as shown in figure 10.

*BD_*a_R, *BD_*a_A, *BR_*a_R, *BR_*a_A are the control signals from bus controller change from 0→1 respectively. Indicate the flowing through of data form bus driver to bus receiver components. B_{ack} signal changes from 0→1 indicate the completion of data transfer on address phase. Then all signals change from 1→0 for resetting. After that signal D changes from 0→1 for data phase. *BDr_*a_R, *BDr_*a_A, *BRr_*a_R, *BRr_*a_A signals are the control signal for read data change from 0→1 respectively. B_{ack} changes from 0→1 indicates the completion of data transfer. After that all signals change from 1→0 for signals resetting on return to zero phase.

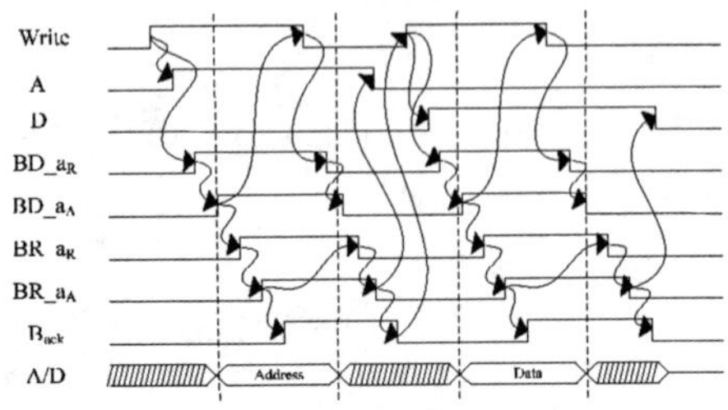

FIGURE 11. Write Transaction

For write transaction, write signal changes from 0→1 then signal A is the address signal changes from 0→1 indicate the address phase as shown in figure 11.

BD_a_R

BD_a_R, BD_a_A, BR_a_R, BR_a_A are the control signals from bus controller change from 0→1 respectively. Indicate the flowing through of data form bus driver to bus receiver components. B_{ack} changes from 0→1 indicates the completion of data transfer on address phase. Then all signals change from 1→0 for resetting. After that signal D changes from 0→1 for data phase.

BD_a_R, BD_a_A, BR_a_R, BR_a_A signals change from 0→1 respectively. B_{ack} changes from 0→1 indicates the completion of data transfer. After that all signal transition from 1→0 for signals resetting return to zero phase.

ASYNCHRONOUS DMA CONTROLLER DESIGN

The asynchronous processor is a master device on the asynchronous system bus, that it is able to initiate read and write transfers to all devices on the bus. The DMA controller is slave device, which is only able to response to read or write requests. The DMA controller is required to initiate read and write transfers to memory and also other I/O on the bus, DMA controller also required to be a master device [6].

DMA Function

The DMA controller waits for *dma_req* line to assert and take over the asynchronous system bus from asynchronous processor. It checks its internal registers to obtained details of the transfer. The DMA controller read data from the source, store it in internal buffer, and then write it out to the memory, until the transfer is complete. The DMA controller then release usage of asynchronous system bus and activate the IRQ line in order to indicate the asynchronous processor that transfer was complete. If any time the asynchronous processor requires usage of asynchronous system bus it is able to take priority over the DMA controller. The DMA controller check it has control of the bus at each stage, if it does not the controller wait for the bus become available, the asynchronous processor has finished using the bus [6].

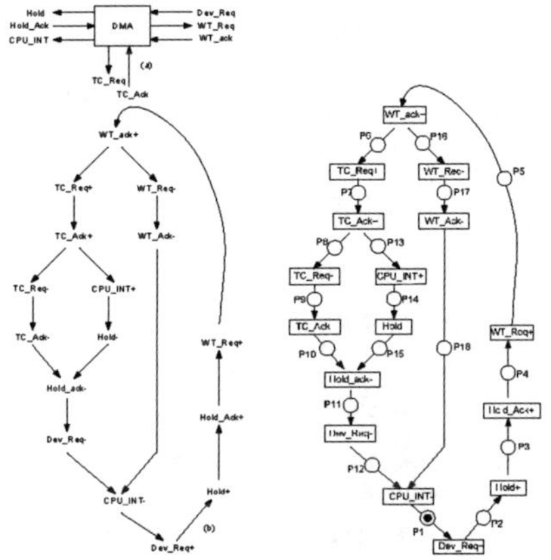

FIGURE 12. Asynchronous DMA controller Specification, STG and structural encoded

DMA Architecture

The internal architecture is very closely based on the classic DMA controller. The architecture proposed of the DMA controller split into above functional units. The most complex of these units is control unit, which consist of a large state transition graph as shown in figure 12.

DMA Controller Synthesizing

This section describes the synthesizing of asynchronous DMA controller based on our Template base method in [7]. The process begins with an encoding STG using Petri-net level in order to form a template STG. Then the projection to each non-input signals from an original STG is done.

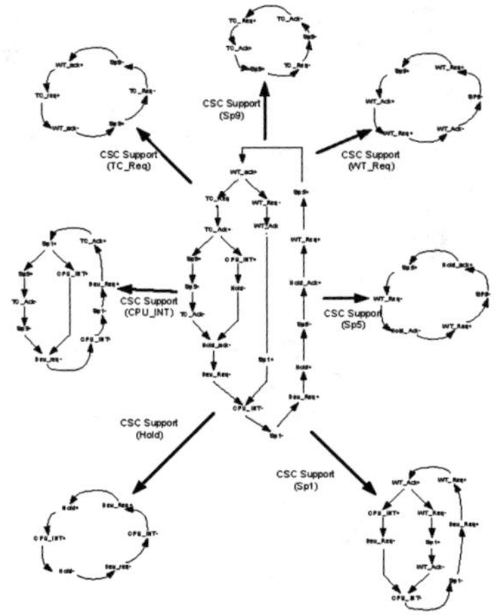

FIGURE 13. Reduced STG and CSC Support of DMA controller

After that, we trace the projection to smaller state space. If the small state space shows conflicts, we have to insert balance signals from template STG. Unbalance signals are inserted after in case the space still shows conflicts. Finally, we can get the STG with appropriate insertion points which is used to be projected for CSC support on each non-input signals as shown in figure 13.

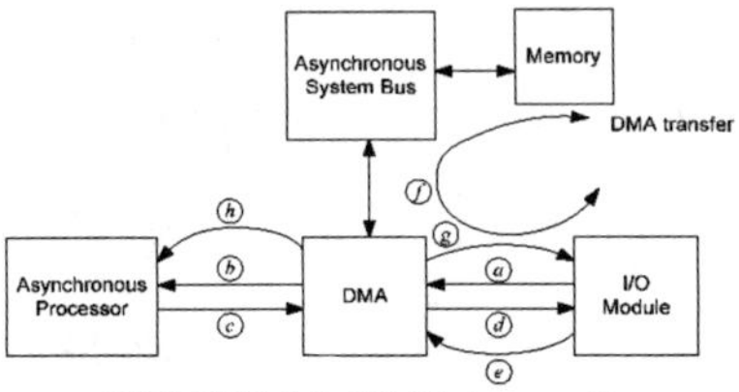

FIGURE 14. DMA transaction

SYNCHRONOUS AND ASYNCHRONOUS INTERFACING

This part is designed for communicating between Synchronous and Asynchronous section, stretchable clock is introduced [5].

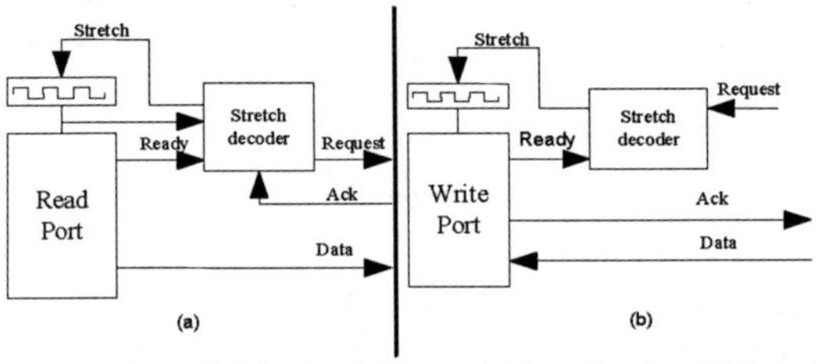

FIGURE 15. Stretchable clock for read (a) and write (b) interfacing

The clock control is generated using the handshaking signals and the internal state of the synchronous region, a block has either a read port or a write port. Read port provides data for an external element while a write port receive data into the module. In a 4-phase protocol, each module must receive two asynchronous transitions to complete a cycle. Each synchronous module must go through to two clock cycles to allow for two possible stretch signals as shown in figure 15.

FULL SYSTEM INTEGRATION

The full system integration is shown in figure 16, consider; only RAM and I/O section are synchronous section, otherwise it's fully asynchrony operation.

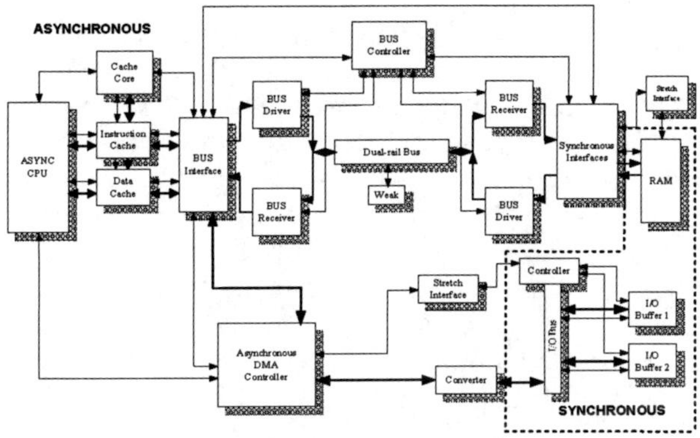

FIGURE 16. Full System Integration

EXPERIMENTAL RESULT

TABLE 1. Timing Summary

TIMING SUMMARY @- Memory 100 MHz							
INSTRUCTION TYPE	**TIME**	**INSTRUCTION TYPE**	**TIME**	**INSTRUCTION TYPE**	**TIME**	**INSTRUCTION TYPE**	**TIME**
	(ns)		**(ns)**		**(ns)**		**(ns)**
Setup Time	32	LD (Direct)	65	AND (immediate)	55	SUB (immediate)	90
Interrupt Response Time	42	LD (Indirect)	58	AND (Register)	60	SUB (Register)	100
Bus Read Cycle	16	ST (Register)	39	AND (Direct)	80	SUB (Direct)	144
Bus Write Cycle	15	ST (Direct)	60	AND (Indirect)	80	SUB (Indirect)	130
DMA Read Cycle	28	ST (Indirect)	57	OR (Immediate)	80	ADD (Immediate)	90
DMA Write Cycle	30	JMP	45	OR (Register)	80	ADD (Register)	110
LD (Immediate)	42	CALL	50	OR(Direct)	120	ADD (Direct)	130
LD (Register)	40	RET	45	OR (Indirect)	120	ADD (Indirect)	130
		RETI	43				

TABLE 2. Device Utilization Summary

Device Utilization Summary			
	Used	**Available**	**Utilization**
Slice Flip Flops	2	9312	1%
4-inputs LUTs	1,662	9312	17%
Occupied Slices	880	4,656	18%
Bonded IOBs	82	232	35%
Block RAMs	2	20	10%
GCLKs	1	24	4%
Gate count	141,063		

CONCLUDING REMARK

This paper presents FPGA implementation of 8-bit almost completely asynchronous computer. The processor employs an asynchronous system bus, asynchronous cache and asynchronous DMA controller. All designed circuits are high level specification, and it operates correctly under FPGA combinational logic propagation delays, the system uses much of function generator on FPGA, Only 2 flip-flop slices are consumed to prepare an interfacing with synchronous one. The designed has been implemented on Xilinx Spartan-3E FPGA no. 3S500EFG320, partitioning each module to prevent place and routing conflicts.

ACKNOWLEDGMENTS

The research is supported by TJTTP-OECF (Thailand - Japan Technology Transfer Project – Japanese Overseas Economic Cooperation Fund)

REFERENCES

1. Davis. A, Nowick.S.M "An Introduction to Asynchronous Circuit Design" , 1997
2. T.Nanya., Y.Ueno., H.Kagotani., M.Kuwako., A.Takamura. TITAC : Design of a Quasi-Delay-Insensitive Microprocessor. IEEE Design & Test of Computers., Vol.11., No.2, pp.50-63, Summer 1994.
3. P.Ruangsilsap. "Design of 8-bit scalable-delay-insensitive microprocessor using FPGA". Master thesis, Chlalongkorn University, Bangkok, Thailand, 2001
4. J. Carmona and J. Cortadella. State encoding of large asynchronous controllers. *In* Proc. ACM/IEEE Design Automation Conference, *pages 939-944*, July 2006.
5. S.Sudeng, A.Thongtak:" A Design of System Bus for Asynchronous Circuits" International Technical Conference in Circuits/Systems Computer and Communication (ITC-CSCC2006), Pang Suan Kaew hotel, Chiangmai, Thailand on July 10-13, 2006.
6. S.Sudeng, A.Thongtak:" Asynchronous System Bus Enhancement by Interrupt and DMA Technique" International Annual Conference in Electrical Engineering/ Electronics, Computer, Telecommunications and Information Technology (ECTI-CON2007), Mae Fah Luang University, Chiangrai, Thailand on May 9-12, 2007.
7. S.Sudeng, A.Thongtak:" Signal Transition Graph Based Logic Synthesis for Asynchronous Control Circuits Using Template Based Method" IEEE Tencon 2007, IEEE Region 10 conference (IEEE Tencon2007), Taipei International Convention center, Taipei, Taiwan on Oct 30- Nov 2, 2007.

Exploitation of Multi-beam Directional Antennas for a Wireless TDMA/FDD MAC

Sedat ATMACA[a], Celal CEKEN[b] and Ismail ERTURK[c]

Kocaeli University, Technical Education Faculty
Electronics and Computer Education Department
41380 Kocaeli, Turkey[a,b,c]
{sedatatmaca[a], cceken[b], erturk[c]}@kou.edu.tr

Abstract. The effects of the multi-beam directional antennas on the performance of a new wireless TDMA/FDD MAC system are presented. Directional antennas intrinsically enable development of the SDMA systems and allow transmitting and receiving signals simultaneously at the same time slot. Employing a dynamic slot allocation table at a base station with 4 or 8 sector directional antennas and holding the wireless terminals' location information, a new SDMA/TDMA/FDD frame structure has been developed for wireless communications. The simulation studies realized using OPNET Modeler show that the proposed SDMA/TDMA/FDD system has substantially increased the traditional TDMA/FDD system capacity and provides 1.37 to 4 times better mean delay results when the number of users is increased from 4 to 32 under the same load in the wireless network models.

Keywords: SDMA/TDMA, Directional Antennas, MAC.
PACS: 07.50.Qx, 07.50.-e.

INTRODUCTION

Together with the developments in high performance wireless computers and other mobile devices, the importance of wireless/mobile data communication has been increased, and the use of directional antennas has gained more importance to enhance the network capacity. Recently, there has been a remarkable interest in developing MAC protocols for wireless networks which are equipped with directional antennas.

Smart antennas generally known as directional antennas are one of the most promising technologies that enable a higher capacity in wireless networks by effectively reducing multi-path and co-channel interference and allowing multi-sector implementations. This is achieved by focusing the radiation pattern only in the desired direction. Smart antennas employ a set of radiating elements arranged in the form of an array. The signals from these elements are combined to form a steerable or switchable beam pattern that follows the desired user or a group of users. In a smart antenna system the arrays by themselves are not smart; it is the digital signal processing that makes them smart. The process of combining the signals and the focusing the radiation pattern in particular direction is often referred to as digital beamforming [1 and 2].

CP1007, *Current Themes in Engineering Technologies*, edited by S.-I. Ao, M. A. Amouzegar, and S.-S. Chen

In this work presented, a new MAC based on SDMA/TDMA/FDD (Space Division Multiple Access / Time Division Multiple Access / Frequency Division Duplexing) using directional antennas has been designed, simulated and compared to a classical TDMA/FDD MAC counterpart employing omni-directional antennas. The metrics used for performance analyses are the end-to-end delay and throughput. In the proposed MAC, WTs (Wireless Terminals) use omni-directional antennas and communicate through a BS (Base Station) which is equipped with multi-beam directional antennas containing beamforming modules for both receiving and transmitting, each of which is capable of directing a beam at an intended sector in the space. The system model uses FDD as the dublexing technique and employs a DSAT (Dynamic Slot Allocation Table) at the BS that is especially utilized for determining to which WT the slot will be assigned in the uplink (WT to BS) direction. By means of the DSAT at the BS, an SDMA/TDMA/FDD frame structure is constructed. This approach ensures an increased wireless system capacity, meanwhile reducing end-to-end delays.

Section 2 begins with a brief explanation of directional antenna systems, multi-beam directional antennas and antenna models used. Section 3 provides an overview of the common TDMA/FDD MAC protocols utilizing directional antennas, which is used for performance comparisons. Overall properties and design stages of the proposed MAC protocol together with algorithms developed are explained in detail in Section 4. Section 5 includes an example wireless networking scenario employing the proposed MAC technique, which has been modeled and simulated under different traffic loads and varying number of users, with OPNET Modeler including the Radio Module. The simulation results obtained are compared with those of another TDMA/FDD MAC based model utilizing omni directional antennas under the same networking conditions as proposed model. And, the last section gives our summary with final remarks.

DIRECTIONAL ANTENNA SYSTEMS

Recently, the need for providing high quality wireless access and the great demand on high speed wireless links have increased dramatically. Consequently researchers have been motivated to enhance the wireless network capacity to meet the growing subscriber and different application requirements. As a result, new sources have been focused on increasing the overall network performance. The use of directional antennas, especially to achieve the throughput expansions and to reduce the communication delays, is of much importance [2, 3, 4, 5 and 6].

There are several techniques of implementing directional antennas. Basically, they are described in two categories: switched beam systems and adaptive array systems [2]. The switched beam antenna system has a switching mechanism enabling it to select and then to switch the right beam which gives the best directivity and gain to the terminal under consideration. This technique does not steer or scan the beam in the direction of the desired signal. Switched beam systems can be divided into two groups: single beam and multi-beam directional antennas. In single beam directional antennas, only one beam is active at a given time. Simultaneous transmissions are not allowed as there is only one transceiver in this system. On the other hand, in multi-beam directional antennas, there are several beam patterns and each beam is directed to a different user or a sector. Therefore, simultaneous transmissions are allowed at the same time and frequency. As a result SDMA is achieved as utilized in this research study. Here, the number of beams is equal to the number of transceivers.

The adaptive array system tracks the mobile user continuously by steering the main beam towards the user and at the same time forming nulls in the directions of the interfering signal. This antenna system is out of the scope of the works carried out and presented here, and therefore is not explained in detail.

WIRELESS MAC TECHNIQUES USING DIRECTIONAL ANTENNAS

In the last few years, directional antennas have been extensively considered for use in both wireless centralized and ad-hoc network architectures. High costs and complexity of the WTs with directional antenna systems lead to difficulties in implementation of ad-hoc network architectures. On the other hand, a centralized architecture, where a base station has directional antennas while the WTs are equipped with omni-directional antennas, is considered more suitable and reasonable than the ac-hoc network architecture. Considering this point, the work presented exploits the wireless centralized architecture with the idea of further development including quality of service support for real-time multimedia applications [6 and 7].

In [8], multiple beam adaptive arrays (MBAA) were used to improve the performance of a slotted ALOHA packet radio network. When used in a packet radio system, an MBAA can successfully receive two or more packets simultaneously. Each beam captures a different packet by automatically pointing its pattern toward one packet while nulling other contending packets. As a result, the capacity of the system can be relatively increased compared to the conventional ALOHA due to a remarkable reduction in the effect of collisions. Differing mainly from this approach, the proposed MAC using SDMA/TDMA/FDD in this work overcomes collisions resulting in high system performance.

In [9], a MAC protocol was proposed to achieve throughput multiplication of wireless LANs, employing a slot reservation mechanism combined with SDMA. The most important benefit of this protocol is that it can adaptively adjust its partition time-space slots based on the network traffic to optimize the throughput and delay performance. As designed for mobile terminals, it exploits an adaptive array antenna with a complex and expensive signal processing unit in the BS. It differs from the proposed simple and easy MAC approach to implement system with sector antennas in that the latter is much more appropriate and cost-effective for stationary wireless terminals considering most of their applications, which are hardly mobile in time, require only fixed-wireless services, also providing foundations for quality of service supported real-time multimedia applications.

The performance of slotted ALOHA when the BS receiver uses multi-beam directional antennas capable of steering the beams selectively on smaller sectors was analyzed in [10]. Under high load conditions, steered beams with long beam service times offer better performance, whereas under light load conditions, static coverage patterns are better. The use of multi-beam directional antennas in the BS to realize sectors in this research work is partially benefited in the proposed MAC protocol.

A TDD-SDMA/TDMA system with a multi-slot assignment for downlink in packet-switched wireless network was presented in [11]. It could assign more than one time slots for downlink to each terminal so as to deal with the asymmetric traffic in packet-switched services. In the proposed system, an efficient algorithm was reported to decide the priority of time slots for downlink, by which the maximum of potential transmission rate for downlink could be obtained.

A NEW TDMA/FDD MAC PROTOCOL USING DIRECTIONAL ANTENNAS

In the proposed MAC protocol, utilizing the TDMA/FDD MAC approach presented in [12, 16], TDMA is preferred as a multiplexing technique due to its simplicity, easy implementation and appropriateness for further development for real-time multimedia applications requiring quality of service support [9 and 13]. By virtue of TDMA, radio spectrum is divided into time slots that can be assigned to different connections where a user can send data only in its own dedicated time-slot. Due to the FDD duplexing technique utilized in the proposed TDMA MAC protocol, two distinct carrier frequencies are used for the uplink and downlink channels. For the system under consideration, the BS is equipped with an N-element antenna array for uplink and downlink communications. In this way, the space has been divided into N sectors, thus SDMA has been exploited. There are also M wireless terminals which are equipped with omni-directional antennas [13, 14 and 15].

Wireless terminals are not able to communicate directly with one another in the proposed method as opposed to the Ad-Hoc networks. They communicate through a centralized base station. When a WT needs to communicate with any other terminal, initially it asks for a transmission time slot from the BS (using the CT slots shown in Figure 1). Having its time-slot reserved (using the CB slots), the WT starts transferring data.

Data packets generated in a wireless terminal are first destined to the BS. The BS is responsible for queuing these incoming packets and arranging their transmission, using its DSAT algorithms, to the destination WTs in the reserved time slot.

If there is not any slot available for a request, a connection can not be established and the WT may later attempt to connect again sending another connection request to the BS. Figure 1 shows the slotted frame structure of the proposed MAC protocol.

The proposed MAC protocol is divided into two main complementary parts operating at the WTs with an omni-directional antenna and BS with directional antennas. Data transmissions can be provided when these two MAC parts work synchronized. The following sub-sections explain in detail all functions and algorithms developed for both parts together with their translation into the modeling and simulation environment in the OPNET Modeler.

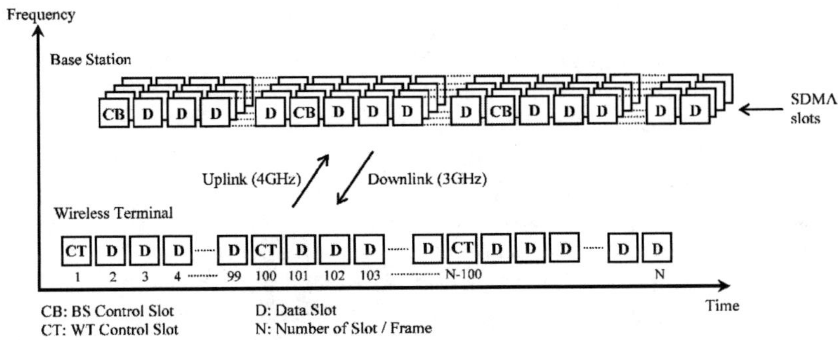

FIGURE 1. SDMA/TDMA/FDD frame structure of the proposed MAC protocol.

The WT MAC Model

The proposed MAC protocol includes three main processes to implement the WT functions and algorithms. The first is to request a connection establishment/termination from the BS, the second is to obtain its own time slot from the BS utilizing the DSAT and the last is to send data in the allocated time slot.

Before sending/receiving data to/from other terminals, a terminal must first inform the BS about its time slot requirement. In this regard, it creates a control packet called *con_req_packet* (Figure 2.a) containing its own terminal address and sector number indicating which beam needs to be activated by the BS to communicate with this terminal. Then this packet is sent to the BS in the first available uplink control slot (CT). After a connection is established, the BS knows the WT's sector information recorded into the 'SectorNo' field in the DSAT (Figures 2 and 5), and the WT is acknowledged about its assigned time-slot to transfer data (Figure 2.b and 2.c). The WT creates a control packet called *con_end_packet* to terminate the active connection and sends it to the BS in the first available uplink CT again (Figure 2.d). After that, the BS releases this data slot (D) for use by other WTs.

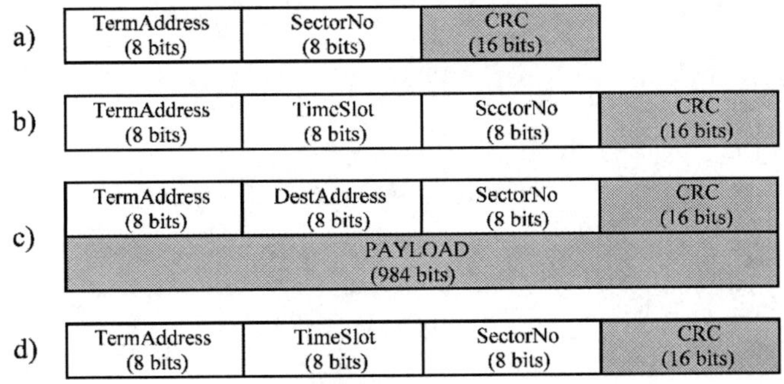

FIGURE 2. a) Connection request packet (*con_req_packet*),
b) Connection response packet (*con_res_packet*),
c) SDMA/TDMA data packet (*data_packet*),

d) Connection terminate packet (*con_end_packet*).

The WT node model designed using OPNET Modeler is shown in Figure 3.a. The *SOURCE* module used in the WT is responsible for packet generation according to the packet size and arrival time parameters defined. The *SINK* module collects statistics of the arrived packets and then destroys them.

The *MAC* module manages the allocation of existing bandwidth shared by all WTs. The physical layer components used in the model are *rx*, *rx_ant*, *tx* and *tx_ant*. They provide the interface between the node module and air (Figure 3.a).

In Figure 3.b, the process model of the *MAC* module used in WT node model is illustrated. The *con_req* state machine creates connection request packets and sends them to the BS. Using the *from_src* state machine, data packets received from the *SOURCE* are sent to destination (through the send state machine) in the connection time slot allocated. The *req_resp* state machine obtains the connection time slot that is assigned by the BS using its DSAT scheduling algorithm. The *con_end* state machine creates the connection termination packet and sends it to the BS. The *from_rx* state machine delivers any arrived packets destined to this WT through its *MAC* layer.

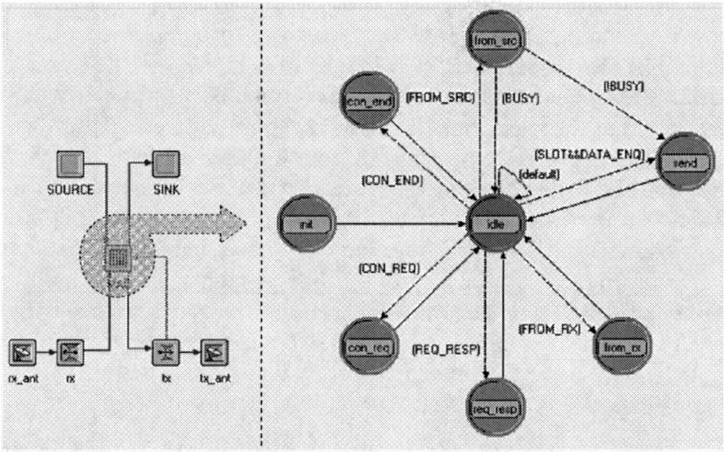

a) b)

FIGURE 3. a) The WT node model, **b)** The WT MAC module process model.

The BS MAC Model

The BS node model of the proposed MAC protocol includes three main modules. These are namely the *Antenna Controller*, the *Antenna Pointer* and the *BS MAC*. The *Antenna Controller* module gets packets from its antennas each directed to a different sector in the space and transfers them to the *BS MAC* module. It also gets packets from the *BS MAC* module, and having extracted 'SectorNo' information, it transmits the packets to the related sector. *Antenna Controller* module has also capabilities getting and sending multiple packets simultaneously.

Antenna Pointer module has predefined individual beams and each beam covers a specified sector in the space. These beams are assumed to have almost ideal shape for receiving and transmitting in the desired direction and rejection in the other directions. For transmission, this module points its main beams to the desired directions in the space. In the antenna module, 4 beams and 8 beams are used. When used a 4-beam antenna, each of these 4 beams is directed to a different sector: North-East, North-West, South-East and South-West. For reception, antenna controller module monitors the signal levels on all beams after that the packets having the signal with the power over the predefined threshold level are forwarded to the *BS MAC*, while the others are discarded as interference.

The *BS MAC* includes three main processes. These are namely, assigning a time slot for a connection using its DSAT scheduling algorithm, forwarding any arrived data packets to their destinations and terminating any active connection. The most important and primary function of the proposed MAC protocol is the efficient management of the DSAT.

Figure 4.a shows the BS node model realized using OPNET Modeler. The MAC module functions in the BS node are used to allocate available time slots to the WTs requesting connection and to downlink actual data packets to the intended WTs.

The *BS MAC* module process model designed is given in Figure 4.b. The *from_rx* state machine delivers arriving packets to the next state machine according to the packet formats. The *con_req* state machine executes the scheduling algorithm managing the DSAT. The *data* state machine delivers data packets and finally the *con_end* state machine handles connection termination requests.

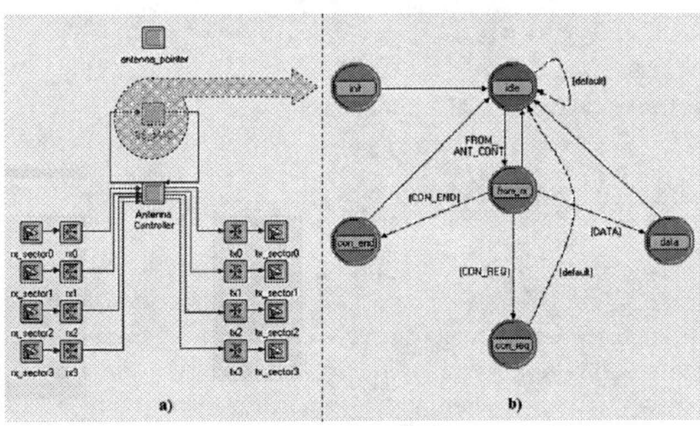

FIGURE 4. a) BS node model, **b)** BS MAC module process model.

The DSAT and its scheduling algorithm running in the BS are the most vital parts of the proposed MAC protocol. The structure of DSAT is shown in Figure 5.

For a new connection request, a WT sends a *con_req_packet* to the BS. After that the BS looks up its DSAT table whether there is any available time slot to be allocated. If so, the BS assigns a time slot to the WT using the WTs position information (SectorNo) contained in the *con_req_packet* and updates the DSAT accordingly. Then BS creates *con_res_packet* and sends it to the destination WT. A connection may not be established if there is not any available empty time slot. The WT repeats this connection request procedure until it succeeds.

For a connection termination request, the BS gets *con_end packet* from the WT and extracts the time slot number from it. It then releases this time slot for a new connection, updating the DSAT. Finally, a *con_end_packet* is created and sent to the related WT.

For a data packet, the BS extracts the sector number information from the 'SectorNo' field and determines in which sector the related WT is located. After that the beam used for this sector is activated to transfer data.

Certain number of slots (i.e. 1, 100, 200, ..., N-100) in the DSAT called control slots (CT) are reserved for connection request and connection termination packets. When a WT wants to send a control packet for connection request or termination, it uses the first empty control slot.

Time slots →

Number of Sectors ↓	1	2	3	4	5	6	7	9	10	......	N
1	1	4	14	16	NU	NU	NU	NU	NU	......	CB
2	2	3	13	15	NU	NU	NU	NU	NU	......	CB
3	5	7	9	11	NU	NU	NU	NU	NU	......	CB
⋮	⋮	⋮	⋮	⋮	⋮	⋮	⋮	⋮	⋮		⋮
M	6	8	10	12	NU	NU	NU	NU	NU	......	CB

NU : None Used Slot, CB : Base Station Control Slot.
M : Number of Sectors, N: Number of Time Slots.

FIGURE 5. Structure of the Dynamic Slot Allocation Table (DSAT).

The BS Directional Antenna Model

This study considers WTs operating in omni-directional mode and the BS in directional mode. The BS has a switched-beam directional antenna comprised of multiple beam antenna arrays for both reception and transmission. Each antenna array with M elements forms non-overlapping sectors covering an angle of *360/M* degrees so as to collectively cover the entire space as shown in Figure 6. The beam shape is assumed as conical. In the computer model developed, OPNET Antenna Pattern Editor is used to model the antenna pattern. In this editor, the main beam and side beam coverage range are set by the transmitting power and the antenna gain. Given a gain value g_m, the antenna pattern for this gain consists of main beam of bandwidth θ_m, and a side beam of gain g_s of bandwidth $(2\pi - \theta_m)$. We have used two different antenna patterns. The first one has 4 beams, conical shape of beamwidth $2\pi/4$ radians and gain of 10 dBi for main beam and -20 dBi for the others. The other has 8 beams, conical shape of beamwidth $2\pi/8$ radians and gain of 10 dBi for main beam and -20 dBi for the others. In Figure 6, the antenna pattern with $\pi/4$ radians bandwidth is shown.

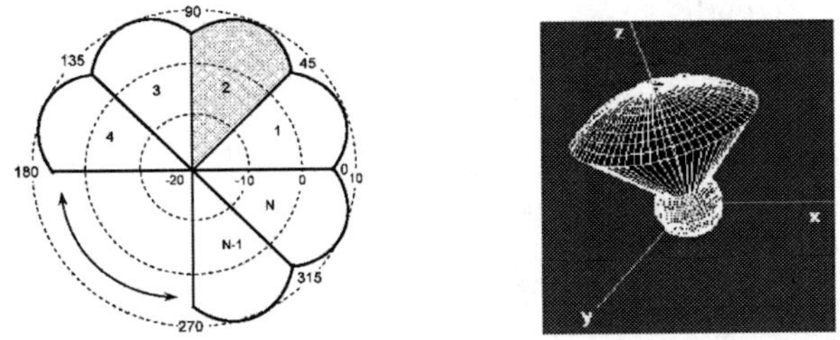

a) b)

FIGURE 6. a) Antenna model used in the base station MAC with 8-Beam, **b)** Antenna model designed in OPNET Antenna Pattern Editor.

COMPUTER SIMULATION OF THE PROPOSED APPROACH

The stationary (fixed wireless) WTs and the BS in the example scenario implemented using OPNET Modeler (Figure 7), employ the proposed SDMA/TDMA/FDD based MAC protocol, which has been explained in the previous section, to communicate with each other in the same wireless environment. Diameter of the cell which constructs the network topology is chosen 10 kilometers. The simulation models consider firstly the TDMA/FDD with 4-Beam directional antennas, secondly TDMA/FDD with 8-Beam directional antennas, and finally TDMA/FDD with an omni-directional antenna (where spatial reuse is not exploited) for consistent performance comparisons, under the same networking conditions. The simulation environment is divided into 4 equal sectors for 4-Beam and 8 equal sectors for 8-Beam directional antenna models. Each BS antenna is directed towards one of these sectors. The sectors contain equal number of WTs to be able to compare the simulation results fairly.

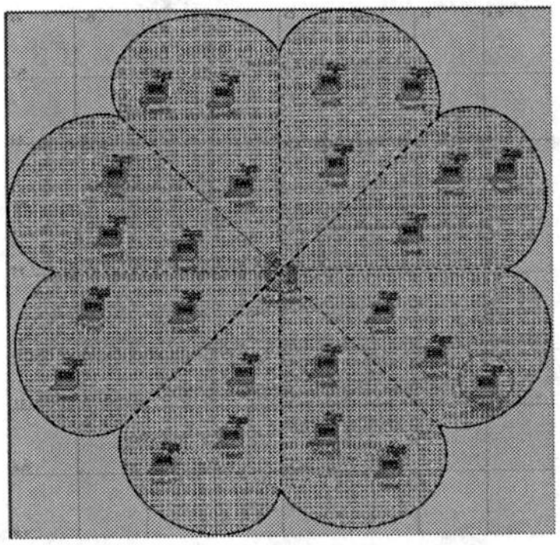

FIGURE 7. The simulation scenario of the proposed MAC with 8-Beam directional antennas.

In the example models shown in Figure 7, for the TDMA/FDD with 8-Beam directional antennas, there are 24 WTs on which data are generated, transferred and received. The data traffic introduced to the network by any WT is randomly destined to another WT. The simulations are repeated for both different number of users and increasing offered loads.

The Simulation Results and Performance Evaluation

Simulation results of both a classical TDMA/FDD model and SDMA/TDMA/FDD models with 4-Beam and 8-Beam DAs described above are presented under varying network load conditions followed by performance comparisons and analysis. In the simulation environment a free space channel propagation model is used. To avoid the transient effects, the simulation statistics are flushed after approximately 10 seconds. For the simulation models, a slot length of 200 μseconds which has been determined considering that a 25 Mb/s data rate has been chosen. The simulation parameters used are given in Table 1.

TABLE 1. Simulation parameters.

Uplink/Downlink Data Rate	25 Mbps
*Data Packet Size**	1024 Bits
TDMA frame length	200 μsn
Frequency Band	Uplink = 3 GHz and Downlink = 4 GHz
Transmitter Power	BS = 10^{-2} W and WTs =10^{-2} W
Modulation Scheme	QPSK
Directional Gain	10 dBi (main beam) and -20 dBi (side beams)
Packet Reception Power Threshold	$1.2x10^{-12}$ W
Simulation Time	1 Hour
**Generated using Exponential Distribution Function Exp(Mean).*	

The performance metrics concerned in this research work are the average end to end delay (EED), normalized average end to end delay and throughput.

In Figure 8, average EED results of the proposed models as a function of the simulation time are presented. Considering the number of users is 16 and each introducing 10000 Bytes/s data traffic to the network, the EED results vary between 0.46 and 0.41 seconds for the classical TDMA/FDD model. This figure has been improved for the worst case (i.e., corresponding to the simulation time of 1800 seconds) to 0.18 seconds and 0.14 seconds using the proposed approach with 4-Beam and 8-Beam DAs, respectively. The graphics also imply that the average EED results of the TDMA/FDD model with 8-Beam DAs are not relatively better, considering the trivial improvement in the results of the TDMA/FDD model with 4-Beam DAs compared to those of the classical TDMA/FDD model.

Figure 9 shows the average EED results for the proposed MAC models with 4 and 8-Beam DAs, normalized with those of the standard TDMA/FDD MAC based counterpart as a function of the increasing number of users. As seen from the figure, the proposed TDMA/FDD model with 4-Beam and 8-Beam DAs with different number of users outperform the classical TDMA/FDD model. The simulation models employing the TDMA/FDD with 4-Beam and 8-Beam DAs provide up to 4 times and 6 times better average EED results, respectively. In addition, the more the number of users is, the more the proposed MAC presents improved results.

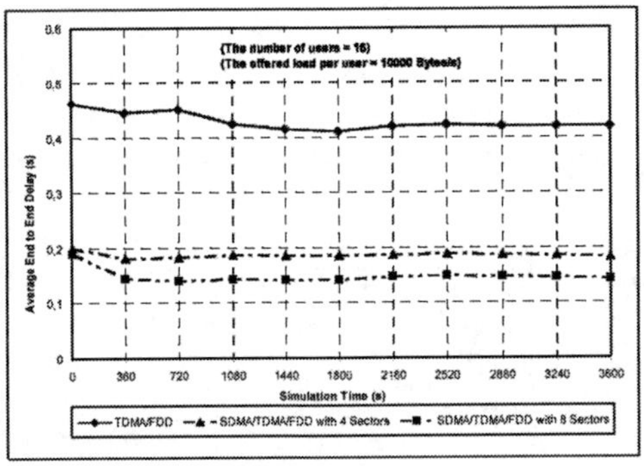

FIGURE 8. Average EED results for the proposed MAC models.

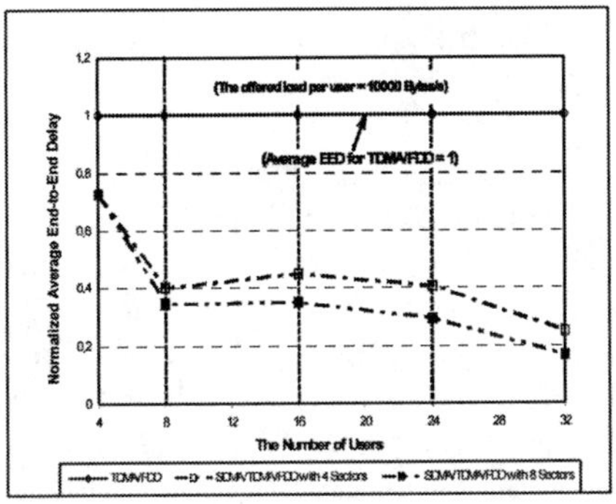

FIGURE 9. Normalized average EED results for the proposed MAC models.

In Figure 10, the system throughput results for the proposed and classical TDMA/FDD MAC models are presented. As the number of users is increased, for the TDMA/FDD MAC with 4-Beam and 8-Beam DAs, the system throughput rises linearly while saturates for that of the classical approach when the number of users is over 16. Referring to the Figures 8 and 9, these results also clarify the reasons both for the increasing average EEDs for the classical TDMA/FDD MAC model with above 16 users and for better average EEDs for the models using the proposed approach.

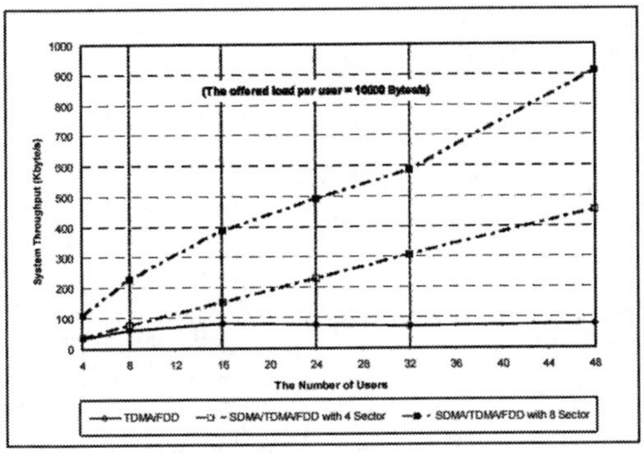

FIGURE 10. System throughput for the proposed and classical TDMA/FDD MAC models.

CONCLUSIONS

This work introduces an SDMA/TDMA/FDD based MAC protocol proposal for wireless data communications. Its design stages and employment in different wireless networking systems with 4-Beam and 8-Beam DAs are outlined, and a detailed performance analysis compared with a classical TDMA/FDD based MAC counterpart is presented. Employing the proposed DSAT algorithms, the BS allows a time slot to be allocated for wireless users in different sectors. As a result of separating the space into different areas using switched-beam directional antennas, the user capacity has been enhanced depending on the number of sectors for the models using the proposed MAC. Thus, multiple users in different sectors are able to utilize the same time slot to access the BS simultaneously (for uplink) and after that the BS directs the data traffic to the related sectors (for downlink).

The simulation results obtained conclude that especially system capacity, average EED and system throughput of the proposed SDMA/TDMA/FDD based MAC models are much better than those of the classical TDMA/FDD based MAC model, as a consequence of new algorithms introduced in the BS. The simulation models employing the proposed TDMA/FDD MAC with 4-Beam and 8-Beam DAs improve the average EED results up to 4 times and 6 times, respectively. As the number of users is increased, for the TDMA/FDD MAC with 4-Beam and 8-Beam DAs, the system throughput rises linearly while saturates for that of the classical approach when the number of users is over 16. The comparative simulation studies realized verify that the use of proposed MAC substantially increases the traditional TDMA/FDD system capacity, which is one of the most crucial issues in wireless networking environments.

REFERENCES

1. E. Ulukan, Wireless LANS with Smart Antennas, M.Sc. Thesis, University of Sabanci, 2005.
2. Lehne, P. H., Pettersen, M.: An overview of smart antenna technology for mobile communications systems, *IEEE Communications Surveys*, Vol. 2, No. 4, (1999) 2–13.
3. Haardt, M., Spencer Q.: Smart antennas for wireless communications beyond the third generation, *Commputer Communications,* Vol. 26, (2002) 41–45.
4. Chryssomallis, M.: Smart Antennas, *IEEE Antennas and Propagation Magazine*, Vol. 2, No.3, (2000) 129–136.
5. Stine, J. A.: Exploiting smart antennas in wireless mesh networks using contention access, *IEEE Wireless Communications*, Vol. 13, No. 2, (2006) 38–49.
6. Winters, J. H.: Smart antenna techniques and their application to wireless ad hoc networks, *IEEE Wireless Communications*, Vol. 13, No. 4, (2006) 77–83.
7. Delicado, F. M., Cuenca, P., Barbosa L. O.: QoS mechanisms for multimedia communications over TDMA/TDD WLANs, *Computer Communications*, Vol. 29, (2006) 2721–2735.
8. Ward, J., Compton, R. T.: High throughput slotted ALOHA packet radio networks with adaptive arrays, *IEEE Transaction on Communication*, Vol. 41, (1993) 460–470.
9. Xu, G., Li, S. Q.: Throughput multiplication of wireless LANs for multimedia services: SDMA protocol design, *IEEE Global Telecommunications Conference*, Vol. 3, (1994) 1326–1332.
10. Chockalingamt, A., Rao, R. R.: MAC layer performance with steerable multibeam antenna arrays, *IEEE PIMRC,* (1998) 973–977.
11. Jial, Y., Hara, Y., Hara, S.: Performance of TDD-SDMA/TDMA system with multi-slot assignment in asymmetric traffic wireless network, *IEEE PIMRC*, (2002) 1–5.
12. Ceken, C., Erturk, I., Bayilmis C.: A new MAC protocol design for WATM networks, *Lecture Notes in Computer Science*, LNCS 3261, (2004) 564–574.
13. Vergados, D. D., Vergados, D. J., Douligeris C.: QoS-aware TDMA for end-to-end traffic scheduling in ad hoc networks, *IEEE Wireless Communications*, (2006) 68–74.
14. Shad, F., Todd, T. D.: Capacity of S-ALOHA protocols using a smart antenna at the base station, *The 9th IEEE Symposium on Indoor and Mobile Radio Communications*, Vol. 3, (1998) 1101–1105.
15. Oikonomoua, K., Proniosa, N., Stavrakakis I.: Performance analysis of topology-unaware TDMA MAC schemes for ad-hoc networks with topology control, *Computer Communications*, Vol. 28 (2005) 313–324.
16. Atmaca, S., Ceken, C., Erturk, I.: Improving wireless TDMA/FDD MAC performance using multi-beam directional antennas", *ICCST 2007*, San Francisco, USA, 24–26 October, 2007.

Traffic Adaptive Energy Efficient and Low Latency Medium Access Control for Wireless Sensor Networks

Rajesh Yadav[a], Shirshu Varma[b], N. Malaviya[c]

[a] *LRDE, Defense R&D Organization, Bangalore, India*
[b] *Indian Institute of Information Technology, Allahabad, India*
[c] *Institute of Engineering & Technology, Lucknow, India*

Abstract. Medium access control for wireless sensor networks has been a very active research area in the recent years. The traditional wireless medium access control protocol such as IEEE 802.11 is not suitable for the sensor network application because these are battery powered. The recharging of these sensor nodes is expensive and also not possible. The most of the literature in the medium access for the sensor network focuses on the energy efficiency. The proposed MAC protocol solves the energy inefficiency caused by idle listening, control packet overhead and overhearing taking nodes latency into consideration based on the network traffic. Simulation experiments have been performed to demonstrate the effectiveness of the proposed approach. The validation of the simulation results of the proposed MAC has been done by comparing it with the analytical model. This protocol has been simulated in Network Simulator ns-2.

Keywords: Wireless Sensor Network, Medium Access Control, Energy Efficiency, MAC Protocols

INTRODUCTION

Wireless Sensor Networks (WSNs) have become very popular in this current decade (2000-2010) due to their wide range of applications in different fields such as military and civilian use. These WSNs can be used for different purposes such as target tracking, intrusion detection, wildlife habitat monitoring, climate control and disaster management [1]. A typical node in the WSN consists of a sensor, embedded processor, moderate amount of memory and transmitter/receiver circuitry. These sensor nodes are normally battery powered and they coordinate among them selves to perform a common task.

Unlike standard wireless network, these WSNs have severe resource constrains and energy conservation is very essential. The radio in the sensor node consumes a significant amount of energy. Substantial research has been done on the design of low power electronic devices in order to reduce energy consumption of these sensor nodes. Because of hardware limitations further energy efficiency can be achieved through the design of energy efficient communication protocols.

Medium access control (MAC) is an important technique that enables the successful operation of the network. One of the main tasks of the MAC protocol is to avoid collisions from interfering nodes. The classical IEEE 802.11 MAC protocol for

CP1007, *Current Themes in Engineering Technologies*, edited by S.-I. Ao, M. A. Amouzegar, and S.-S. Chen

wireless local area network wastes a lot of energy because of idle listening. This *Idle Listening* problem in wireless sensor networks can be minimized by putting the radio into sleep mode.

Apart from the Idle Listening problem, the other source of overhead is *Collisions*. The collisions occur when two nodes transmit at the same time. The packets can get corrupted and it may be required to be retransmitted. So a lot of time and energy gets wasted during this transmission and reception. The other major problem is the *Control Packet Overhead*. These Control Packets does not contain any application data but are essential for the communication. The transmission and reception of these packets is overhead on the sensor network. The last problem is *Overhearing* in which a sensor node may receive packets that are not destined for it. This node could have turned off its radio to save its energy.

In order to design a good medium access control protocol for the wireless sensor networks, it is very important to consider the energy efficiency. As the sensor nodes are mostly battery powered so it is difficult to change or recharge the batteries. In the coming years we expect the sensor nodes to be cheap enough so that the nodes can be discarded rather than recharged. These sensor nodes are battery powered and it is often very difficult to change or recharge batteries for these nodes. After few years we may expect some nodes to be cheap enough so that they can be discarded rather than recharged.

The other important attributes of the wireless sensor network are fairness, latency, throughput and bandwidth. The medium access control scheme proposed in this chapter is good for applications where apart from energy efficiency there is need for low latency.

The rest of the chapter is organized as follows. First we discuss the related work. Then the design description of the proposed MAC protocol is presented. This is followed by the experimental results and finally the conclusions are presented.

RELATED WORK

Medium access control for the wireless sensor network is one of the active research areas for the researchers [9]. The existing wireless MAC protocols such as IEEE 802.11 is not suitable for the sensor network application because these sensors are battery powered and recharging is expensive & also not possible. The medium access control protocols for the wireless sensor networks can be classified broadly into two categories: Contention based [2, 3, 4] and Schedule Based [10, 11, 12].

The schedule based protocol can avoid collisions, overhearing and idle listening by scheduling transmit & listen periods but have strict time synchronization requirements. The contention based protocols on the other hand relax time synchronization requirements and can easily adjust to the topology changes as some new nodes may join and others may die few years after deployment. These protocols are based on Carrier Sense Multiple Access (CSMA) technique and have higher costs for message collisions, overhearing and idle listening.

IEEE 802.11 is a contention based MAC protocol in which collisions are avoided by carrier sensing and randomized back off [14]. The power saving mode (PSM) of IEEE 802.11 reduces the idle listening by putting the sensor node in sleep mode periodically. This PSM mode is for single hop network and may not be suitable for multi hop communication networks.

Sensor S-MAC [2] a contention based MAC protocol is modification of IEEE 802.11 protocol specially designed for the wireless sensor network in 2002. In this medium access control protocol sensor node periodically goes to the fixed listen/sleep cycle. A time frame in S-MAC is divided into to parts: one for a listening session and the other for a sleeping session. Only for a listen period, sensor nodes are able to communicate with other nodes and send some control packets such as SYNC, RTS (Request to Send), CTS (Clear to Send) and ACK (Acknowledgement). By a SYNC packet exchange all neighboring nodes can synchronize together and using RTS/CTS exchange the two nodes can communicate with each other. A lot of energy is still wasted in this protocol during listen period as the sensor will be awake even if there is no reception/transmission.

Timeout T-MAC [3] is the protocol based on the S-MAC protocol in which the Active period is preempted and the sensor goes to the sleep period if no activation event has occurred for a time 'Ta'. The event can be reception of data, start of listen/sleep frame time etc. The time 'Ta' is the minimal amount of idle listening per frame. The interval Ta > Tci + Trt + Tta + Tct where Tci is the length of the contention interval, Trt is the length of an RTS packet, Tta is the turn-around time (time between the end of the RTS packet and the beginning of the CTS packet) and Tct is the length of the CTS packet.

The energy consumption in T-MAC is less as compared to S-MAC. However the latency of T-MAC is more as compared to S-MAC. The proposed scheme takes in to account both energy consumption and latency.

PROPOSED MAC PROTOCOL DESIGN

In the proposed MAC protocol, the sensor duty cycle is changed based on the network load. If the traffic is more than the duty cycle will be more and for low traffic the duty cycle will be less. The network load is identified based on the messages in the queue pending at a particular sensor. The control packet overhead is minimized by reducing the number and size of the control packets as compared to those used in the S-MAC protocol.

Synchronization of Sensor Nodes

The synchronization of the neighboring sensor nodes is done using the SYNC_RTS packet as is done in S-MAC [2] protocol. This SYNC_RTS packet contains the time of its next sleep. After deployment a sensor node starts by waiting and listening. If it hears nothing for a certain amount of time, it chooses a frame schedule and transmits a SYNC_RTS packet. If the node, during startup, hears a SYNC_RTS packet from another node, it follows the schedule in that SYNC_RTS packet and transmits its own SYNC_RTS accordingly. The synchronization table is maintained by all sensors for its neighboring nodes. Upon reception of the SYNC_RTS packet the synchronization table is updated and the timer is adjusted accordingly.

Modified Data and Control Packets

The data and control packets in the wireless sensor networks are broadcasted. These packets apart from the frame control, duration, cyclic redundancy check contains the source & destination address [6] which have been removed as every node is the recipient. The removal of the source and destination address minimizes the control packet overhead in the sensor network communication.

Moreover some of the control packets such as SYNC and RTS has been combined in to one control packet SYNC_RTS (FIGURE 1 (d)) generated by combining SYNC packet FIGURE 1(b) and RTS packet FIGURE 1(c). This reduces the packets overhead and finally contributes in the reduction of the energy consumption and latency.

Bytes 2	2	6	6	6	2	6	0-2312	4
Frame Control	Duration	Addr1	Addr2	Addr3	Seq. Control	Addr4	Data	CRC

(a)

Bytes 2	6	7	2	4
Frame Control	From Addr	Sleep Time	Duty Cycle	CRC

(b)

Bytes 2	2	6	6	4
Frame Control	Duration (NAV)	To Addr	From Addr	CRC

(c)

Bytes 2	2	6	6	7	2	4
Frame Control	Duration (NAV)	To Addr	From Addr	Sleep Time	Duty Cycle	CRC

(d)

Bytes 2	2	6	4
Frame Control	Duration (NAV)	To Addr	CRC

(e)

Bytes 2	2	6	4
Frame Control	Duration (NAV)	To Addr	CRC

(f)

FIGURE 1. (a) IEEE 802.11 MAC Frame Format, (b) SYNC Packet, (c) RTS Packet, (d) SYNC_RTS Packet (Combined from (b) and (c)), (e) CTS Packet, (f) ACK Packet

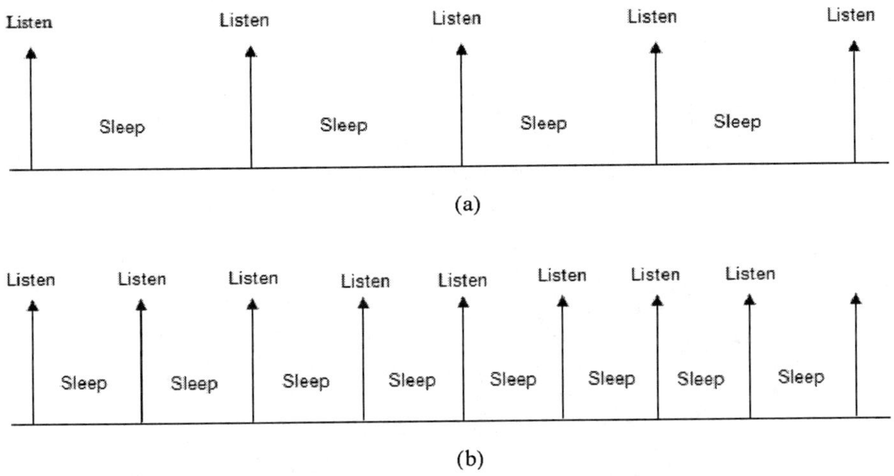

FIGURE 2. (a) Original Duty Cycle, (b) Increased Duty Cycle

Adaptive Duty Cycle

The S-MAC protocol which has fixed duty cycle is energy efficient but this efficiency is achieved at the expense of compromise in the latency. In the proposed protocol each sensor keeps track of the traffic load based on the number of messages in its queue. When a message is received, the counter is increased and when it is transmitted the counter is decreased. If the message counter is greater than a threshold (COUNTthres), then the duty cycle is increased and the changed duty cycle is reported to the neighboring sensor in the SYNC_RTS packet. The neighboring sensor on the reception of SYNC_RTS packet checks if its queue also contains message more than the COUNTthres, it also increases it duty cycle. If not, it simply updates the synchronization table and continues with the original duty cycle. When the traffic is less and so when message counter is less than COUNTthres, the duty cycle is decreased. The sensor node intimates the changed duty cycle to its neighboring nodes.

In this proposed scheme it is not necessary for all the sensor nodes to maintain the same duty cycle. The initial/ previous duty cycle is still valid as the new duty cycle is multiple of the initial/previous duty cycle. If the initial duty cycle is 15%, the increased duty cycle will be 30%.Similarly when the load is less the duty cycle will be reduced from 30% to 15%.

The synchronization of the sensor nodes in the S-MAC protocol is done by maintaining the common sleep-listen cycle i.e fixed duty cycle. In the proposed MAC protocol all the sensor nodes on deployment have the basic duty cycle. Once the traffic on a sensor becomes high it automatically increases it duty cycle by listening again during sleep time as shown in FIGURE 2 (b). So the sensor gets more time to receive the packets. The original listen time remains unchanged. When the traffic is low, its duty cycle is decreased and the sensor stops sensing the channel during sleep interval. So it starts maintaining its original listen-sleep cycle as shown in FIGURE 2 (a).

TABLE 1. Parameters of MAC Implementation on NS-2

Serial Number	Parameter Name	Value
1.	Channel Bandwidth	20 kbps
2.	Control packet length	10 bytes
3.	Data packet length	Up to 250 bytes
4.	MAC header length	10 bytes
5.	Contention window for SYNCrts	31 slots
6.	Contention window for data	63 slots
7.	Slot Time	1 ms
8.	Power consumption for transmission	36 mW
9.	Power consumption for reception	14.4 mW
10.	Power consumption for sleep	15 μW

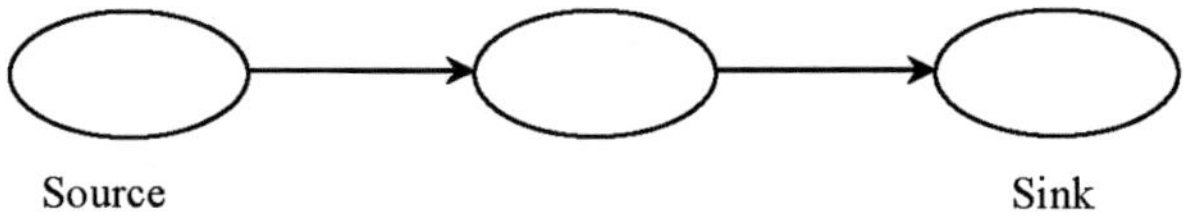

FIGURE 3. Two-hop Network Topology Used in Experiment

Experimental Results

We have simulated the above proposed scheme in the Network Simulator ns-2 [8] version 2.30 under Linux environment. The performance of the proposed MAC protocol was evaluated based upon energy consumption, latency and throughput. In the simulation, no mobility is assumed. In our experimental simulation we performed test on a simple two-hop network topology as illustrated in FIGURE 3. We vary the traffic load by changing the packet inter-arrival time on the source node. The packet inter-arrival time changes from 1 Sec to 12 Sec. Under each traffic condition, the simulation is independently carried out for 10 times. In our simulations we evaluate the performance of our scheme and compared it with the standard S-MAC with 15% duty cycle & T-MAC protocol. The parameters for the implementation of MAC protocol on network simulator (ns-2) are given in TABLE 1.

Energy Consumption

The comparative aggregate energy consumption of nodes is shown in FIGURE 4. The energy consumption of a node is computed by measuring the time the sensor node spends in the different modes: transmit, receive, idle and sleep. The total power consumption is then obtained by multiplying the time with power consumed by the radio in that mode. It is evident from the results that the energy consumption of the proposed MAC protocol is less than S-MAC protocol. This is because of the reduction in the size and number of the control packets.

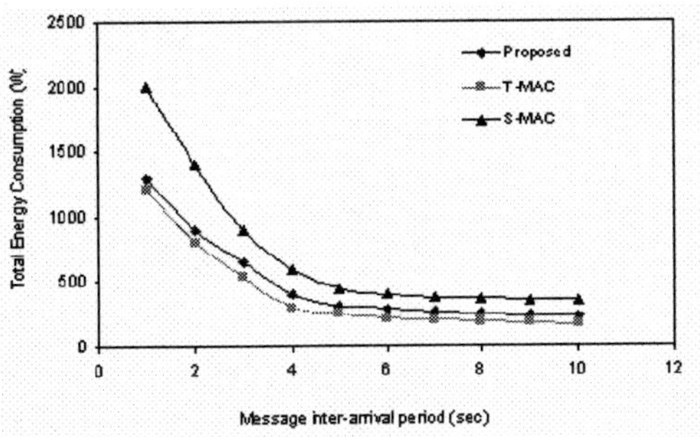

FIGURE 4. Aggregate Energy Consumption

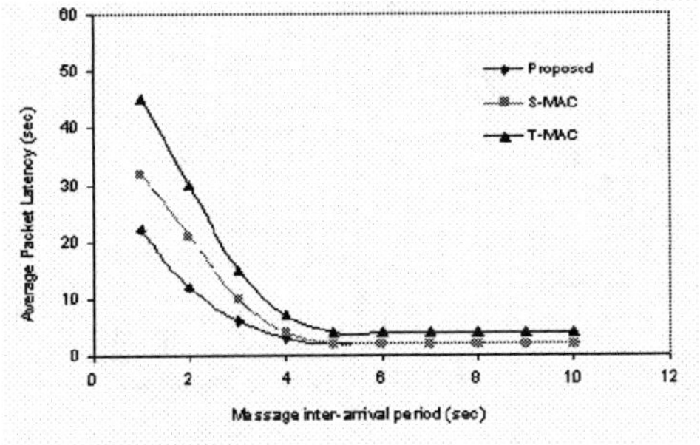

FIGURE 5. Latency Behavior

The reduction in the size of the data and control packets is obtained by eliminating the source & destination address. This reduces the overhead by 1% for IEEE 802.11 data packet, 40% for the SYNC_RTS and 43% for the CTS and ACK packets. The overhead is computed as ((number of bytes in source & destination address) / (total number of bytes in the packet) * 100).

This overhead is further reduced because of transmission of one combined SYNC_RTS packet instead of two separate packets SYNC & RTS apart from CTS and ACK. In the experiment the total number of control packets transmitted and received was also recorded and is shown in FIGURE 6. As expected the total number of control packets in the proposed scheme is lesser than both S-MAC and T-MAC protocols.

These modifications in the control packets contribute to the lower energy consumption in the proposed MAC scheme as compared to S-MAC. The energy consumption of the proposed MAC protocol is however comparable to that of T-MAC protocol in which there is premature termination of listen interval when no event is expected to occur.

Latency Behavior

The latency behavior of the different MAC protocols is shown in FIGURE 5. The latency of the proposed MAC protocol is less than both S-MAC and T-MAC protocol because of adaptive adjustment of the duty cycle based on the network traffic.

When the load is high (i.e., message inter arrival period is small), the number of messages in a queue pending at the sensor node increases resulting in adjustment to higher duty cycle. So the latency of the proposed MAC is low as sensor node gets more time to receive the packets.

However when the traffic is low, the duty cycle is again changed from higher to lower. So the latency of proposed MAC protocol follows the pattern of the S-MAC protocol. As expected the latency of T-MAC protocol is higher as compared to Proposed MAC and S-MAC protocols. This is because T-MAC trades off latency for energy savings under heavy traffic conditions.

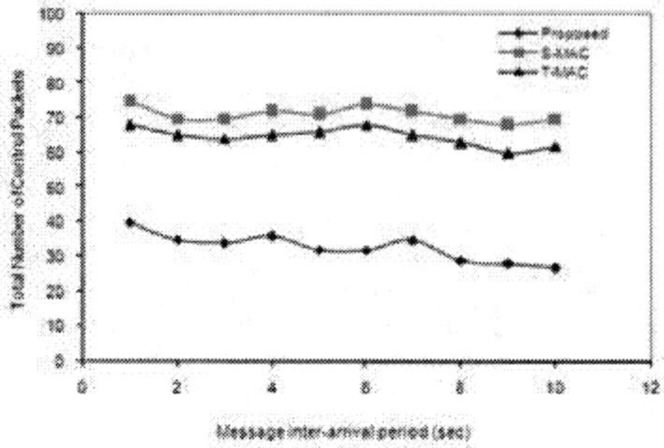

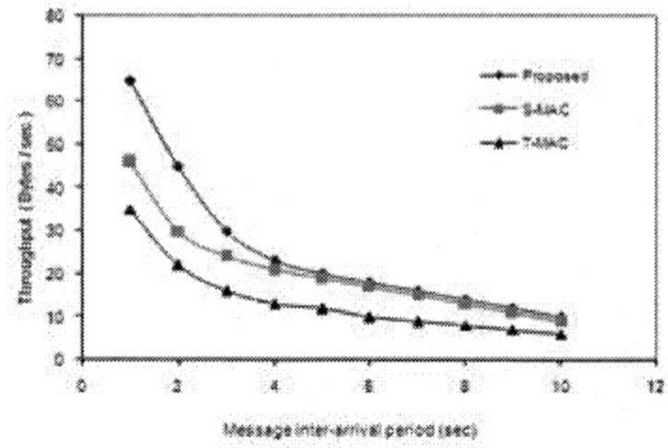

FIGURE 6. Total Number of Control Packets

FIGURE 7. Throughput Comparison

Throughput Comparison

The throughput measured for the different MAC protocols for the different message inter-arrival time is shown in the FIGURE 7. The throughput is measured as the total number of data bytes received at the destination node in unit time. When the traffic is high, more messages will be queued at the source node and so the throughput is also high. Under low traffic conditions the throughput is low as the number of messages to be delivered are less.

The throughput of the proposed MAC scheme is higher than both S-MAC and T-MAC under heavy load when the packet inter-arrival time is less than 4 sec. This is because of the sensor nodes adjustment to the higher duty cycle when the message count becomes greater than threshold. Under the low traffic when the packet inter-arrival time is greater than 4 sec, the throughput of the proposed MAC follows the pattern of the S-MAC because of adjustment to the low duty cycle. The throughput of the T-MAC is however always lower than both S-MAC and proposed MAC scheme because of trade off for the energy saving.

Validation of Results

The simulation results of the proposed MAC protocol for energy consumption have been validated by comparing with the analytical model. The proposed MAC scheme is based on the concept of S-MAC protocol, so the energy consumption analytical model for S-MAC protocol presented in paper [7] has been modified for adaptive duty cycle of 15% and 30% based on the network traffic conditions.

The parameters used for the analytical model are same as those used in the simulation of the proposed MAC scheme under ns-2 network simulator. The power consumption for receive and idle are 14.4 mW, transmit is 36 mW and sleep is 15 μW as given in TABLE 1.The traffic load is varied by source node 0 by changing the message inter arrival time between 1 to 12 sec. The results of analytical model for energy consumption quite match with the simulation results as shown in FIGURE 8 and the percentage difference between the two results is about 8%.

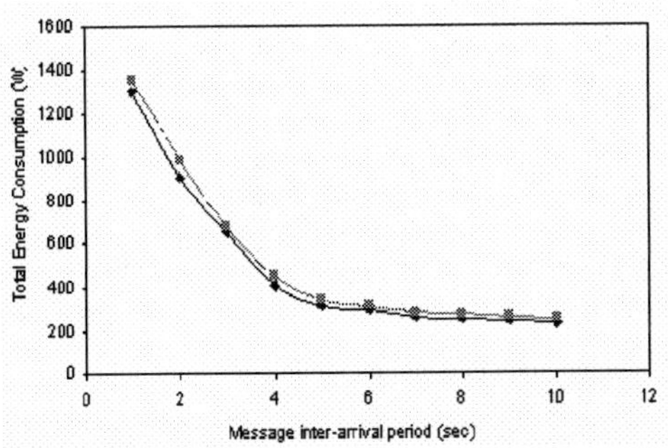

FIGURE 8. Comparison of Analytical & Simulation Results

CONCLUSIONS

In this paper we proposed a Medium Access Control (MAC) scheme for the wireless sensor network. This medium access control scheme is based on the concept of listen/sleep mode cycle like S-MAC. The S-MAC protocol achieves the energy efficiency at the expense of sensor latency and may not be suitable for the delay sensitive medical and defense applications.

The proposed scheme is good for applications where apart from energy efficiency there is need for lower latency. In the proposed scheme the latency is less because of the adaptive adjustment in the sensors duty cycle based on the network traffic conditions. The energy consumption has been reduced because of the reduction in the number and size of the control packets.

REFERENCES

1. I. Akyildiz, W. Su, Y.Sankarasubramaniam and E. Cayirci, " A Survey on Sensor Networks", IEEE Communication Magazine, Aug. 2002 p.p. 102-114
2. W. Ye, J.Heidemann and D.Estrin "An Energy-Efficient MAC Protocol for Wireless Sensor Networks",IEEE INFOCOM,New York, NY, June 2002, Vol.2, p.p. 1567-1576
3. Tijs van Dam, Koen Langendoen, "An Adaptive Energy Efficient MAC Protocol for Wireless Networks" In Proceedings of the First ACM Conference on Embedded Networked Sensor Systems, Nov 2003.
4. Changsu Suh, Young-Mi Song,Young-Bae Ko, We Duke Cho "Energy Efficient & delay Optimized MAC for Wireless Sensor Networks" In Proceedings of the Workshop in the Seventh International Conference on Ubiquitous Computing (Ubicomp'05),Sep 2005.
5. S. Park, A. Savvides and M. B. Srivastava, "SensorSim: A Simulation Framework for Sensor Networks, "Proc. of MSWiM 2000, Boston, MA, August 11, 2000.
6. Brenner, Pablo "A Technical Tutorial on the IEEE 802.11 Protocol", Breezecom Wireless Communications, July 1996.
7. H.Tseng . Yang, P. Chuang, H. Wu, and G. Chen, "An Energy Consumption Analytic Model for a Wireless Sensor MAC Protocol," in Proceedings of the IEEE Vehicular Technology Conference Spring (VTC 2004- Fall), vol.6,pp 4533-4537
8. The Network Simulator –ns-2, http://www.isi.edu/nsnam/ns/

9. M. Ali, Saif, A. Dunkels, T. Voigt, K. Romer, K. Langendoen, J. Polastre, Z. A. Uzmi, "Medium Access Control issues in Sensor Networks", ACM SIGCOMM Computer Communication Review, Volume 36, Number 2, April 2006.
10. V. Rajendran, K. Obraczka and J.J. Gracia-Luna-Aceves "Energy Efficient, Collision Free Medium Access Control for Wireless Sensor Networks", International Conference on Embedded Networked Sensor Systems, Nov. 2003, p.p. 181-192
11. K. Sohrabi, J.Gao, V.Ailawadhi and G.J.Pottie, "Protocols for Self Organization of a Wireless Sensor Network", IEEE Personal Communication, volume 7, issue 5, Oct 2000, p.p. 16-27
12. Zhihui Chen, Ashfaq Khokhar "Self Organisation and Energy Efficient TDMA MAC Protocol by Wakeup For Wireless Sensor Networks" In Proceedings of the IEEE Conference SECON 2004, August 2004.
13. J. Reason and J.M. Rabaey "A Study of Energy Consumption and Reliability in a Multi-Hop Sensor Networks" ACM SIGMOBILE Mobile Computing and Communication Review, volume 8 Issue 1, Jan 2004, p.p. 84-97
14. IEEE Standard 802.11. Wireless LAN Medium Access Control (MAC) and Physical Layer (PHY) Specifications, 1999.

Authors Info

Rajesh Yadav completed his B.Tech (Hons) in Computer Science and Engineering from Bundelkhand Institute of Engineering and Technology, JHANSI (U.P) in 1993. He obtained his M.Tech from Dayalbagh University AGRA (U.P) in 1998. He joined Electronics and Radar Development Establishment, DRDO, Bangalore in 1998. Presently he is working as Scientist 'D' and his areas of interest are wireless sensor networks and real time embedded systems for radar applications. Before joining DRDO, he has worked as Lecturer in Computer Sc.& Engg. Department in Kumaon Engineering College, ALMORA (Uttranchal) from 1995 to 1998.

Dr. Shirshu Varma graduated in Electronics & Communication Engg. from Allahabad University and postgraduated in Communication Engg. from BIT Mesra Ranchi. He completed his Ph.D in Optical Communication from University of Lucknow. He has served many organizations like BIT Mesra Ranchi, IET Lucknow, C-DAC Noida in the capacity of lecturer, Sr. lecturer & IT Consultant. Presently he is working Assistant Professor in IIIT Allahabad. Dr.Varma has published about 27 papers in international and national journals and conferences of repute. He is a member of IEEE and life member of ISTE. He has been a recipient of many national awards in this area. His areas of interest are intelligent sensor network, wireless sensor network, Optical wireless communication, Wireless communication & network.

Dr. N. Malaviya is working as Prof & Head Electronics Department at Institute of Engineering & Technology, Lucknow (U.P). He completed his Ph.D and M.Tech from Indian Institute of Technology, Roorkee. He has over thirty years of teaching and research experience. He has guided 10 Ph.D students and several M.E and B.Tech students. He is also Dean Research in U.P Technical University, Lucknow (U.P)

Functional Security Model: Managers and Engineers Working Together

Edward Paul Guillen[a] and Rulfo Quintero[b]

[a] *Telecommunications Engineering, Military University "Nueva Granada", Cra. 11 No. 101-80, Bogota, Colombia, e-mail: edward.guillen@umng.edu.co*
[b] *Telecommunications Engineering, Military University "Nueva Granada", Cra. 11 No. 101-80, Bogota, Colombia, e-mail: gissic@umng.edu.co)*

Abstract. Information security has a wide variety of solutions including security policies, network architectures and technological applications, they are usually designed and implemented by security architects, but in its own complexity this solutions are difficult to understand by company managers and they are who finally fund the security project. The main goal of the functional security model is to achieve a solid security platform reliable and understandable in the whole company without leaving of side the rigor of the recommendations and the laws compliance in a single frame. This paper shows a general scheme of the model with the use of important standards and tries to give an integrated solution.

Keywords: Security Models, Management. Web technologies. Security platforms.
PACS: 89.20.Ff 89.20.Hh.

INTRODUCTION

Most of people think that information security is a relevant subject only for engineers and security architects inside an organization. Nevertheless, security goes beyond technical resources, operators, security tools, and indeed it includes a cooperative technical, operative and management effort that relates the whole organization through controls application and security policies implementation.

Nowadays information security is more relevant for organizations than ever before and because of the legislation compliance, some medium-sized and international companies already have been started to working on it, but it is important that everybody tries to do it. In fact, strategic security suggests to apply and to recognize the informatics security as a key business component which must be present, and proactively adopted inside all company levels and departments. To achieve this, tools and mechanisms are required in order to plan, to execute and monitoring a functional model. For small industries it is very important to optimize resources, and it has been proved to be more efficient under better planning environments. To achieve a functional security model we will show a legal framework as an initial requirement of the model and after that we will cover some security recommendations and standards including ITIL, COBIT, ISO and best practices. With these references, we will show the steps to develop the functional security model and finally the proposed model is shown. The implementation of the model and our actual work on it is analyzed to give some points to future work.

CP1007, *Current Themes in Engineering Technologies*, edited by S.-I. Ao, M. A. Amouzegar, and S.-S. Chen

LEGAL FRAMEWORK AS AN INITIAL REQUIREMENT OF THE MODEL

Security is an evolving area which in its beginnings was not regulated by law. But due to transparency problems, cyber terrorism, online business, partners and stakeholders connectivity, it was necessary a strong legislation development like HIPAA (The Health Insurance Portability and Accountability Act, which covers privacy rules for people handling healthcare information), SOX (Management's assessment of internal control over financial reporting), GLBA (The Gramm-Leach-Bliley Act, which describes privacy protection rules for the banking and financial services industry), COPA (The Children's Online Protection Act, which covers a broad area of information access and protection for children), and others, which has been the triggers to start to work in the security field across the entire organizations.

When SOX was released, it was not expected that IT staff would be so important to be SOX compliant, but when statistics showed that 70% of time was used to solve deficiencies on IT controls, and 70% of the security budget was spent identifying the rights controls to apply on IT, the subject became really relevant [3]. And by the other side the numbers revealed a very expensive and inefficient first years of work to be SOX compliant.

Companies are trying to apply too many IT controls during these first SOX years, without selecting them carefully. The security designers should have in mind the multidisciplinary essence of the security with special emphasis in the active management participation and in the realization of educational programs.

Due to language between management and engineering is so different, it is very important that engineers recognize management concepts like risk management as managers the security concepts, then we can put them inside an unified framework to join efforts to fulfill a complete strategic security.

It is very significant to identify the right way to motivate all people belonging to the organization, to commit with regulation compliance. Next, we will show five related factors that were analyzed by security researchers in North American experiences [3].

1. Deliver continuous training to employees while ensuring accountability with policy: Being strategic and relate everybody inside the organization is more likely to achieve and maintain compliance.
2. Restructure the risk management function, internal controls, and IT security: risk involves integrate multiple disciplines and departments across organization. Experience shows that many resources were used to solve IT security controls deficiencies, policies, procedures and their management.
3. Reallocate IT expenditures by shifting spending from consultants and contract labour to automated tools: The control deficiencies decrease as the company increase the investment on IT security. In the other hand it was revealed that is more useful invest on automated IT based continuous control and monitoring tools, than on contractors and services.
4. Automate IT measurements, reporting, controls, change management processes, and IT security policies: focused on documenting procedures, making changes to both business and technical procedures, and automating these processes as much as possible.

5. Focus on managing risk to improve IT controls, information collection, and reporting.

TABLE 1. Responsibilities Distribution as a Result of Lessons Learned

Lesson No.	CEO	IT Admin	Internal Auditor	Corporative legal
1	X			X
2	X	X	X	
3		X		
4		X	X	X
5	X	X	X	X

The suggested responsibilities distribution between the different roles inside any organization is shown on Table 1.

The most successful applying control to be SOX compliant where companies which invested to identify the IT controls needed to satisfy SOX and then put procedures and tools in place to monitor, automate, and report on these controls.

There is an emerging issue: "data custody", which involves data protection, data privacy, data retention, data destruction, and data discovery. The research shows that custody of sensitive data was more important for many organizations in the last 12 months. Because most of this information resides on IT systems, "the question of general IT controls and information security controls designed to safeguard sensitive data has come to the forefront" [3].

EXISTING RECOMMENDATIONS AND STANDARDS

In the security world, there are documents which guide organizations from diverse point of view to get a better security level from management to daily operations. However these documents are so extensive to be analyzed by organizations without lose their main goal. A functional security model suggests being a practical help to work with this documents.

IT security professionals have read, during the last years, the CSI and FBI security reports (http://www.gocsi.com/forms/fbi/csi_fbi_survey.jhtml) which report security behaviour during the year. But W. Baker y L. Wallace show that the way surveys questions are done, are not the best kind of question to extract more valuable information of their participants therefore this report cannot show more realistic information [4].

Including such tips inside a functional security model, it is shown that recollect and associate information is not enough and the necessity of design better strategies to react to security challenges is more relevant because this can affect their applicability results.

IT governance and management related with IT security operations can't work without IT itself, in the way, COBIT–Control Objectives for Information and related Technology– and ITIL–Information Technology Infrastructure Library– support IT at the organization and ISM3–Information Security Management Maturity Model– represents the quality vector.

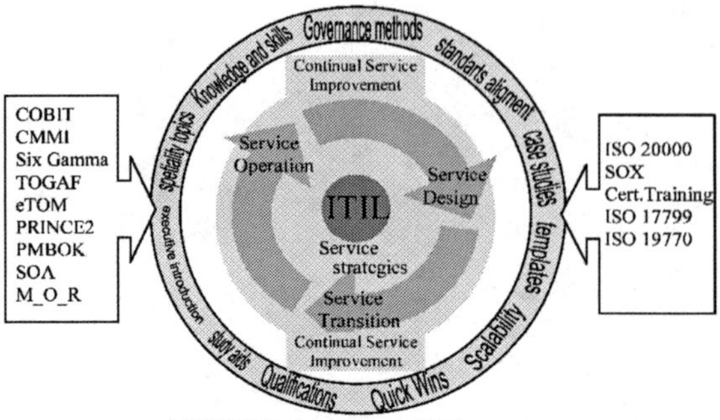

FIGURE 1. ITIL 3.0 Diagram.

Putting all the Documentation Together

The first step is to get a copy of SOX, COBIT, ISM3 which extends ISO 9001 quality management principles to information security management (ISM) systems and is distributed under Creative Commons license, and CBK (if available), because CBK aims to reconcile industry and academia needs and reflect information security's multidisciplinary nature [6], as starting IT security frameworks, and other important documents like manufactures guidelines as shown on Fig 2 without forget its unique strategic position and scope.

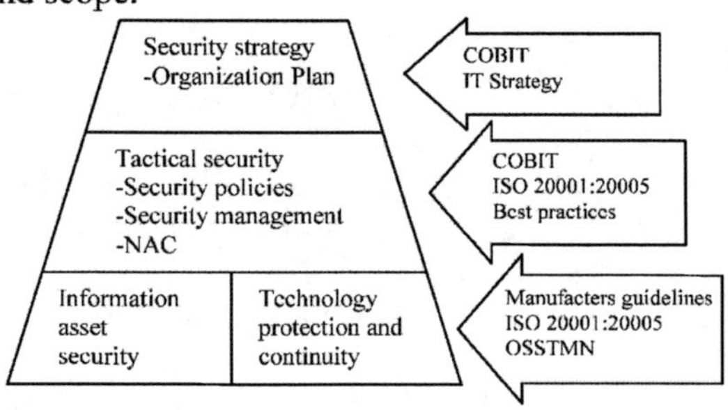

FIGURE 2. Guides and Their Strategic Scope.

Maintain the idea that any organization has some kind of policies, even written or not, applied or not, and start with an audit to determine the initial conditions of the organization, in a first stage, considering all IT targets as black box and lately as a white box.

The next step is to Document, catalogue and centralize the audit results in a corporative portal under a restricted area protected for any authentication mechanism. This process allows the group to develop a self assessment and found the key security controls and even more general IT controls. The GISSIC ethical hackers did the audits and used the portal http://gissic.umng.edu.co to handle all the resulting information.

The risk could be identified if the vulnerabilities identification is made in an early stage with information that must be real and potentially manipulated. In that way, key controls can be easily identified and applied, optimizing resources and effort in the planning and execution of tasks.

STEPS TO DEVELOP THE FUNCTIONAL SECURITY MODEL

It is very clear that security issue goes further technical implementations, and it means to think in a whole process integrated into the core business function, where find the right reporting structure for security is between looking at it as an information technology function or as a control function.

Security is an evolving area, and learning while it evolves is important for executive managers.

This commitment is wrote in mission statements and transfers this ideals internally and to their customers, as the results of applying strategic information security models, and giving trust to other companies as a good business partner.

Information security strategic role understanding must include at least the next basic rules:

1. Information is an asset that is critical to the success of the business. As a result, the information assets of the company must be protected against relevant risks.
2. Information security must be integrated with business plans.
3. Customers should expect that the company will respect their privacy and protect their information with the same diligence that will be used to protect its own information.

It is important to understand the necessity of safeguard the information in nowadays global interconnected market. To achieve this change are important the next steps.

1. Made a deep understand of the problem and the risks a business faces
2. Use the correct steps to mitigate those risks.
3. Measure the success of the security program and monitor it to ensure that it continues to function at the desired level.

All CIOs need to know security reasons based on information triad CIA (Confidentiality, Integrity and Availability), and more detailed Parkerian Hexad (Confidentiality, Possession or Control, Integrity, Authenticity, Availability and Utility). This concept helps him to understand the global implications of security in organization behavior, specially its business core.

Information security professionals point of view is a further definition of CIA and Parkerian Hexad, to accomplish this CIOs support on CBK (Common Body of Knowledge), COBIT, ISO 27001:27005, OSSTMN, SDL, best practices, and others security frameworks and guidance documents that have been integrated to security and are operated, sometimes from similar point of view and others from a complementary one.

The next example shows the different points of view about a new service deployment, when business man and security agent meet to talk about it. a) For business man, whom technical things are easy and ROI (Return of Investment) and cost effective solutions important. b) Information security professional will be focused on solutions, product features, and implementation issues. Therefore his conversation will be reflected on many technical questions of huge importance.

The conclusion in this well known circumstance inside any organization, is the importance of keep in mind both of them, and create a life cycle approach of information assurance, to a solution that will be good for both of them: business and engineering. Putting together both technical and business view is essential to move security into a strategic role. This ideas support the developed model which were validated in [7] and [8].

CEOs are familiar with risk management and this is an important advantage, because security behaves in a similar way and they don't need to know about the technology to face risk management. However they will be careful to who delegate such responsibilities given that CIO use to be 2 levels down, giving the impression that security is less important.

"The business manager will set lofty goals and have challenging aspirations. The security manager must learn that those goals are his goals too, but his job is to ensure the goals are met with a minimum of risk. That is the real challenge of defining strategic security" [1].

Monitoring

Strategic security depends of monitoring of its strategy to succeed. A Proactive, flexible, updated, documented strategy, which tests constantly the new rules to react to novel challenges and deploying the computer security incident response team CSIRT (outsourced or not), with policies, procedures and escalation plans to communicate with management, are a must.

This will be monitored proactively in the form of Security Information Management (SIM) products which runs a console to analyze information and agents to recollect it in all the devices to be monitored, It is clear that monitoring is beyond a single product. It is a hard work take one of these products and adapts it to internal policies, but It is very important too.

Information security management can be achieved in three ways:

1. Using existing tools for monitoring and reporting in consoles, like HP OpenView http://h20229.www2.hp.com/ and IBM Tivoli http://www-306.ibm.com/software/tivoli/.
2. Consolidating information through product consoles. It is more possible since multi-function appliances appearance. According to Gartner, an appliance is "a computing entity that delivers predefined service(s) through an application-specific interface, with no accessible operating software" [2].
3. Implementing a **S**ecurity **I**nformation **M**anagement –SIM– product: This approach is in theory the best solution, it deals with diversity to do correlation, but It is the most expensive.

It is highly recommended open and secure standards usage, which reduce the likelihood of monopolies and transparency doubts inside security platforms, on processes such as data gathering, monitoring, management, unified storage, correlation and presentation.

Management Communication Channels Importance

The collected information must be shown in easy terms to management. This will support the key finding of IT security deficiencies on technology, training, processes, policies, management and decision making areas. Communicative skills are a must for security team.

Outsourcing

When enterprises thought on outsourcing, usually it is associated with the necessity of reducing operational (staff) and acquisition costs (technology), looking for a bigger ROI, and trying to transfer some responsibilities to the service provider.

However, it could be about sign with more specialized security companies and understand the agreement terms, and the kind of offered services such as operations, monitoring, design, management, decision making with or without value-add.

Any third party will not only being protecting, monitoring and comply with policies and requirements but this service provider has the potential to handling, analyzing, exposing, thieving and even selling company's more sensitive information.

Experiences report, the best equilibrium is reached when organization IT personal guides the service provider inside unique organization characteristics and eventually, the provider drives his work aligned with organization objectives

Involve Suppliers Into The Process

In a functional security model application, it is clear that decisions done by the organization must be implemented not only internally but in the whole production chain. If some deficiencies directly related with the core of the business on suppliers resources are found, the group suggest that third party uses some automated methodologies like OWASP Top Ten Project [15] or Automated Security Functional Testing (NIST) [16], to solve this deficiencies together.

People require continuous training and awareness plans for day to day operations against new and ever changing threats, in this field creative ways to work with groups are highly recommended, it doesn't matter how difficult a subject could be as shows COBIT training program at www.cobitgames.com.

PROPOSED MODEL

The model has a schematic structure which covers: governance and management as the highest goals, which rules over IT assets and procedures with IT and IT security policies of the organization; and all this applied by people who must be trained and informed periodically.

Fig 3 shows the proposed functional security model. This model integrates and put in strategic way some of the most effective methodologies and frameworks as a quick guide inside the IT and security management world.

MODEL IMPLEMENTATION

As a result of information analysis, we suggest a working plan through wizards based on open source software which characterize for reduce acquisition costs, increase flexibility, dynamism, information sharing and even enhance security under certain circumstances compared with closed source developing model [9],[10]. Open source tools include: risk management, project management, knowledge base, bugs tracking, control tools, massive training channels, policies and strategic plans organization facilities for multidisciplinary teams.

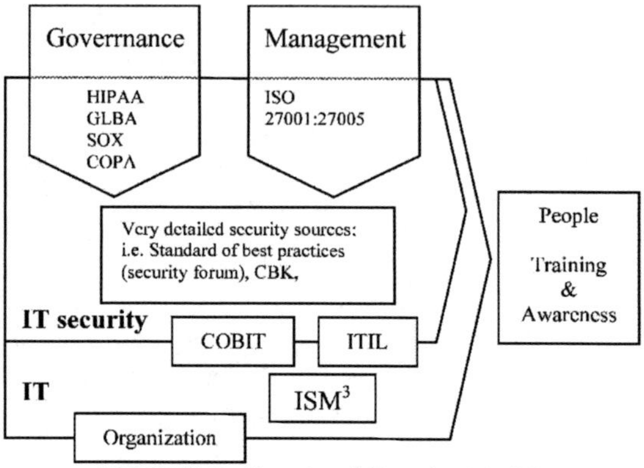

FIGURE 3. Functional Security Model.

Functional Security Model Wizards

This research has been materialized into web wizard modules that guide users into security questions. These wizards help CEOs (Chief executive officer), CIOs (Chief information officer), and CISOs (chief information security officer), in the large process of being compliant with some legislation at many levels of the organization, in a self explanatory vocabulary and inner skills for each created wizard role, as shown on Fig 4.

Wizard's modules have been catalogued in the next domains, inspired on CBK for the reasons explained above:

1. Access control and methodology
2. Telecommunications Network and Internet Security
3. Security Management Practices
4. Applications and systems development security
5. Cryptography
6. Security Architecture and Models
7. Operations Security
8. Business continuity planning (BCP) and disaster recovery planning (DRP)
9. Law, Investigations and Ethics
10. Physical Security

The designed wizards have report interfaces and its outputs can be shown on management and decision meetings to show results and levels of IT security in order to resolve differences obtained on wizard answers and actual policies statements. The approach can be used to update and adapt policies and procedures into a more realistic and effective way.

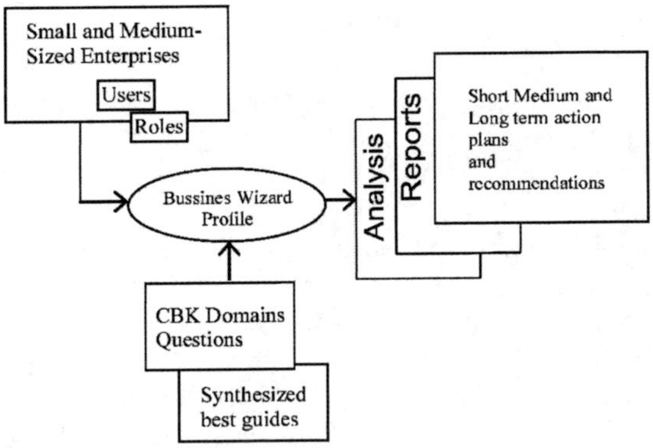

FIGURE 4. Functional Security Model's Wizards.

Using Open Source Tools

During acquisition process, measuring advantages of open source software and determining more pros than cons, GISSIC group is using open source software tools with reporting capabilities for monitoring, like the promising Cobia, its Strata IPS and OSSIM, which are fundamentals to keep control and think in adaptability, flexibility and governance as a reachable challenge on real time [11].

Some solutions and appliances [12] which simplify monitoring and integrate multiple protection mechanism with enterprise quality, does not fit neither to security policies nor attack prevention strategies on some organizations. There are projects that try to integrate tools under unified and adaptable platforms, which by definition must be open to do not monopolize and constraint the final result. One of this security platforms is called HEKA, a Military University "Nueva Granada" open source project for designing security modules, for adapting them to organization necessities and focused on functional security [13],[14]. HEKA platform is divided in three parts. The first one is a designer application where a security architect develops the network security architecture according to the organization policies without the use of commercial tools. This kind of design guarantee an unique design and reduce the risk of being vulnerable by security holes discovered in commercial platforms that are equal for different organizations. The second part is the use of exportable software produced by the architect design that is installed in the necessary equipment according to the original design. The software products are unique in each equipment because the original designer decides the necessary applications to be implemented, making difference with commercial tools where there are some applications that are not in use and you can't disable.

The third part of HEKA is the final user and its applications that can be monitoring for special agents in order to achieve a dynamic tool. This manager applications audit the whole design in the network looking for security holes and ensuring the law compliance.

FUTURE WORK

With some enterprises being beta users during the first months, many results has been collected and are being used as a feedback to the next year's development process, enforcing automation, effective self assessments and integration with platforms and appliances.

A remarkable point is about web technologies. As stats suggest, the web is the first target of 2007. GISSIC research security group is developing data-mining techniques to correlate all risks events, thinking in a preventive and proactive approach against insiders with the capability of use search engines with hacking purposes. With this information, it will be possible to create risk profiles per web user, and determine criminal profiles and take isolating and quarantine countermeasures against potential high risks and suspicious activities.

During our beta tests, it has been very interesting find incoherencies on the wizard results, between CEOs, CIOs and CISOs mind to the questions. And often it is directly related with holes in policies and poor communication between all of them. These communication problems have been reduced after data recollection and these have been communicated to the participants as a feedback capability of the model and platform implementation.

Security platform HEKA is not integrated with the functional security model and a wide range of applications must to be developed. The GISSIC investigation group web portal has information related with the necessary programming work.

CONCLUSIONS

The development of a functional security model allows to link technological affairs with company's management, putting different languages in the same framework without forget law compliance and following recognized security standards.

The results of the functional security model can be implemented with the use of a security platform based on open source technologies to achieve a better security level. Nevertheless, the results of the model can be implemented with any security application since this application complies with regulation and minimum security tests.

Asynchronous and collaborative wizards and portals enhance and take advantage of people available time into the organization, and well designed graphical wizards with different roles and questions capabilities can set a common starting point for management, IT groups and security experts in the way to identify the security weaknesses and implementation methodologies.

Although a wide variety of security models have been proposed, the election of one of them is a matter of experience, education and application. The experts decide but moving away the knowledge of most people at the companies because of their specialized concept. However it is possible to propose software based wizard that helps the inexpert management to achieve a reasonable security level according to the company security policies.

The functional security model characteristics demand to involve the whole company process according to the designed security policies. Even the suppliers and their processes must be followed.

REFERENCES

1. J. Wylder, Strategic Information Security Auerbach Publications, 2004.
2. N. MacDonald, T. Bittman, M. Reynolds, B. Gammage. "Findings: Not all appliances are appliances". Gartner. White paper, September 2006.
3. Symantec Corporation. "Lessons Learned for SOX Compliance and Other Regulatory Challenges". White paper, May 2007.
4. W. Baker, L. Wallace "Is Information Security Under Control? Investigating Quality in Information Security Management". IEEE Security and Privacy, Vol. 5 No. 1, February 2007, pp. 36-44.
5. B. Clacy, B. Jennings "Service Management: Driving the Future of IT". IEEE Computer, Vol. 40 No. 6, May 2007, pp. 98-100.
6. M. Theoharidou, D. Gritzalis. Common Body of Knowledge for Information Security". IEEE Security and Privacy, Vol. 5 No. 2, April 2007, pp. 64-67.
7. E. Guillén, R. Quintero. "Functional Security Model: A Way to Link Technological Affairs with Companies Management". Proceedings of World Congress on Engineering and Computer Science - WCECS 2007. San Francisco. USA. October 2007. IAENG Newswood Limited 2007. pp 368-373.
8. E. Guillén, R. Quintero. "Strategic Functional Security Model: An Initial Approach". Proceedings of Second International Conference on Informatics (Informatics 2007) " Research Excellence and Knowledge Enrichment in ICT". Kuala Lumpur. Malasia. November 2007.
9. Trend Micro: Open source is more secure. Available: http://news.zdnet.com/2100-1009_22-6083490.html and http://news.zdnet.co.uk/security/0,1000000189,39274646,00.htm
10. Why Open Source Software / Free Software (OSS/FS, FLOSS, or FOSS)? Look at the Numbers! Available: http://www.dwheeler.com/oss_fs_why.html
11. L. Herreros. "NAC y Gestión de Vulnerabilidades. Desarrollo de Negocio-Área de Seguridad". Sociedad de la información. February 2007. Available: www.socinfo.info/contenidos/pdf34feb07/p12-34datos.pdf
12. Sophos. " Managed appliances, security solutions that do more". White paper, May 2007. Available: http://whitepapers.zdnet.co.uk/0,1000000651,260281023p,00.htm
13. E. Guillén, J. Constante, M. Hernandez. "Self-Management Security Design Platform HEKA". Proceedings of Mexican Conference on Informatics Security 2006
14. E. Guillén, R. Quintero. "Hacia un modelo de seguridad funcional: casos prácticos que sugieren su utilización". Proceedings del Simposio Internacional de Tecnologías de Información y Comunicaciones- TIC. March 2007.
15. Owas top ten Project, available: http://www.owasp.org/index.php/Category:OWASP_Top_Ten_Project
16. Automated Security Functional Testing (NIST) Project, available: http://csrc.nist.gov/auto-func-test/index.html

Characterizing Nanometer Sized Platinum Films for Temperature Measurements

Fred Lacy

College of Engineering, Electrical Engineering Department, Pinchback Hall, Room 415
Southern University and A&M College, Baton Rouge, LA 70813, USA
(e-mail: fredlacy@engr.subr.edu)

Abstract. Sensors for temperature detection can be achieved with platinum films. After these metallic films have been fabricated, measurements of the electrical resistance will correspond to the temperature of the film. Although metals have a well characterized temperature-resistance relationship, this relationship changes if one of the dimensions of the material is nanometer sized. Laboratory experiments were performed to determine this relationship when thin films (of nanometer thickness) of platinum are sputtered on a glass substrate. Results of these experiments indicate that the resistance of the thin films is linearly related to temperature below a certain temperature, however, above this temperature, the resistance remains relatively constant as temperature increases. Based on these experimental results, calculations were performed to determine resistivity as a function of film thickness and an equation was determined that represents the resistance as a function of both temperature and film thickness.

Keywords: detector, nanomaterial, resistance, resistivity, temperature, sensor, temperature coefficient of resistance, thin film
PACS: 72.15.Eb, 73.61.At, 73.50.Lw

INTRODUCTION

Background

Fabricating thin films has become routine due to the well-established photolithographic processing involved. These films can be conductors, semiconductors, or insulators. Thin films find uses in various fields such as optics, semiconductor electronics, micro-electro-mechanical-systems (MEMS), corrosion, nanotechnology, biomedical sciences, and sensors.

It is being recognized that materials exhibit different properties or characteristics when their dimensions decrease to submicron size. Characteristics such as wavelength absorption and scattering, electrical and thermal conductivity, bandgap spacing, and melting temperature are some of the material properties that are affected by reducing dimensions. Metals or conductors have well characterized bulk resistive properties that are predictable and follow a standard equation (the resistance depends on the material dimensions and the resistivity which is a proportionality 'constant'). Research is showing that if the metal is made sufficiently thin, the electrical resistance of the conductor does not have the predictable response because the material's resistivity is

CP1007, *Current Themes in Engineering Technologies*, edited by S.-I. Ao, M. A. Amouzegar, and S.-S. Chen

not constant, but dependent on the conductor's thickness.

Barnat et al. have shown that the electrical resistivity of thin films of copper has a dependence on the thickness of the film provided that the thickness is below the bulk mean free path for electrons [1]. Similarly, Jen et al. studied metallic films composed of palladium, gold, and copper and found that mean free path of electrons as well as surface roughness could play a role in determining the resistivity of the metallic film [2]. They showed that the resistivity for gold and copper significantly increases if the film thickness falls below about 30 nm. Additionally, Rossnagel et al. found that the resistivity of thin copper films becomes dependent on film thickness as the dimensions approach the electron mean-free-path [3]. The key contributions are from electron–surface scattering, grain boundary scattering, and surface roughness-induced scattering. W. Zhang et al. also show that the resistivity of thin metal films exhibit a high dependence upon their thickness due to the mean free path of electrons [4]. However, they notice that the fabrication or deposition method also have an effect on the resistivity at these dimensions.

Fan et al. developed a model in which includes electron scattering from the metallic film surfaces and grain boundaries [5]. Their theoretical results seem to correlate quite well with experimental data. Yarimbiyik et al. developed a computer simulation program to model various parameters that could affect the electrical resistance of thin metal films [6]. They show that a change in resistivity with temperature begins to increase as dimensions approach the bulk mean free path of the electrons in the metal. Kawamura et al. showed that the temperature coefficient of resistance (TCR) for sputtered platinum films is a function of temperature [7]. Specifically, they showed that as the film thickness increases from 0.4nm – 600nm, the TCR increases as well. X. Zhang et al. provide experimental results showing that the electrical conductivity and TCR of nanofilms are greatly lower than the corresponding bulk values for temperatures from 77 to 330 K [8, 9].

It is known that the electrical resistance, of a material is proportional to the length and resistivity of the material and inversely proportional to the cross sectional area of the material. In equation form, this is given by Eq. 1.

$$R = \rho \frac{l}{A} \tag{1}$$

Furthermore, the resistivity of a material is a function of temperature as given by Eq. 2

$$\rho = \rho_0 [1 + \alpha (T - T_0)] \tag{2}$$

where α is the TCR, T is the temperature, and ρ_0 is the resistivity at temperature T_0. So, it is easy to see that the change in resistance of a material is proportional to its change in temperature as given in Eq. 3.

$$\frac{\Delta R}{R_0} = \alpha \Delta T \tag{3}$$

Thus, based on these equations and the aforementioned research results, for a given temperature change, the change in resistance of thinner films is expect to be smaller than for thicker films. Recent results have shown that thin film electrical resistance remains relatively constant if the temperature is above a certain level [10].

Device Information

A functional block diagram of the temperature sensing device is shown in Fig. 1. It consists of the temperature sensing structure, electronic circuitry (which could consist of a current generator, amplifiers, wireless communications circuitry, etc.), and an electrical power source. This power source would be a micro-battery and would be used to supply power to the appropriate components on the chip. This device would ultimately consist of components that are constructed on the surface of the chip using microfabrication (and/or nanofabrication) techniques. Thus, one of the keys to manufacturing a good device is to keep the power consumption of the sensor and electronic circuitry to a minimum.

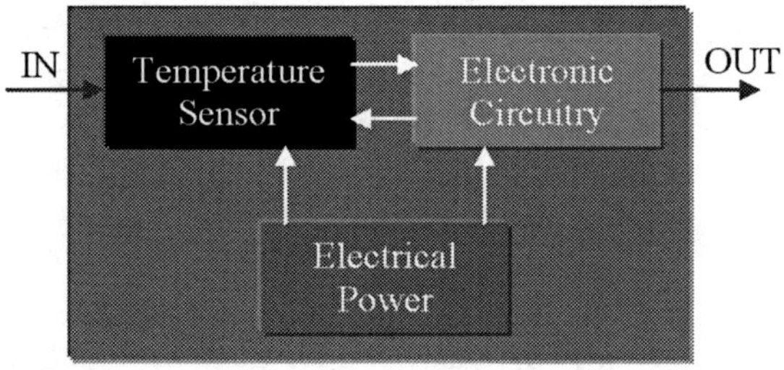

FIGURE 1. Block diagram of the temperature sensing device.

The research results provided in this paper will describe the fabrication of thin films of platinum, show the resistance of these films as a function of temperature, and provide an analysis of the temperature coefficient of resistance as well as the electrical resistivity. The film thickness ranges from 46.3nm to 92.6nm and the temperature ranges from 25°C to 150°C.

PROCEDURES

Experimental Device Fabrication

The thin film platinum temperature sensors were fabricated using the following procedure (as illustrated in Fig. 2). A glass substrate was first cleaned with acetone and isopropyl alcohol. Then S1805 photoresist was deposited on the substrate by spinning at 3550 r.p.m. for 20 seconds. Then, these substrates were soft baked for 4 minutes in an oven at a temperature of 100° C.

Then the substrate was patterned using photolithography. A photomask was placed over the sample and then exposed to 22mJ/cm^2 of ultraviolet radiation for 8 seconds. Next, the unwanted photoresist was removed by immersing the substrate in MF-CD-26 developer for approximately 2 minutes. Then the substrate was rinsed with water to remove the developer.

Subsequently, DC sputtering was used to deposit the desired thickness of platinum onto the substrates. After the platinum was deposited over the entire surface of the substrate, the unwanted platinum was removed through liftoff by rinsing the substrate with acetone for approximately 10 seconds until only the patterned platinum remained. Again, the fabrication steps are shown schematically in Fig. 2.

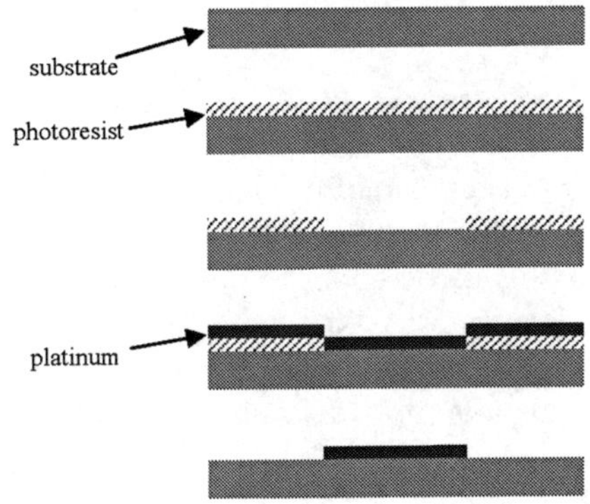

FIGURE 2. Diagram of the fabrication processing steps. (Reprinted from [10]).

This diagram shown in Fig. 2 is a side view of the structure during the processing steps. After the processing steps have been completed, the top view of the serpentine shaped structure that appears on the glass substrate is shown in Fig. 3. This shape was selected to achieve a substantial length of the temperature sensor while still maintaining a reasonable surface area for the other components of the device.

FIGURE 3. Top view of the temperature sensor pattern on the substrate. (Reprinted from [10]).

Again, this structure was fabricated and three different film thicknesses, 46.3nm, 74.0nm, and 96.2nm, were created. For each of these thicknesses, the width of each of the serpentine lines was 0.1mm, and the length of the structure (i.e., the distance that electrical current would have to flow through the structure) was 10.655cm.

Experimental Measurements

To calibrate and characterize the sensors, the fabricated structures were placed in an oven/furnace and the temperature was set to a desired level. Upon stabilizing at this temperature, the electrical resistance of the sensors was measured with a digital multimeter. The oven/furnace temperature was varied from 25°C to 150°C and electrical resistance values were recorded over this temperature range. Thin films of 46.3nm, 74.0nm, and 92.6nm were used to characterize the sensors during this procedure.

RESULTS AND DISCUSSION

Electrical Resistance

The electrical resistance vs. temperature for the 46.3nm thin film is shown in Fig. 4a. The graphs provide the experimental measurements of the resistance as well as the resistance that would theoretically be obtained for platinum in bulk form.

Likewise, results are shown in Figs. 4b and 4c for temperature sensing thin film of 74.0nm and 92.6nm, respectively. Again, these figures provide the measured electrical resistance for the thin films as well as the ideal electrical resistance for platinum in bulk form.

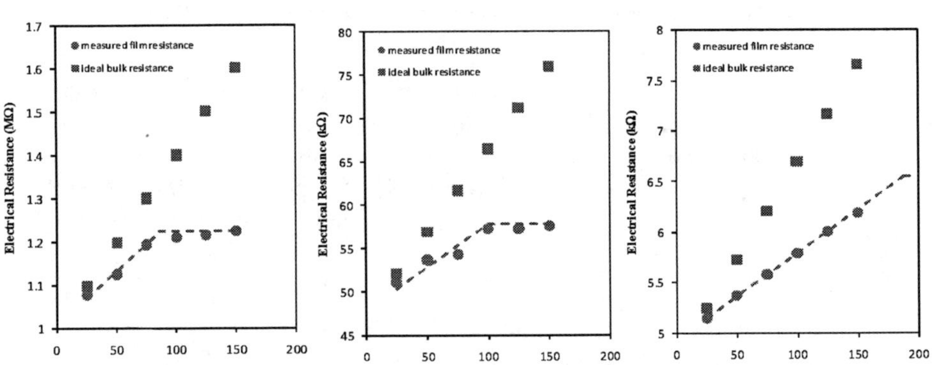

FIGURE 4. Graph of the electrical resistance vs. temperature for (a) 46.3nm platinum film, (b) 74.0nm platinum film, and (c) 92.6nm platinum film. These graphs show the 'saturation' level for a linear curve fit of the measured data along with the ideal resistance that would be obtained for bulk platinum. (Reprinted from [10]).

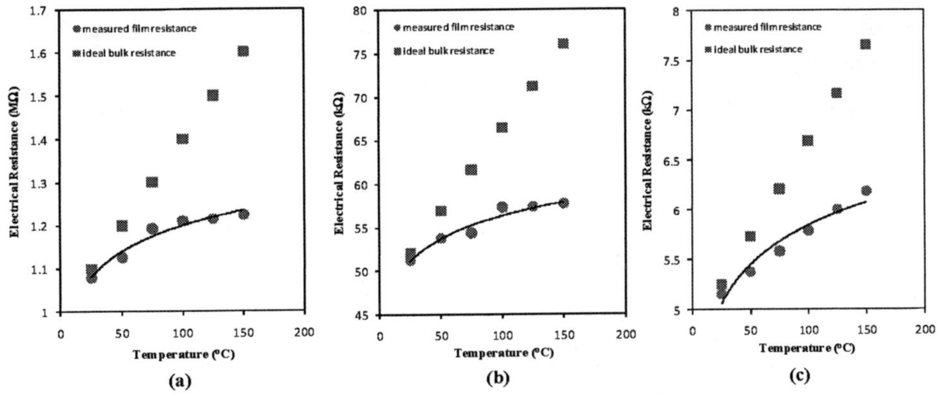

FIGURE 5. Graph of the electrical resistance vs. temperature for (a) 46.3nm platinum film, (b) 74.0nm platinum film, and (c) 92.6nm platinum film. These graphs show a logarithmic curve fit for the measured data along with the ideal resistance that would be obtained for bulk platinum. (Reprinted from [10]).

It is seen from the resistance vs. temperature graphs that all of the films exhibit a resistance below the ideal bulk resistance over the entire temperature range. Figure 4a shows that for a film thickness of 46.3nm, there is very little change in the resistance once the temperature exceeds 85°C. Similarly, from Fig. 4b, a film with a thickness of 74.0nm will have a significantly smaller change in resistance after the temperature rises above 100°C. Whereas the 'transition temperature' (or the temperature where the resistance does not significantly increase for further increases in the temperature) can be clearly recognized from the graphs in Figs. 4a and 4b, this value for the 92.6nm platinum film cannot be easily discerned from the graphed data in Fig. 4c. Undoubtedly, this transition temperature occurs beyond the maximum temperature level used in this research (i.e., 150°C), so an estimated saturation level at 190°C is depicted in Fig. 4c.

TABLE 1. Correlation and Saturation Information for the Thin Films

	46.3nm film	**74.0nm film**	**92.6nm film**
Linear fit correlation	$R^2 = 0.989$	$R^2 = 0.952$	$R^2 = 0.999$
Saturation temperature	85°C	100°C	190°C
Logarithmic fit correlation	$R^2 = 0.955$	$R^2 = 0.948$	$R^2 = 0.946$

Using the same data, logarithmic curve fits provided a good and logical correlation between the electrical resistance and temperature for the three film thicknesses as shown in Figs. 5a, 5b, and 5c. A summary of this linear fit correlation, logarithmic fit correlation, and saturation information is given in Table 1. Although the data for the 92.6nm thin film looks "linear" over the 25°C to 150°C temperature range, since the logarithmic curve fit provides a similar correlation coefficient as 46.3nm and 74.0 nm film, the saturation effect for the 92.6nm film should manifest itself if temperatures beyond 150°C are used.

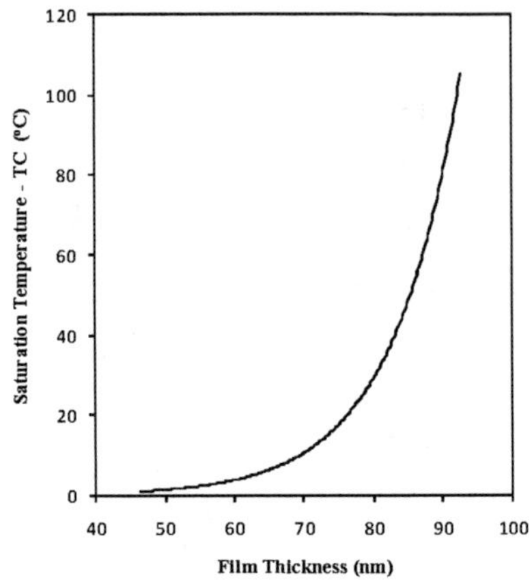

FIGURE 6. Graph showing the relationship between the linear curve fit 'saturation' temperature and the film thickness.

To determine where the saturation temperature occurs for the 92.6nm film (as well as for other film thicknesses), the experimental data was used. Figure 6 shows that the saturation temperature is exponential with regard to the film thickness. When the temperature constant (TC) is subtracted from the saturation temperature (ST) for each data point, the relationship between this variable and the film thickness is given by

$$ST - TC = 0.009e^{0.1t} \quad (4)$$

where the temperature constant TC for the platinum film is 84°C, and t is the film thickness (in nm). This expression given in Eq. 4 and represented by the graph in Fig. 6 yields a correlation coefficient of $R^2 = 1$.

One of the most interesting results to come from this research is that the electrical resistance of thin films can be modeled 'linearly' (i.e., the electrical resistance increases directly with temperature) as long as the temperature is below a certain level; above this temperature level, the electrical resistance does not change as rapidly and can be modeled with a horizontal line of constant electrical resistance. It is also seen that as the film thickness becomes smaller, the transition point between the linear and the horizontal regions also becomes smaller (i.e., occurs at a smaller or lower temperature). This suggests that the thin film structures cannot respond to increases in temperature changes because they have a certain energy capacity that is dependent upon the size or thickness of the film.

This work provides results that are consistent with the work from other research groups. Several groups have shown that the resistivity (and thus the resistance) of metallic thin films will be different from their 'bulk' values if the films have dimensions on the order of the electron mean free path (i.e., in the nanometer range). The platinum films used in this work were in the nanometer range and it was found that the resistance of each film was always less than its corresponding ideal bulk value.

From the logarithmic equations that are used to fit the electrical resistance vs. temperature data, a general equation can be found that represents the electrical resistance as a function of temperature and film thickness. It is noted that each of the electrical resistance vs. temperature curves has a general equation form of $R = A\ ln(T) + B$ where R is the electrical resistance (in Ω), T is the temperature (in °C) and A and B are constants. After some mathematical analysis, the parameters for A and B can be found to obtain one general equation for the electrical resistance. After performing this, it is found that the resistance as a function of film thickness and temperature is

$$R(t,T) = 10^8 e^{-0.1t}[0.1\ln(t)+1][-0.009t+1.273] \qquad (5)$$

where t is the film thickness in nm. Thus, with Eq. 5, sensor structures can be manufactured with a specified thickness and the desired electrical resistance will be achieved. Additionally, for a sensor that has been manufactured with a certain film thickness, the electrical resistance can be calculated at a given temperature.

Temperature Coefficient of Resistance

Using a logarithmic curve fit for the measured electrical resistances [as shown in Figs. 5a, 5b, and 5c], the temperature coefficient of resistance (TCR) can be calculated for each of the thin films (using Eq. 3). The results are graphed and the curves are shown in Fig. 7.

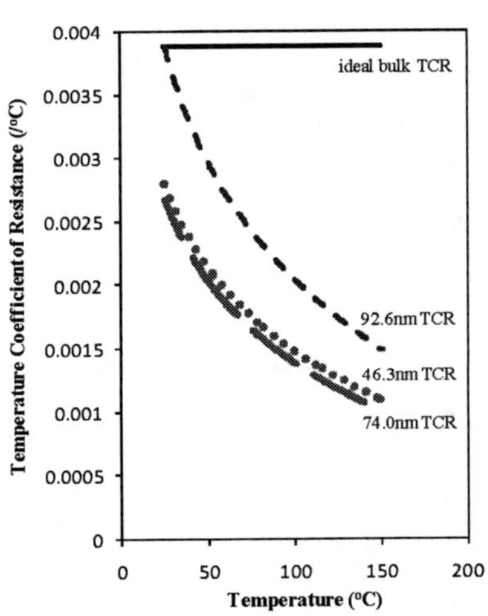

FIGURE 7. Graph of the temperature coefficient of resistance vs. temperature for the thin films and for the ideal case of bulk platinum. (Reprinted from [10]).

This figure shows the calculated TCR for the three thin films (along with the ideal bulk TCR). All three of these temperature coefficients follow logarithmic curves. For thin films of thicknesses 46.3nm and 92.6nm, the correlation coefficient is $R^2 = 0.998$, and for film thickness 74.0nm, the correlation coefficient is $R^2 = 0.997$. Thus, the logarithmic curves fit the measured data really well.

From Fig. 7 it is seen that at a particular temperature the two thinnest films have a noticeably lower TCR than the largest film. It is believed that the two thinnest films have about the same TCR (due to reasonable measurement error), and thus as the film thickness increases, the TCR increases and will eventually equal the ideal bulk TCR. The other observation is that for a particular film thickness, the TCR decreases as the temperature increases. In other words, a fabricated film that is designed to measure temperature through its resistance will eventually reach a 'saturation point' and not respond (or not respond very well) to further changes in temperature.

These results are consistent with the work from other research groups. Research findings from others have shown that the TCR of metal thin films will vary if the temperature or film thickness changes. Results are provided in this paper that support this referenced work. Figures 4 and 5 show clearly that the resistance of the films does not change as dramatically if the temperature goes beyond a certain point. Figure 7 shows directly that the TCR is a function of temperature and film thickness.

Electrical Resistivity

From these electrical resistance measurements, the electrical resistivity for each of the three film thickness can be calculated. These results, for the 46.3 nm, 74.0nm, and 92.6nm films, are shown in Figs. 8a, 8b, and 8c, respectively. These resistivities are plotted as a function of temperature.

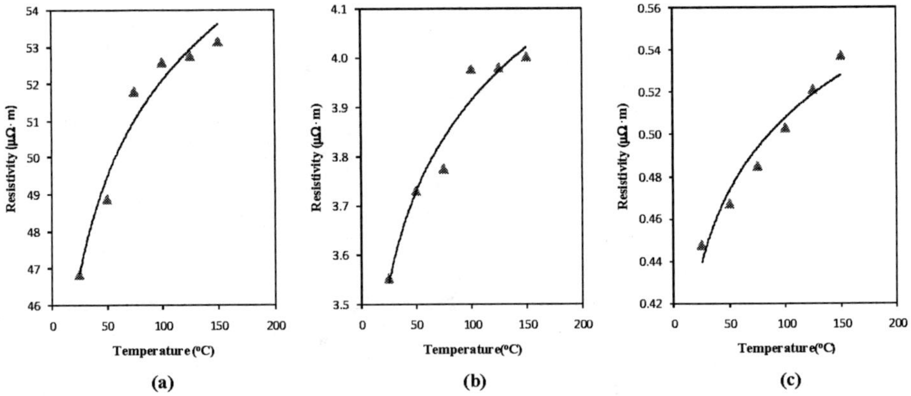

FIGURE 8. Graph of the electrical resistivity vs. temperature using a logarithmic curve fit for (a) 46.3nm platinum film, (b) 74.0nm platinum film, and (c) 92.6nm platinum film.

It is well known that the resistivity of a metal has a linear relationship with temperature as given in Eq. 2. However, this relationship applies to metals in 'bulk' form and does not hold for metal films that have one or more dimension that is smaller than (or on the order of) the mean free path of the conduction electrons. Again, just as the electrical resistance of the thin films exhibits a saturation effect, the electrical resistivity also displays this same effect.

The data used to generate Figs. 8a, 8b, and 8c as well as the electrical resistance as given in Eq. 5 can be used to determine the electrical resistivity as a function of film thickness and temperature. This resistivity can be modeled with the following equation

$$\rho = 93.85e^{-0.1t}[0.1\ln(t)+1][-0.009t^2+1.273t] \tag{6}$$

where ρ is the resistivity (in $\mu\Omega\cdot$ m), t is the thickness of the film (in nm), and T is the temperature (in °C).

It is noted that platinum has a 'bulk' electrical resistivity of $1.06\text{x}10^{-7}$ $\Omega\cdot$m or 0.106 $\mu\Omega\cdot$m. Thus, as the film thickness becomes larger and larger, it is expected that the resistivity will approach and reach the bulk value. It is seen from these results that the resistivity of these platinum films does approach the 'bulk' value as the film thickness increases.

An example of the electrical resistivity and film thickness relationship is shown in the graph in Fig. 9. The plot shown is for a temperature of 25°C. It is noted that all of the graphs of resistivity vs. film thickness will resemble the one shown in Fig.9 regardless of the temperature (i.e., they will all have the same form of an exponential decay).

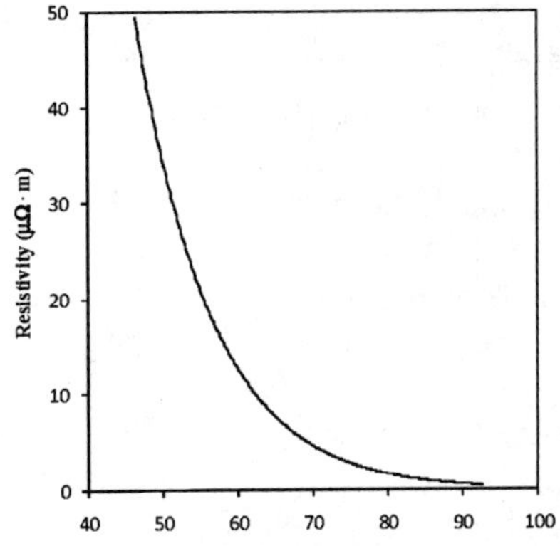

FIGURE 9. Graph of the electrical resistivity vs. film thickness for a temperature of 25°C.

These results are consistent with the preceding results reported here as well as the results from other researchers. The results show that the electrical resistivity increases as the film thickness increases. From Figs. 8a – 8c, it can be seen that at lower temperatures, the resistivity could be modeled linearly with respect to temperature. However, just as in the case for the electrical resistance, as the temperature increases, the slope becomes much smaller and could be modeled with a horizontal line. Additionally, results from several research groups have shown that the resistivity of thin metal films increases in an exponential manner when one of the dimensions of the metal decreases to a sufficiently low level (i.e., below the electron mean free path). As shown in Fig. 9, the electrical resistivity of the nanometer platinum films is consistent with these results.

CONCLUSIONS

This work characterized nanometer sized platinum films for use as temperature sensors. Structures were fabricated using photolithography techniques and then laboratory experiments were performed to characterize their performance. Experimental results showed that a saturation level exists for these thin films that will limit the sensitivity and thus the detection limit for these sensors. Linear curve fits can be applied to the measurement data (along with saturation limits) or logarithmic curve fits can be applied to the data. A logarithm based equation was found that represents the electrical resistance as a function of temperature and film thickness. Both the linear and logarithmic curve fits have very good correlation coefficients.

The temperature coefficient of resistance (TCR) was calculated from the measured resistance values and it was determined that the TCR was not constant as is the case for bulk platinum. All of the TCRs for the platinum films follow logarithmic form and decrease as temperature increases.

The electrical resistivity of the thin films was determined from the laboratory measurements of the electrical resistance. The electrical resistivity was determined as a function of temperature and film thickness and is very similar to the equation for electrical resistance. It was found that for a given film thickness, graphs for the electrical resistivity have the same form as the electrical resistance. The electrical resistivity was also plotted as a function of film thickness. There is an exponential relationship between the electrical resistivity and the film thickness and this is consistent with results from other researchers.

This work shows that thin platinum thin films can be used as temperature detectors by determining the electrical resistance of the film. However, when the films have nanometer thicknesses, care must be taken in 'predicting' the temperature that caused/created that measured resistance. Additionally, if the sensors will be used in a place where the temperature measurement range is known, proper design can be performed to achieve sufficient output from the thin film sensor.

ACKNOWLEDGMENTS

The author would like to thank Dr. Ernest Walker for his guidance and support on this project. The author would also like to express gratitude to Ms. April Page for her hard work and dedication in obtaining the data. The author would also like to thank Dr. Pradeep Bhattacharya for use of the equipment in the Device Processing and Characterization Laboratory and for discussions regarding this subject matter. Finally, the author would also express appreciation to Dr. Habib Mohamadian for continued support for this project and toward laboratory development for future research projects.

REFERENCES

1. E. V. Barnat, D. Nagakura, P. I. Wang, T. M. Lu, "Real time resistivity measurements during sputter deposition of ultrathin copper films," J App Phys, vol. 91 (3), Feb 1, 2002, pp. 1667-1672.
2. S. U. Jen, C. C. Yu, C. H. Liu, G. Y. Lee, "Piezoresistance and electrical resistivity of Pd, Au, and Cu films," Thin Solid Films, vol. 434 (1-2), Jun 23, 2003, pp. 316-322.
3. S. M. Rossnagel, T. S. Kuan, "Alterations of Cu Conductivity in the size effect regime," J Vac Sci & Tech B, vol. 22 (1), Jan-Feb 2004, pp. 240-247
4. W. Zhang, S. H. Brongersma, O. Richard, B. Brijs, R. Palmans, L. Froyen, K. Maex, "Influence of the electron mean free path on the resistivity of thin metal films," Microelectronic Engineering, vol. 76 (1-4), Oct. 2004, pp. 146-152
5. P. Fan, K. Yi, J. D. Shao, Z. X. Fan, "Electrical transport in metallic films," J App Phys, vol. 95 (5), Mar 1, 2004, pp. 2527-2531.
6. A. .E. Yarimbiyik, H. A. Schafft, R. A. Allen, M. E. Zaghloul, D. L. Blackburn, "Modeling and simulation of resistivity of nanometer scale copper," Microelectronics Reliability vol. 46 (7), Jul 2006, pp. 1050-1057
7. M. Kawamura,T. Mashima, Y. Abe, K. Sasaki, "Formation of ultra-thin continuous Pt and Al films by RF sputtering," Thin Solid Films vol. 377, Dec. 1, 2000, pp. 537-542.
8. X. Zhang, H. Q. Xie, M. Fujii, H. Ago, K. Takahashi, T. Ikuta, H. Abe, T. Shimizu, "Thermal and electrical conductivity of a suspended platinum nanofilm," App Phys Letters, vol. 86 (17), Apr 25, 2005, pp. 171912-1 – 171912-3
9. X. Zhang, H. Q. Xie, M. Fujii, H. Ago, K. Takahashi, T. Ikuta, H. Abe, T. Shimizu, "Thermal and electrical properties of a suspended nanoscale thin film," Int J Thermophysics, vol. 28 (1), Feb 2007, pp. 33-43.
10. F. Lacy, "Investigating Thin Films for Use as Temperature Sensors" in *World Congress on Engineering and Computer Science 2007*, edited by S. I. Ao et al., WCECS 2007 Conference Proceedings, Newswood Limited, International Association of Engineers, Kwun Tong, Hong Kong, 2007, pp. 441-444.

An Approach to the Lightning Overvoltage Protection of Medium Voltage Lines in Severe Lightning Areas

Omidiora M.A. and Lehtonen M.

Power Systems and High Voltage Engineering Laboratory, Electrical Engineering Department, Helsinki University of Technology, Otaniemi, Espoo, FINLAND

Abstract. This paper deals with the effect of shield wires on lightning overvoltage reduction and the energy relief of MOV (Metal Oxide Varistor) arresters from direct strokes to distribution lines. The subject of discussion is the enhancement of lightning protection in Finnish distribution networks where lightning is most severe. The true index of lightning severity in these areas is based on the ground flash densities and return stroke data collected from the Finnish meteorological institute. The presented test case is the IEEE 34-node test feeder injected with multiple lightning strokes and simulated with the Alternative Transients Program/ Electromagnetic Transients program (ATP/EMTP).The response of the distribution line to lightning strokes was modeled with three different cases: no protection, protection with surge arresters and protection with a combination of shield wire and arresters. Simulations were made to compare the resulting overvoltages on the line for all the analyzed cases.

Keywords: Direct strokes, Lightning flash, MOV arresters, Lightning Overvoltages, ATP/EMTP, ATPDraw, Test feeder, shields wire.

INTRODUCTION

The inland part of southern Finland is an area where the rate of lightning strikes is very high as compared to other locations in the country, based on the statistical occurrence of lightning and other lightning related incidents. Medium voltage circuits for all system voltage levels are exposed to lightning, and outage and damage occur frequently due to this natural phenomenon. Distribution lines existing in this severe lightning area will be more culpable to direct strokes than other network areas. Therefore, the assessment of overvoltage due to direct strokes and augmentation of surge protection to lines in the area are the subjects of this paper.

BACKGROUND

Lightning data around the world have shown that more than 80% of cloud-to-ground lightning flashes contain more than one stroke [1]-[3]. Outage due to lighting is associated with the frequency of thunderstorm activity or Ground Flash Density

CP1007, *Current Themes in Engineering Technologies*, edited by S.-I. Ao, M. A. Amouzegar, and S.-S. Chen

(GFD) in a given power distribution area. This can be assessed through information from lightning location systems (LLS), such as thunderstorm or GFD maps (Figure1). In Finland, lightning flash counter results from 1998-2007 give an average of 117,530 cloud-to-ground flashes per year [4]. Multiple strokes from these flashes range from 1-15 strokes with a mean of 2 strokes per flash. In both the northern and coastal parts of the country, fewer flashes are detected than in the inland part of the southern region (inset of Figure 1). The average thunderstorm day is 12 days/year.

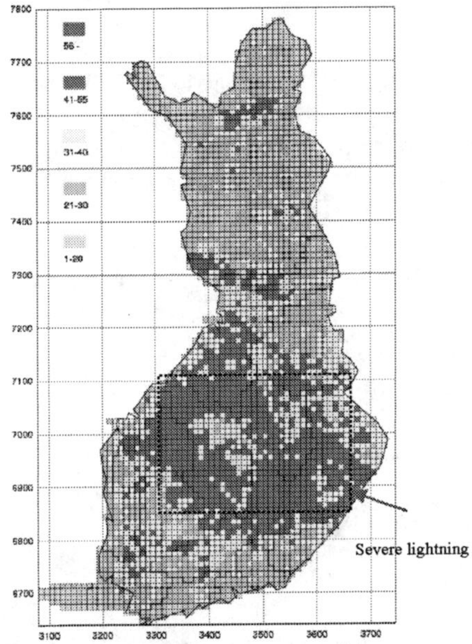

FIGURE 1. Average Ground Flash Densities (GFD) per 100 km^2 in Finland from 1998-2007. There are 5 sensors plus 3 supporting sensors from Sweden, Norway and Estonia to boost the sensitivity of detection. Inset is the area with highest number of flash densities since 1998.

Lightning data (1998-2007) collected from the Finnish Meteorological Institute (FMI) on thunderstorm days have shown good correlation with Eq.1 which is a slight modification of Macherras' equation out of the examined equations shown in Table 1.

$$N_g = 0.01T_d^{\ 1.4824} \qquad (1)$$

where N_g= the ground flash density (flash/ square km/year)
T_d= the thunderstorm days (days/year)

TABLE 1 : Ground Flash Density (N_g) [5]	
Reference	**N_g (flashes/km^2/year)**
N. Klairuang *et al*	$6.5*10^{-5}T_d^{\ 2.277}$
R.B Anderson	$0.023T_d^{\ 1.3}$
D.Macherras	$0.01T_d^{\ 1.4}$
A.J. Eriksson (IEC)	$0.04T_d^{\ 1.25}$

In Finland, the average GFD was estimated to be 0.786 strokes/ km^2/year. Field data analysis has shown that the cumulative distribution of lightning stroke magnitudes $P\ (I_p)$ can be represented as shown in Figure 2. The probability density function given by Klairuang et al [5] fits the curve very well. This is given as;

$$P(I_p) = \frac{1}{1+\left(\frac{I_p}{I_{50}}\right)^{3.09}} \qquad (2)$$

where;

$P(I_p)$ = Probability that any peak return-stroke in any given flash will exceed I_p (kA),

I_p = Lightning peak current (kA),

I_{50} = 50% lightning current. (15kA is the average negative lightning peak current for Finland).

Therefore, a typical stroke in any flash is expected to comply with Figure 2.

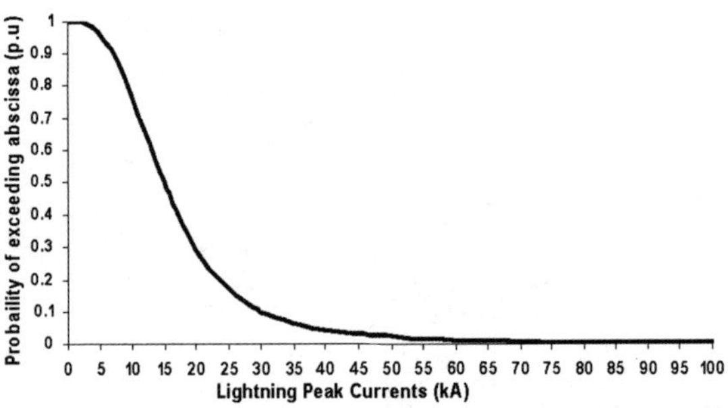

FIGURE 2. Cumulative distribution of Lightning peak current in Finland

SIMULATION PROCEDURE

Test system

In this study, the IEEE 34-node radial distribution test feeder was tested with direct multiple strokes with the test data in [6]. The test feeder diagram is shown in Figure 3. It was modeled with ATPDraw, as shown in Figure 4, and all data available in [6] were used for the simulations with slight modifications. Brief descriptions of the model are as follows;

The test system has an ac source of 69kV, 60Hz from the utility. It was model with a 3-phase infinite source.

- The 2.5MVA (69/24.9kV) substation transformer (T_1) was increased to 25MVA with its per unit impedance, in order to compensate for losses due to the inclusion of additional loads. The 5kVA in-line transformer (T_2)

(24.9/4.16kV) was used with its original parameters. The transformers were simulated in a saturable 3-phase delta-wye format based on the standard ATP GENTRAFO model.

- The two regulating transformers at bus feeder points 814 and 856 were excluded from the simulation as their impact would be negligible in this kind of analysis.
- All line parameters except line 816-824 were included as lumped or pi equivalents based on the IEEE 34-node specifications.
- A set of 20 distributed and unbalanced loads together with 6 spot and balanced loads were modeled as constant impedance loads.
- Line 816-824 of length 3.112 km, was modeled as a 10 pole distribution line of 9 equal spans. The whole line was simulated with the ATP line and cable constant (LCC) subroutine using the physical configuration of Finnish distribution lines, shown in Figure 5.
- A lightning flash of multiple strokes was assumed to terminate on phase B (the middle conductor) at pole number 6 along line 816- 824.

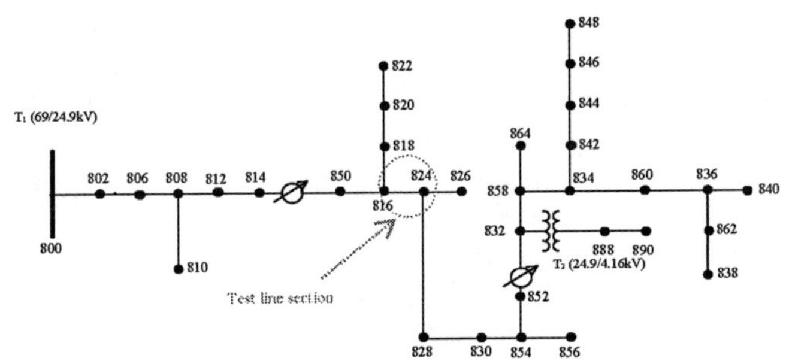

FIGURE 3. IEEE 34-node Test Feeder

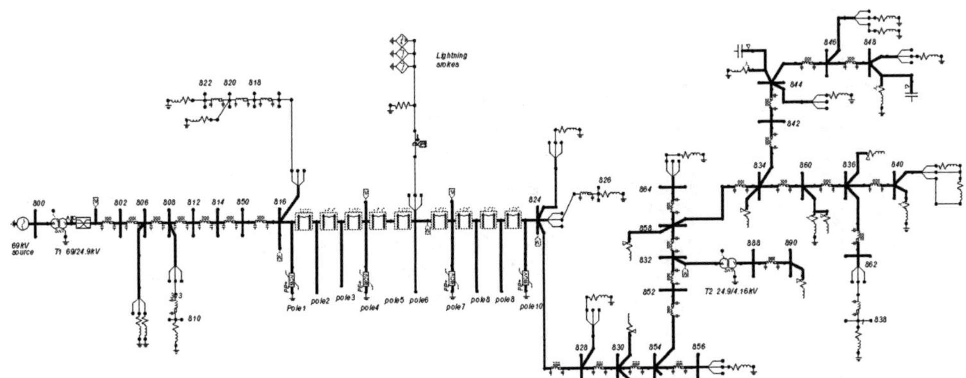

FIGURE 4. ATPDraw Model of the IEEE 34-node Test Feeder with line model modification from 816 to 824

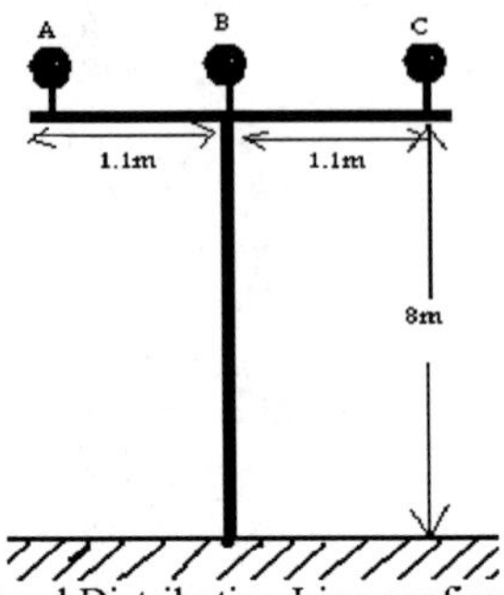

FIGURE 5. Finnish Overhead Distribution Line configuration (diameter -12.7mm)

Multiple Lightning Strokes

A selection of multiple lightning strokes of Figures 6 (a) and 6(b) were based on the mean values of positive lightning strokes observed in Finland in 2006 [4] with amplitudes of 18.6kA, 15kA and 12kA for the 1st, 2nd and 3rd strokes, with 1ms intervals. In the simulation, three ideal sources were used for the strokes, with time durations of 0.6ms for the 1st and 0.3ms for the 2nd and 3rd strokes. The 1st stroke was modeled with a *Heidler* ideal source (1.2/50 μs), Figure 6a, and the 2nd and 3rd strokes were modeled with the two slope ramp *Type 13* of ATPDraw, based on the characteristic of the lightning strokes in the lightning literature [7]. Figure 6(b) gives the waveform of the multiple strokes simulated.

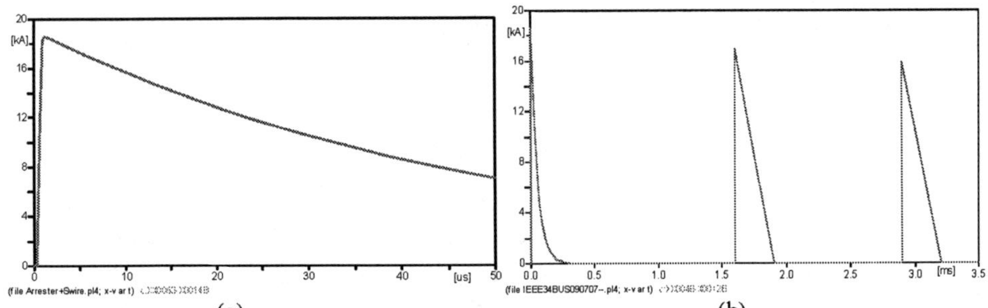

FIGURE 6. (a) Waveform of the 1st injected lightning stroke with amplitude of 20kA (1.2/50 μs). (b) Waveforms of the 1st, 2nd and 3rd positive lightning strokes modeled with ATPDraw

MOV arresters were employed for the protection of lightning surges at various locations on the test system. Obtained from the manufacturer's datasheet [8], the V-I characteristic curve of the arrester used for the simulation is shown in Figure 7.

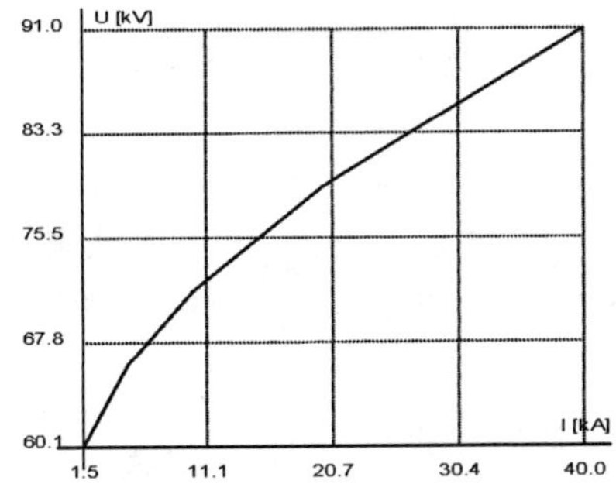

Figure 7. V-I Characteristic curve of the 30kV MOV arrester (8/20 μ s)

Case Descriptions and Simulation Results

All cases considered in this study were analyzed with the ATP/EMTP program based on fast transient studies of lightning surges [7]. Three cases were studied to observe the suppressing effect of lightning overvoltages from direct strokes due to the combination of shield wire and surge arresters. As shown in Figure 4, direct strokes were terminated on phase B, at pole number 6

Case 1

No protection was considered on feeder 816-824. The strokes terminated on phase B of pole number 6. Table 2 summarizes the result obtained from the simulations. Overvoltages obtained at points of observation are far above the flashover voltage and Basic Impulse Level (BIL) ratings of the feeder equipment (25kV/150kV BIL). Overvoltages at the terminals of the two transformers surpassed their BILs (350kV for 69kV & 150kV for 24.9kV). Figure 8 shows the overvoltages on all the phases at pole number 6. It can be observed here that the overvoltages in phase B resulted in induced overvoltages to the other phases due to mutual coupling effect of the line, with phase C's overvoltage more than half of the overvoltage in phase B and phase A's overvoltage slightly higher than the overvoltage in phase C at pole number 6. Overvoltages recorded at others poles and the transformers follow almost the same pattern.

TABLE 2. Peak Overvoltages at Poles -No Protection, Direct Multiples Strokes on Phase B, at Pole 6

	Phase A (MV)	Phase B (MV)	Phase C (MV)
Pole 1	1.86	4.25	1.46
Pole 4	3.31	7.16	3.49
Pole 6	5.92	9.01	5.80
Pole 7	4.92	8.00	4.67
Pole 10	3.39	4.18	2.14
Transformer 1	1.52	2.74	0.93
Transformer 2	2.22	1.59	2.16

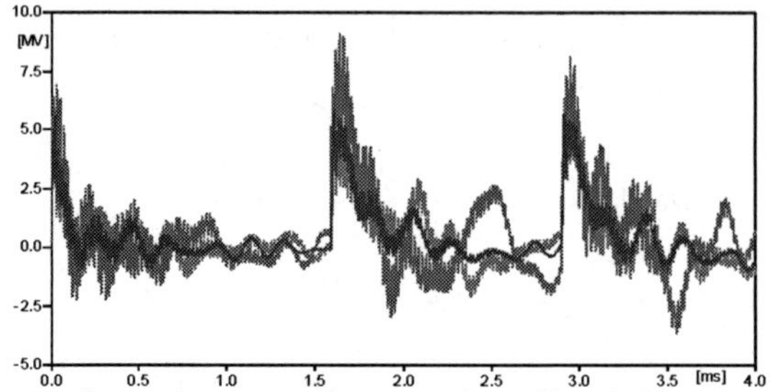

FIGURE 8. Case 1- Overvoltage waveshape at pole number 6 - No protection on the line

Case 2

Three MOV surge arresters were installed on phases A, B and C on each of the poles 1, 4, 7 and 10, but no arresters were installed at the terminals of the substation transformer terminal (T_1) and the in-line transformer terminal (T_2) transformers. It was assumed here that the strategic location of surge arresters at three pole intervals, starting from pole 1, would suppress the overvoltages due to the strokes simulated in this study. Table 3 summarizes the results and Figure 9(a) shows the remaining overvoltages clamped by the arresters at pole number 6. Only the highest peak overvoltages from multiple strokes are displayed for clarity. Also shown in Figures 9(b) and 9(c) are the overvoltages at the terminals of the two transformers, from the whole flash. Only the energies absorbed by the arresters at poles 4 and 7 are displayed in Figures 10(a) and 10(b) due to space limitation.

TABLE 3. Peak Overvoltages at Poles –With Arresters, Direct Multiples Strokes on Phase B, at Pole 6

	Phase A (kV)	**Phase B (kV)**	**Phase C (kV)**
Pole 1	43.50	54.90	31.70
Pole 4	67.87	79.36	66.74
Pole 6	2910	5230	3000
Pole 7	69.33	80.11	68.22
Pole 10	46.60	59.05	47.09
Transformer 1	20.32	73.73	56.40
Transformer 2	43.00	33.46	32.92

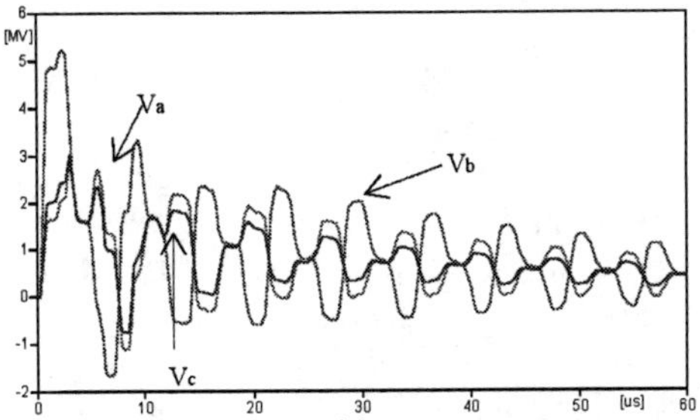

(a)

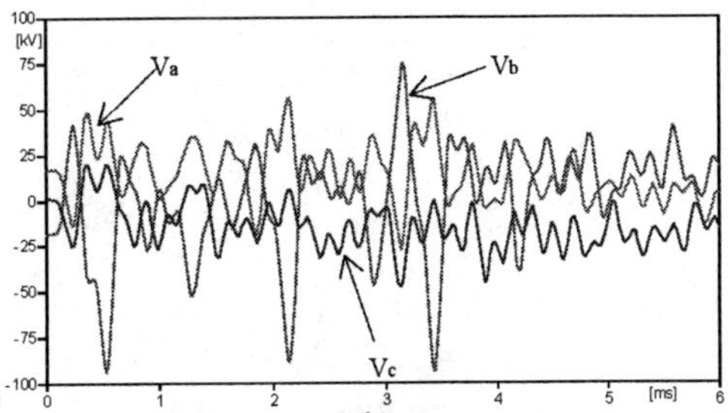

(b)

FIGURE 9 (a). Case 2- Remaining overvoltages from 1st stroke measured at Pole 6. Phase B has the maximum overvoltage from the multiple strokes.

FIGURE 9 (b). Case 2- Remaining overvoltages from the strokes measured at substation transformer (T_1). Phase B has the maximum overvoltage from the multiple strokes

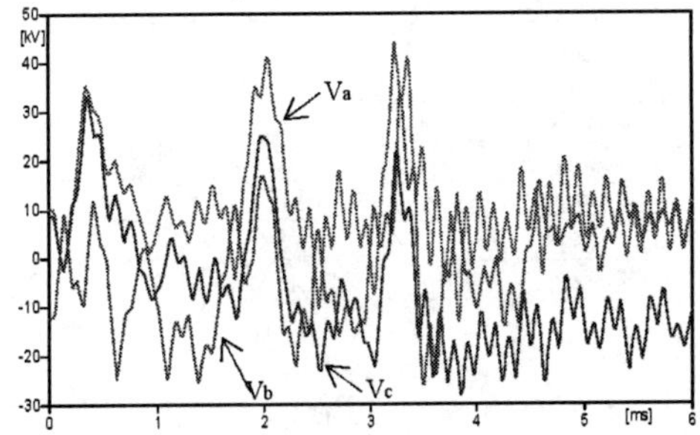

FIGURE 9 (c). Case 2- Remaining overvoltages from the strokes measured at in-line transformer (T_2). Phase A has the maximum overvoltage from the multiple strokes

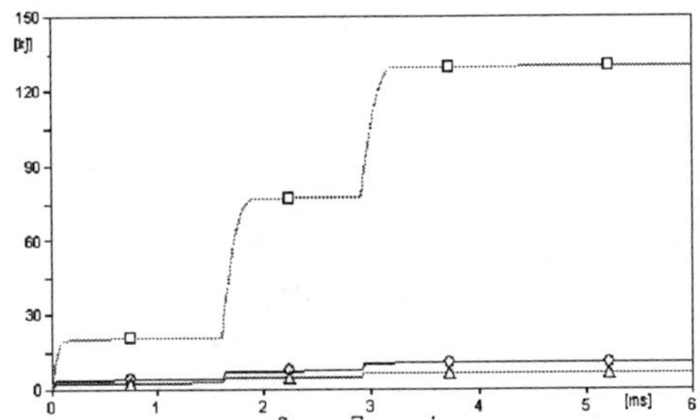

(a)

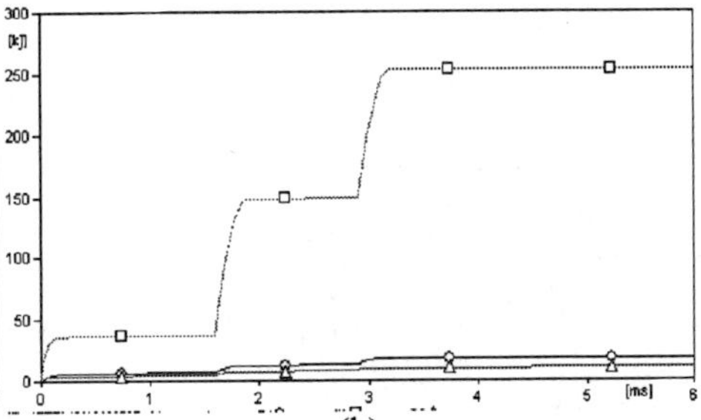

(b)

FIGURE 10 (a). Case 2-Energy absorbed by arresters at pole 4 after operation.

FIGURE 10 (b). Case 2- Energy absorbed by arresters at pole 7 after operation

Case 3

The shield wire was augmented with surge arresters; therefore the pole configuration of Figure 5 was modified with the inclusion of a shield wire 1.5m above phase B on line 816-824. It was assumed that both the shield wire and phase B were directly hit by the strokes as they were positioned in the same vertical plane as the lightning flash. The arresters were installed on all phases as in *Case 2* and the shield wire was grounded at the poles where the arresters were installed. Table 4 summarizes the results of the simulation and Figures 11(a), 11(b) and 11(c) show the remaining overvoltages at pole number 6 and the transformers. The energies absorbed by the surge arresters at pole numbers 4 and 7 are also shown in Figures 12 (a) and 12(b).

TABLE 4. Peak Overvoltages at Poles –with Arresters & Shield wire, Direct Multiples Strokes on Phase B, at Pole 6

	Phase A (kV)	Phase B (kV)	Phase C (kV)
Pole 1	43.29	56.52	33.29
Pole 4	60.50	79.64	58.70
Pole 6	900	4090	1170
Pole 7	62.27	80.46	60.50
Pole 10	47.79	57.82	49.53
Transformer 1	45.70	54.36	16.46
Transformer 2	75.71	50.75	51.99
Pole 1	43.29	56.52	33.29

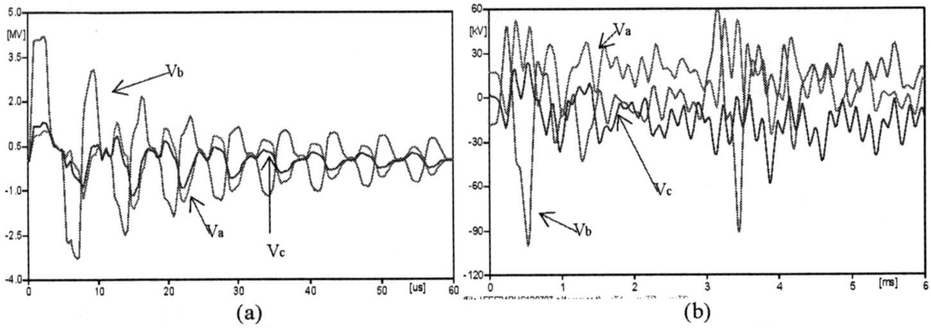

FIGURE 11(a). Case 3- Remaining overvoltages from 1st stroke measured at Pole 6. Phase B has the maximum overvoltage from the multiple strokes

FIGURE 11(b). Case 2- Remaining overvoltages from the strokes measured at substation transformer (T_1). Phase B has the maximum overvoltage from the multiple strokes

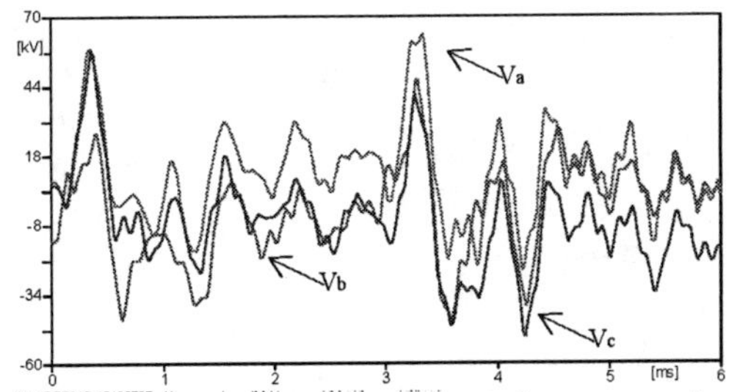

FIGURE 11(c). Case 3- Remaining overvoltages from the strokes measured at in-line transformer (T_2). Phase A has the maximum overvoltage from the multiple strokes

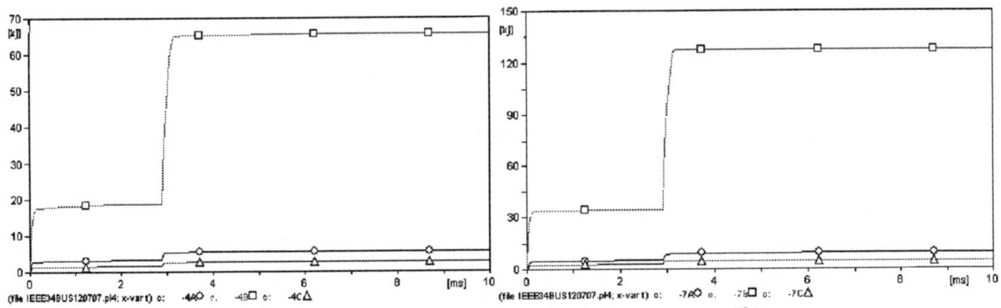

(a) (b)

FIGURE 12(a). Case 3-Energy absorbed by Arresters at pole 4 after operation.
FIGURE 12(b). Case 3-Energy absorbed by Arresters at pole 7 after operation.

DISCUSSION OF THE RESULTS

To quantify the need for the provision of shield wire to MOV-protected networks in severe lightning areas in a realistic simulation, an unbalanced IEEE 34-node test feeder was considered for the simulations made in this study, where the line 816-824 was assumed to be most prone to lightning strokes. All the tested cases showed the effects of direct strokes at the strike point, mutual overvoltages to other phases and the propagation effect at the transformers which are very distant to the strike point.

With the use of only arresters on line 816-824 at the designated poles, the overvoltages were reduced by 59%, 41.95% and 48.28% on conductors A, B and C respectively, at the pole number 6 where the lightning flash was terminated. Overvoltage reduction was observed at every other pole including the two transformers. However, energies absorbed by the arresters at pole numbers 4 and 7 (Figures 10(a) & 10(b)) located at the extremities of the strokes, were enough to damage the arresters even after successful operation, as they may require replacement for guaranteed protection in future. It will be recalled that the energy rating of the 30kV class MOV arrester is 74.8kJ, 3.4kJ/kV, where Uc= 22kV [8].

With the use of a shield wire on the lines and arresters at the designated poles, the overvoltages were reduced by 84.8%, 54.95% and 79.83% on phases A, B and C respectively, at pole number 6. In addition, energies absorbed by the arresters on pole number 4 were reduced by 49.73%, 72.19% and 73.23%, and those of pole number 7 were also reduced by 77.79%, 48.02% and 78% for phases A, B and C respectively (Figure 13). These results have shown clearly the effectiveness of adding shield wires to MOV-protected distribution networks in areas where power distribution lines are much more prone to direct strokes. Therefore, the combination of two protection types is very effective in suppressing to ground all overvoltages from the lightning strokes that the kind of network considered in this study is frequently subject to.

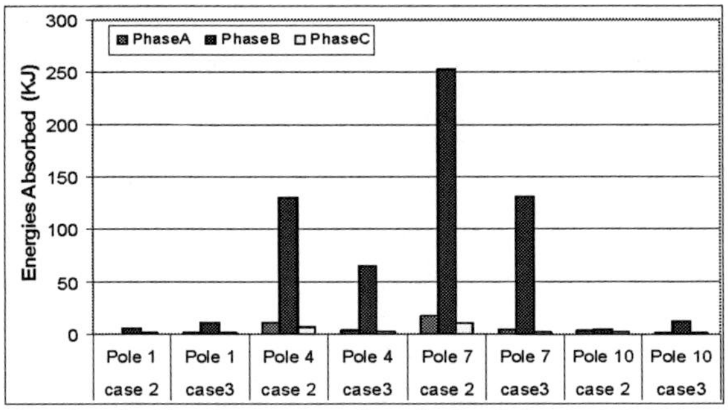

FIGURE 13. Energies absorbed by the Arresters installed at Poles 1, 4, 7 & 10. (Case 2-MOV only, Case 3 - MOV with Shield wire)

CONCLUSION

A distribution line located in a high lightning flash area is much more susceptible to lightning strokes than any other lines. For a given voltage level, surge arresters installed on distribution lines are mostly of the same rating and energy capabilities. However, surge protection will be more effective with the combination of shield wire and surge arresters in situations where there is the incessant occurrence of this natural phenomenon.

In this paper, the responsiveness of two protection schemes were simulated and analyzed, using MOVs only and MOVs with a shield wire. The following conclusions are drawn from the results obtained; Good design of protective measures against lightning strokes in an area solely depends on the knowledge of GFDs of that location, which can be used to determine the severity of any detected flash to distribution circuits. In distribution networks, multiple lightning strokes which occur in real situations are much more severe than a single lightning stroke, even in a situation where the surge arresters are adequately installed into the network. For the stroke magnitudes simulated in this work, surge arresters were not effective in suppressing the overvoltages due to direct strokes. Arresters at the extremities of the struck point may therefore suffer physical and electrical damage even after successful operation. However, shield wires can help relieve MOV-protected lines from overvoltage stress if installed in severe areas, such as were tested in this study. The study successfully establishes the need for an augmented protection scheme against lightning overvoltages in distribution networks where lightning incidence is most severe.

ACKNOWLEDGMENTS

The authors are grateful to Mr. Tapio Tuomi and Mr. M Antti of the Finnish Meteorological center for the provision of lightning data for this work. The discussions with Dr. John Millar are highly appreciated. Financial support from the laboratory of Power Systems and High Voltage Engineering of HUT is well acknowledged.

REFERENCES

1. V. A. Rakov and M. A. Uman, "Some Properties of Negative Cloud-to-Ground Lightning" Proceedings of 20th International Conference on Lightning Protection, Interlaken, Switzerland, 1990. Paper 6.4,
2. V.A Rakov,., M. A. Uman, R.Thottappillil, T. Shindo, "Statistical Characteristics of Negative Ground Flashes as Derived from Electric Field and TV Records" Proceedings of the USSR Academy of Sciences (Izvestiya AN SSSR ser. Energetika Transport), 37, No. 3, 1991, pp. 61-71.
3. V. A. Rakov, Uman, M. A., Thottappillil, "Review of Lightning Properties from Electric Field and TV Observations" J. Geophysics Res., 99, 1994, pp 10,745-10,750.
4. J. T Tapio, M Antti, *Lighnting observations in Finland*, Report of Finnish Metrological Institute, Helsinki, No. 6, 2006
5. N. Klairuang, W. Pobporn and J. Hokierti , "Effect of Electric field Generated by Direct Lightning Strikes on Ground to Underground Cables" International Conference on Power System Technology, POWERCON, Singapore, 2006, No 21-24,
6. W.H. Kersting, "Radial Distribution Test Feeders" Power Engineering Society Winter Meeting, 2001 IEEE. Volume: 2, 28 Jan.-1 Feb. 2001 pp. 908-912.
7. W.Scott Meyer, Tsu-huei Liu, *Alternative Transient Program (ATP) Rule Book*, Canadian/American EMTP User Group, EMTP Center, USA.
8. *Ultrasil Housed VariGap Arresters*, Cooper Power Systems Electrical Apparatus I235-39, April 2003.

A New Mode of Lecturing for Self-Directed Learning – Virtual Classroom on a DVD

Eliathamby Ambikairajah, Julien Epps, Ming Sheng, and Branko Celler

School of Electrical Engineering and Telecommunications, The University of New South Wales, Sydney, NSW 2052, Australia (phone: +61 2 93854002; fax: +61 2 93855993; email: ambi@ee.unsw.edu.au)

Abstract. Results of a large four-year longitudinal study of issues affecting student learning has motivated the development of a new mode of teaching, which takes the context of student learning into account and enhances student understanding of subject material. This new mode of lecture delivery is based on digital capture and student-controlled, user-configurable playback of synchronized lecturer dynamic annotation and video. This approach was tested in a large undergraduate course during which lectures were delivered entirely via pre-recorded lecture material on DVD, and in which the face-to-face teaching time was used instead for focused discussion classes. We present the results of a study of this novel delivery mode, using an electronic whiteboard and DVD capture. Our evaluations show convincingly that students are better able to review and understand lecture material.

Keywords: Self-directed learning, Educational technology, Evaluation, Teaching modes.
PACS: 01.50.ht, 01.40.gb, 01.40.Ha

INTRODUCTION

Students of today balance university study, part-time work and many other commitments. It is important that new and creative teaching methods are used in order to support students in balancing study and work and to help them learn effectively [9]. We have been seeking to understand the complete picture of the student experience in engineering studies, and searching for new ways to address problems in student learning. Having identified student learning issues, we have looked hard for a range of solutions, initiated many programs to address these, and in the process of addressing them, have come up with some radically innovative new approaches to teaching. These new methodologies of teaching that we have developed have proven to be highly effective, and help students to develop critical thinking skills that enable them to take control of their own learning process.

Maintaining student attention span is a perennial challenge in the teaching and learning environment, especially in a large class [1]. This is exacerbated by other difficulties students face, such as fatigue or non-attendance due to balancing part-time work or other commitments, or difficulties with comprehension for students whose first language is not English. As a result, only a small percentage of students are usually able to grasp the key concepts at the time of the live lecture delivery, while the remaining students are left to develop this critical understanding in their own time, with whatever assistance they can find and comprehend [2]. Previous studies [5] have shown that students prefer to study in their own time, when they are less fatigued,

CP1007, *Current Themes in Engineering Technologies*, edited by S.-I. Ao, M. A. Amouzegar, and S.-S. Chen

rather than attending regular structured classes. We believe that it is particularly critical for science and engineering students, where long and/or complex mathematical concepts are often conveyed, and in which the live lecture environment provides little support for any student who loses the train of a particular derivation or explanation, for any reason. For example, even with printed lecture notes and student note-taking, it is still easy for a student to miss a crucial step in the explanation of a difficult but conceptually important problem. In order to address the above problems and in pursuit of optimal student learning, students' individual learning styles need to be taken into account.

Previous attempts to address the problems have included videotaping of lectures [3], but this proved to be somewhat unsuccessful, as large scale equipment set up was required. Multiple cameras were needed and students were not compelled to watch these videotaped lectures on their televisions at home. Another method included a voiceover to accompany PowerPoint slides, which also proved to be ineffective, because students preferred to see a visual image of the lecturer explaining the concepts, rather than listening to a disembodied voice and trying to work out which equation or diagram the lecturer is referring to at a given point in time.

The objectives of revisiting live lectures were to:

- Gain a deeper understanding of the context in which student learning occurs;
- Propose new teaching methods to allow students to exercise greater control over their learning experience, and which address specific problems discovered from student feedback;
- Implement new technologies designed to convey the richness of the classroom experience to students engaged in self-directed study;
- Experiment with the novel technologies in a variety of different teaching modes during the regular teaching semester;
- Evaluate both the new educational technology and the delivery mode it facilitated, over a series of courses.

ISSUES AFFECTING STUDENT LEARNING

In order to understand directly from students the underlying problems and parameters of their learning approaches and behaviors, a survey of engineering students was conducted over four years, between 2003 and 2006 [2]. In each year, approximately 100 students participated, with a total of 400 respondents over the past four years. Participants were mainly (82%) in their third year of study, with the remainder (16%) mainly in the fourth year of a four-year degree program. The aim of the survey was to determine what duration and proportion of students' time was spent on learning activities (lecture, tutorial, laboratory attendance, class preparation, assignment work and study), perceived reasons for poor performance, what types of materials they used for their study, what financial support they received during their studies, and what other commitments competed with study for their discretionary time.

A further objective was to map survey findings over a period of time, to remove the dependence on a particular student cohort, and to identify likely trends.

The survey was conducted as a written questionnaire, comprising a total of thirty questions. Multiple choice style questions were used throughout, and covered student attendance of various types of classes and the reasons behind this, study patterns and approaches outside of class hours, the form and quantity of financial assistance received by students, the effectiveness of lectures versus self-directed study, and the effect of large class sizes.

Of considerable interest were students' ratings of their attendance at lectures and tutorials. As seen in Fig. 1, 100% lecture attendance has been dropping in trend terms, while students are increasingly dependent on both tutorials and self-study. More than 50% of students have part-time jobs. Significantly, the most common reason reported for non-attendance at class was tiredness, ranking above lack of interest and illness.

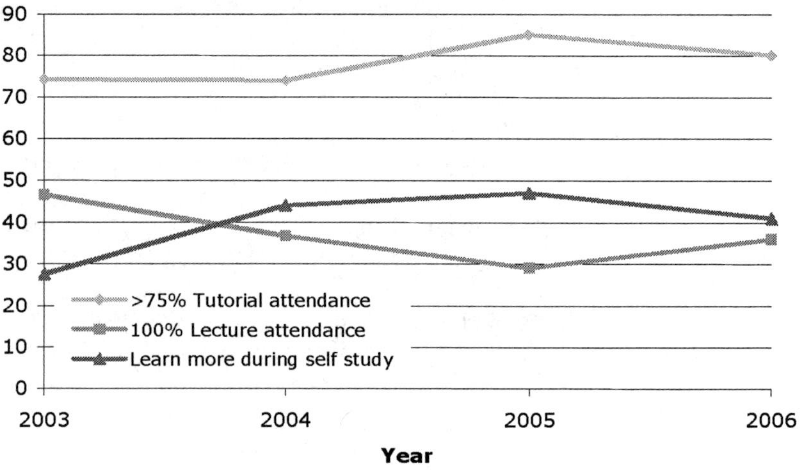

FIGURE 1. Tutorial, lecture attendance and use of self-directed study.

The use of text books as the main study aid was low (approximately 10% of students), while the most commonly used study aids were past exam papers, lecture notes and worked examples (Fig. 2).

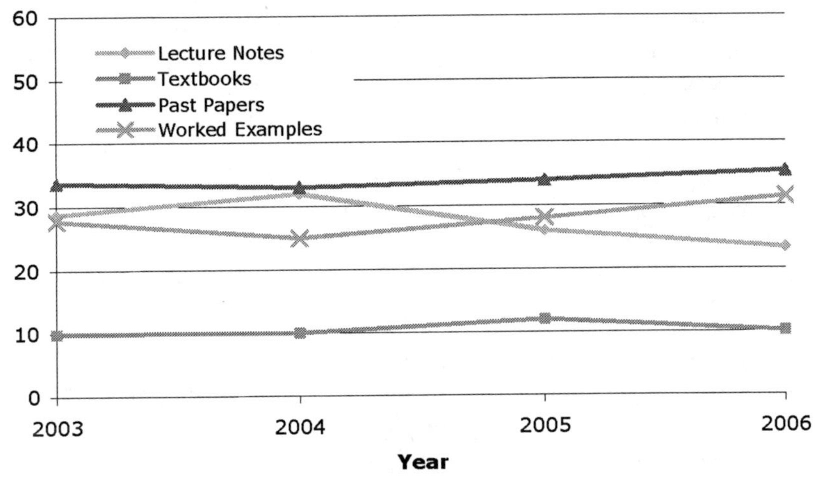

FIGURE 2. Main study aids.

From our survey it was clear that a significant increase in the proportion of students with part-time work was evident in trend terms even during the period of this survey. Clearly, there is an inverse relationship between part-time work and class attendance, with only 30% percent of the students attending 90% of the lectures with about 60% of the students doing part time work. Many students indicated in their responses that one of the reasons they were absent from university was tiredness and work.

The time students spent on study out of class hours (not including assignment work or laboratory preparation or write-up) was particularly revealing, with fewer than 10% (Fig. 3) of students spending more than 20 hours per week on study out of class.

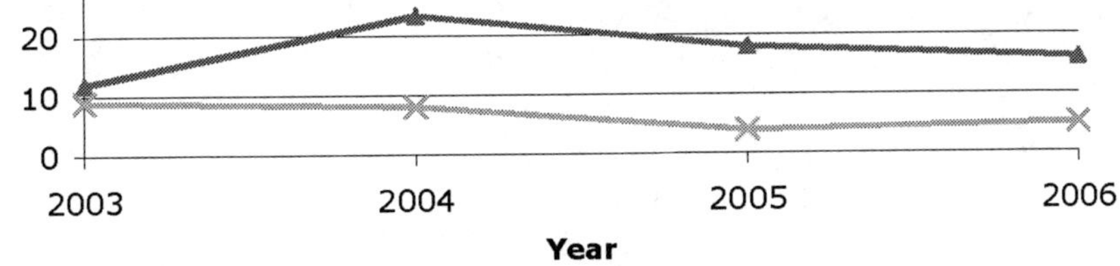

FIGURE 3. Time spent on study after class.

The University of New South Wales typically recommends that students study at least 20 hours per week outside of class hours, indicating that the majority of students are not pursuing self-directed study to the required extent. Many students indicated in their responses that one of the reasons they were absent from university was tiredness and work.

These surveys indicated that the ability to independently review live lecture content at their own pace outside of class hours would be of major benefit to students. Fig. 1 shows that there is a drop in lecture attendance, due mainly to part-time work, and therefore students who miss lectures need to have a way of accessing the live lecture at a later time.

In a large class lecture environment, students experience various problems that range from language difficulties, struggles in picking up difficult concepts and an inability to do ongoing revision throughout the course. Having access to live lectures, with full visual and audio impact, outside of class hours helps to alleviate these problems, encourages independent learning and time spent on study after class, which is low as highlighted in Fig. 3. In response to the survey findings, we developed a new mode of lecture delivery based on DVD capture and student-controlled playback of synchronized lecturer dynamic annotation and video. This gives students the ability to independently review live lecture content at their own pace outside of class hours.

DVD-BASED LECTURES FOR SELF-DIRECTED LEARNING

With the use of the electronic whiteboard and tablet PC [10], we pre-recorded the lectures and provided them to students on a DVD using software we developed (Fig. 4), called VCPlayer [3], [7], [8]. The DVD included the lecturer's annotated slides in digital format (See Fig. 5), along with a video and audio of the lecturer's presentation, which resembles the live classroom situation. Having the lectures on DVD allowed students to watch them as often as required in order for them to acquire the level of understanding needed. In addition, students could watch them in their own time and at their own pace, so their concentration and interest levels were high. Previous studies [4] have shown that in practice most annotations that are used are attention marks (e.g. highlighting in a different color), which provide critical linkage between spoken context and the slide content. We believe that the recording of rich non-persistent information such as video and dynamic annotation can be used to improve self-directed student learning or as preparatory material for more focused lecture/tutorial sessions which offer new possibilities for course delivery. Furthermore, unlike existing reports of some similar tools [6], we have applied the VCPlayer extensively in signal processing teaching, and describe an undergraduate evaluation herein.

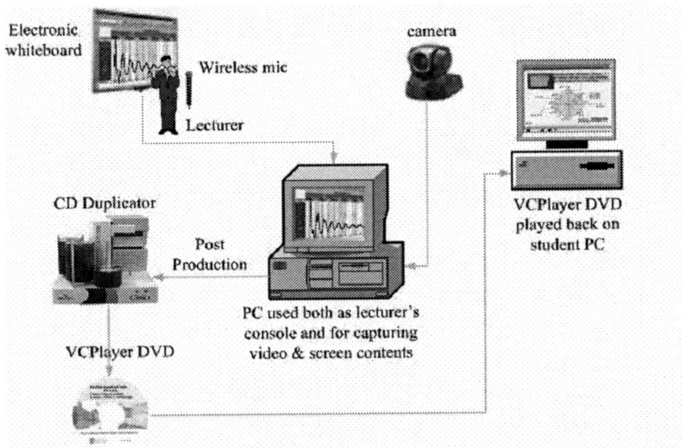

FIGURE 4. Live capture of lecture material using the electronic whiteboard and VCPlayer software.

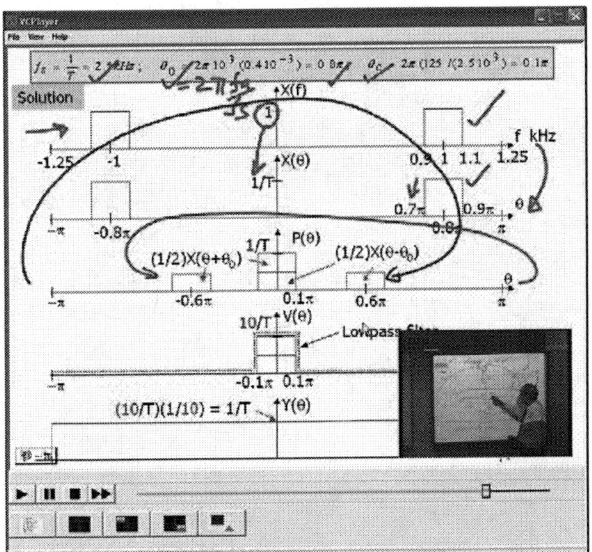

FIGURE 5. An example of captured lecture content.

From our experiences with this mode of teaching, the first lecture of any course should be delivered live, in order to motivate the students while familiarizing them with the delivery mode. In subsequent weeks, students then receive DVDs that capture lecture material in either of two modes:

(i) Lectures are given live, in the traditional format, except that the lecturer's video and annotation of electronic slides are captured using the VCPlayer, and later made available to students; OR

(ii) Lectures are pre-recorded using either electronic whiteboard or tablet PC, and then made available to students one week in advance of their class.

In the case of delivery mode (ii), we suggest that tutorial classes be made compulsory, and that an optional focused discussion class be run in place of the usual lecture time. The purposes of the discussion class are to give the students the face-to-face time needed to ask questions and resolve course-related issues. Obviously there is no reason why a delivery modes (i) and (ii) should not be combined as desired when designing a course. Students should be encouraged to watch the DVDs as many times as they need, in order to acquire the level of understanding they want. Ideally, a large proportion of students would actually watch at least selected portions of the lectures more than once. Assessment methods for this mode of delivery need not change from those employed concurrently with traditional lecture delivery.

RESULTS OF DVD-BASED TEACHING APPROACH

The self-directed learning mode (ii) outlined in the previous section was evaluated for the third-year undergraduate course "Signal Processing I", which ran over 14 weeks at UNSW in 2006, and was attended by 113 students. In this study, after a live introductory lecture was given in the first week, all lecture material was delivered via DVDs pre-recorded in the VCPlayer format (for playback on student PCs). Optional discussion classes were also scheduled during the allocated lecture slots. Compulsory tutorials and laboratories were scheduled and their format was traditional. Assessment of student understanding of the DVD-based lectures was conducted by a series of three surveys, each completed by students at the end of their class exams. The surveys consisted of eight questions, of which seven used a Likert scale (from strongly agree to strongly disagree) and one comprised a free-format comments field. The questions in all surveys were identical. The numbers of participants in the three surveys were 113, 104 and 88 respectively.

As illustrated in Fig. 6, responses to the first seven questions of the surveys showed that the acceptance of the DVD-based lecture delivery mode was supported by about 80% of students. Around 10% of the students were undecided whether DVD-based or traditional lectures were preferable, and around 10% did not support the DVD-based lecture format.

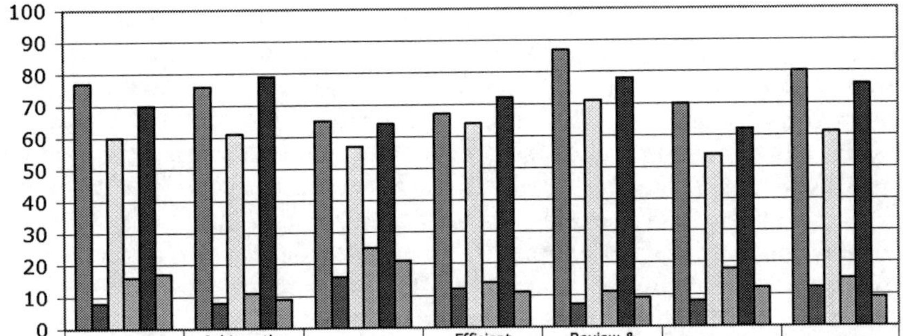

	Acceptable alternative	Achieve the level of understanding	Preferred choice	Efficient Learning Method	Review & understand better	More learning	Recommend to other courses
Agreed April	77	76	65	67	87	70	80
Disagreed April	8	8	16	12	7	8	12
Agreed May	60	61	57	64	71	54	61
Disagreed May	16	11	25	14	11	18	15
Agreed June	70	79	64	72	78	62	76
Disagreed June	17	9	21	11	9	12	9

FIGURE 6. Results of seven questions from a survey of DVD-based self-directed learning mode show that students consistently support this mode of delivery.

While these are positive results, clearly caution must be exercised in balancing the use of the pre-recorded lecture format with other methods of lecture delivery. The remaining (eighth) survey question revealed that 80% of students used the DVD for every lecture, while 34% used it more than once (Fig. 7). Over the duration of the course, only 30-40% of students attended the optional discussion classes, perhaps indicating their preference for the DVD-based lectures.

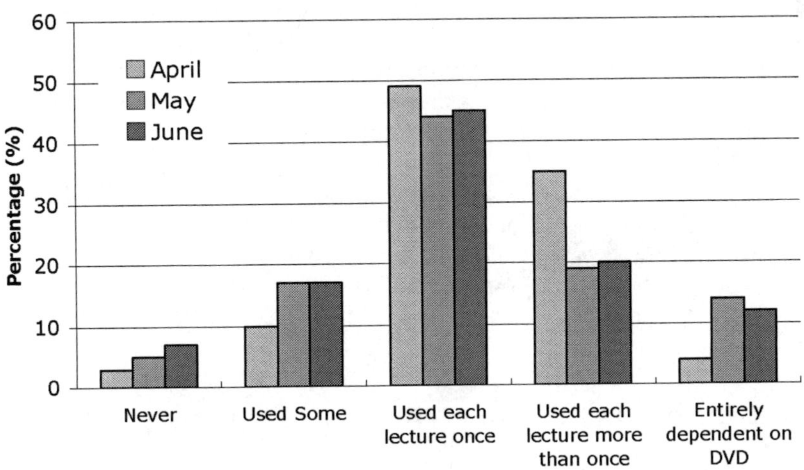

FIGURE 7. Results of the eighth survey question concerning how often students used each lecture recording on DVD.

Furthermore, an analysis of student pass rates over the duration of this study was carried out. Table 1 shows that results from the midterm exams and overall marks for the same course (given by the same lecturer) across four consecutive years. As indicated in this table, week 9 results are traditionally poor (with standard lecture delivery), since students just begin to realize the deficiencies in their understanding of signal processing concepts. Although the assessment scheme (in terms of optional versus compulsory class exams) differed slightly and the group of students was different from one year to another, these results suggest that a selective use of the DVD-based delivery mode combined with traditional course delivery may enable students to achieve higher levels of understanding.

TABLE 1. Comparison of mid-term and final exam pass rates for different lecture delivery modes.

	Students	**Week 4 (April)**	**Week 9 (May)**	**Week 14 (June)**	**Final Pass Rate**
2003 Traditional	121	Not held	48.7%	Not held	81.0%
2004 Traditional plus some AVI tutorials and electronic whiteboard discussions	118	Not held	43.8%	70.1% (compulsory)	86.4%
2005 Mixed traditional (80%) and DVD-based (20%)	129	Not held	37.4%	72.0% (compulsory)	89.9%
2006 DVD-based (100%)	113	74.2%	66.1%	60.0% (optional)	85.1%

COMMENTS FROM STUDENTS

Many of the free-format comments from students in the survey supported the findings in Fig. 5, for example:

"The DVD is a great way to save time and make learning more efficient and personal . . . it is really beneficial that you can skip the bits you already know and concentrate on the more difficult ideas";

"DVDs are better as this eliminates distractions due to people talking in class";

"I can revisit more difficult concepts more than once and develop a better understanding. This could benefit the lecturer as they may not have to re-explain concepts at a consultation";

"If I can't understand a concept early in the lecture, I can go back before trying to understand the later material which could depend on the previous concept, and avoid becoming 'lost' as I might in a live lecture";

"I think it is a good idea and should be used in other courses within the school".

CONCLUSION

This paper presents results from a survey of issues affecting student study behavior and learning. Key implications were that students increasingly had more part-time work commitments competing for both their class attendance and their discretionary study time, and that they preferred lecturer-generated materials such as worked solutions, past exam papers and lectures notes to text books as study aids. It was also found through the survey that students are not spending the recommended time studying after class.

This motivated the introduction of a new mode of teaching, using DVD-based electronic whiteboard and video capture. The proposed mode of delivery enables self-paced, individualized learning for students with a range of different preferences or difficulties, and offers a genuine alternative to the large classroom environment while allowing students to achieve their desired level of understanding. The novel method of course delivery has been highly commended by students, and results presented herein show potential for increasing student understanding of course material beyond the level that would be possible using traditional methods. This teaching methodology is currently being trialed in many other courses within the School. Future work will include progressive evaluations of this novel method of teaching delivery.

ACKNOWLEDGMENT

This project was funded by the Australian Government Department of Science and Training (DEST) Capital Development Pool Program 2004-2005 (Education Delivery) and UNSW School of Electrical Engineering Educational Technology Development funding (2006).

REFERENCES

1. J. Foreman, NEXT-Generation: Educational technology versus the lecture, EDUCAUSE Review, July, 2003, pp. 13-22.
2. E. Ambikairajah, Issues Affecting Student Learning, School of Electrical Engineering and Telecommunications Report, The University of New South Wales, 2006.
3. E. Ambikairajah, J. Epps, M. Sheng, B. Celler, and P. Chen, "Experiences with an electronic whiteboard teaching laboratory and tablet PC-based lecture presentations", in Proc. IEEE ICASSP, vol. 5, 2005, pp. 565-568.
4. R. J. Anderson, C. Hoyer, S. A. Wolfman, and R. Anderson, "A study of digital ink in lecture presentation", in Proc. Conf. on Human Factors in Comp. (CHI), 2004, pp. 567-574.
5. J. A. Brotherton, and G. D. Abowd, "Lessons learned from eClass: Assessing automated capture and access in the classroom", ACM Trans. Computer-Human Interaction, 11(2), 2004, pp. 121-155.
6. M. Ma, V. Schillings, and C. Meinel, "T-Cube: A multimedia authoring system for eLearning", in Proc. AACE E-Learn, 2003, pp. 2289-2296.
7. M. Sheng, B. Celler, E. Ambikairajah and J. Epps, "Development of a virtual classroom player for self-directed learning", Proc. Int. Conf. on Multimedia and Inf. & Comm. Tech. in Education (Caceres, Spain), 2005, pp. 718-721.
8. E. Ambikairajah, J. Epps, M. Sheng, B. Celler, "Signal Processing Education using the TabletPC and Electronic Whiteboard", IEEE Signal Processing Magazine, 2007, pp. 130.1-33.
9. H. A. Latchman, C. Salzmann, D. Gillet, and H. Bouzekri, "Information Technology Enhanced Learning in Distance and Conventional Education", IEEE Trans. On Education, vol. 42, No. 4, 1999, pp. 247-254
10. X. Teng, J. G. Tront, B. Muramatsu, and A. Agogino, "Best Practices in the Design, Development and Use of Courseware in Engineering Education", in Proc. of the 35th ASEE/IEEE Frontiers in Education Conference, 2005.

DCBITS: Distributed Case Base Intelligent Tutoring System

O.P.Rishi[1] and Rekha Govil[2]

[1]O. P. Rishi is with the Birla Institute of Technology, Extension Centre Jaipur, BISR campus, 27 Malviya Industrial Area Jaipur 302017 INDIA phone: + 91 94142 58030; fax: + 91 141 2751601; (e-mail: omprakashrishi@ yahoo.com / oprishi@bitmesra.ac.in).

[2]Rekha Govil is with the Apaji Institute of Mathematics & Applied Computer Technology, Banasthali University, Rajasthan-304022 INDIA (e-mail: grekha@banasthali.ac.in,).

Abstract. Online learning with Intelligent Tutoring System (ITS) is becoming very popular where the system models the student's learning behavior and presents to the student the learning material (content, questions-answers, assignments) accordingly. In today's distributed computing environment, the tutoring system can take advantage of networking to utilize the model for a student for students from other similar groups. In the present paper we present a methodology where using Case Based Reasoning (CBR), ITS provides student modeling for online learning in a distributed environment with the help of agents.

Student modeling is a key component in any Intelligent Tutoring System (ITS). In today's distributed computing environment, the tutoring system can take advantage of networking to utilize the model for a student for students from other similar group. In the present paper we present a methodology where using Case Based Reasoning (CBR), the ITS provides student modeling for online learning in a distributed environment with the help of agents.

The Chapter describes the approach, the architecture, and the agent characteristics for the student modeling in the ITS. This concept can be deployed to develop ITS where the tutor can author and the students can learn locally whereas the ITS can model the students' learning globally in a distributed environment. The advantage of such an approach is that both the learning material (domain knowledge) and students' model can be globally distributed thus enhancing the efficiency of ITS with reducing the bandwidth requirement and complexity of the system.

Keywords: CBR, ITS, Student Modeling, Distributed System, Intelligent Agent, Personal Agent, Teaching Agent, Course Agent.

INTRODUCTION

The major challenge in teaching is to improve both instructional productivity and learning quality for large and diverse population of students under real world constraints such as limited financial resources and insufficient number of qualified instructors [1]. Different researches in the field suggest that students who are engaged in learning through intelligent tutoring processes are more likely to achieve success [2].

Since mid 90's, few educational system models were web based and used ITS with student modeling in distributed style [2][8]. In past decade, researchers from different

CP1007, *Current Themes in Engineering Technologies*, edited by S.-I. Ao, M. A. Amouzegar, and S.-S. Chen

disciplines have come out with systems which define and classify different teaching and learning styles in distributed environment [2][9].

Here we conceptualize a Case Based Distributed Student Modeling (agent based) ITS architecture to support student-centered, self-paced, and highly interactive learning. In this system the first step in building an effective learning environment is building a Case Base where the system maintains a rich set of cases (scenario) of student's learning pattern, and employs an efficient and flexible case retrieval system. This maximizes the interactivity between the ITS and the students; and customizes the learning process to the needs of an individual student [3]. The system must use the student's learning profile such as learning style and background knowledge in selecting, organizing, and presenting the learning material to support case based learning. It also supports personalized and more intensive interaction between the student and the ITS [4]. Distributed CBR based student modeling enables adaptive delivery of educational contents and facilitates automatic evaluation of learning outcomes[1][9]. The system also incorporates a new approach to course content organization and delivery, which can be developed, based on distributed and agent based instructional components. Instructional components represent the customized interactive presentation of any topic of a subject or different subjects.

It is in this context that this research focuses on the design of a *Distributed Case Based Intelligent Tutoring System* architecture abbreviated as *DCBITS*. The chapter describes the approach, the architecture, and the agent characteristics for such system. This concept can be deployed to develop ITS where the tutor can author and the students can learn locally whereas the ITS can model the students' learning globally in a distributed environment. The advantage of such an approach is that both the learning material (domain knowledge) and student model can be globally distributed thus enhancing the efficiency of ITS with reducing the bandwidth requirement and complexity of the system.

PROPOSED SYSTEM FOR DISTRIBUTED CASE BASED REASONING

The DCBITS system is based on finding a case that is a similar to the learning domain of a past student in a distributed environment. A similar learning domain means that when a student enrolls in the system and on presentation of content, questions, examples, and assignments, it responds accordingly. CBR Based student modeling system compares the student's solving process with the patterns of responses stored in the case base. At this time, the inference engine reasons the state of the student's knowledge through the student's solving process. If a new situation occurs then case based system treats this as a new case and records it, termed as machine leaning.

The DCBITS work with the help of a number of specialized agents with different expertise. There is a *personal agent* that manages the student's personal profile, including knowledge background, learning style, interests, courses enrolled in, etc. The *personal agent* talks to other two agents in the system namely t*eaching agent* and c*ourse agent* through different communication channels situated in distributed environment (this is more precisely discussed under architecture).

A web based course is supported jointly by a *teaching agent* and a *course agent* that manage course material and course-specific teaching techniques and student modeling strategy. Multiple *course agents* may exist on distributed sites to provide greater efficiency, flexibility, and availability. The *teaching agents* can talk to any *course agent*, and often choose one nearby for better performance. The *course agents* also act as mediators for communication among students and tutors. A *teaching agent* interacts with a student and serves as an intelligent tutor of a topic or course. From a *course agent*, each *teaching agent* obtains course material and course-specific teaching techniques and then tries to teach the material in the most appropriate form and pace based on the background and learning style of the student [5]. Lecture note, presentations, multiple examples with different difficulty level are used to make difficult concepts and operations easy to understand.

ARCHITECTURE OF DCBITS

In this work we propose and developed the complete software architecture of Distributed Case Base ITS (DCBITS). DCBITS presents an environment for developing an agent based ITS which works in a fully distributed mode. It uses case based reasoning for student modeling. Thus, over World Wide Web on the internet or an intranet, or on a LAN, in a distributed environment tutors can develop content for the curriculum and any student can log in and learn any where any time. The main feature of the system is its student modeling, which stores/ indexes the learner's behavior in a case base and models a student using CBR. All the communication between the tutor and system, and, the learner and system is maintained by agents. The chapter describes the design architecture, network architecture, agent architecture and access architecture of DCBITS.

DCBITS is a web oriented system that can be hosted on both the Internet and an Intranet. In DCBITS one course presents an integrated structure of many learning resources that can be hosted on different web locations. The same resources can be combined with others in different courses. Also, more student groups can learn many courses at the same time. In these conditions, the system must have powerful management features. This means that DCBITS needs specialized ITS properties as well as the capacity to perform the administration, integration and distribution tasks.

Student modeling in DCBITS is based on finding a case that is similar to the learning domain of a past student in a distributed environment. A similar learning domain means that when a student enrolls in the system and on presentation of content, questions, examples, and assignments responds accordingly, the CBR based student modeling system compares the student's solving process with the patterns of responses stored in the case base. At this time, the inference engine reasons the state of the student's knowledge through the student's solving process. If a new situation occurs, then the case based system treats this as a new case and records it, termed as machine leaning. The system updates its knowledge by prompting the authorized tutor to provide content for this new case.

Design Architecture of DCBITS

The general architecture of DCBITS consists of three basic parts: *administration module*, *tutor module,* and *student module*. All the three modules are present on every node. The knowledge of ITS and data on students and tutors are distributed over the network and are managed by agents. The administrative tools support the realization of different management tasks e.g. maintenance of student and teacher records, administration of the domain knowledge and the system security protection.

The Tutor module of the system help teachers to create learning objects, combine them with existing learning objects and compose the courses. Domain experts can design the domain ontology that should describe and structure the knowledge. The tutor module provides the monitoring of student's results that teachers can use to track student sessions with DCBITS. The tutor/teacher tasks are well known. A teacher can create ones own courses. These courses can be about different domains. Like as in the Case base Tutoring System, the teacher can monitor his students' results on individual basis as and where (s)he uses the system. (S)He can modify the learning contents during the students learning. Students are divided into different groups as per their learning and they access the courses according to their group. Their communication with the system (logging into the system, customizing the interface, learning the course chapters, solving the tests and accepting the skill levels and recommendations) runs over any web browser. The system is designed to support changeable navigation possibilities to the student. It provides dynamic creation of the learning material.

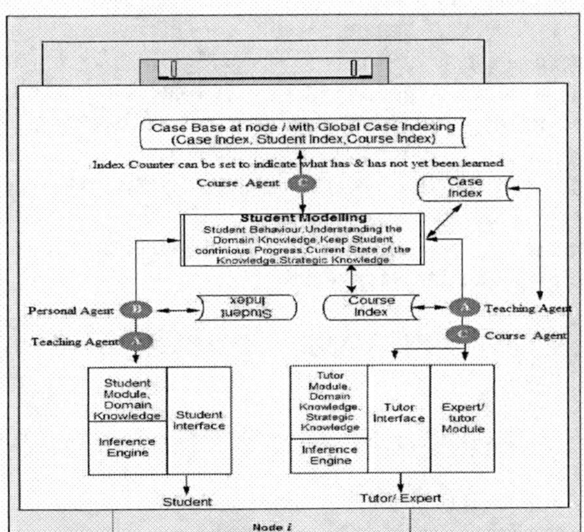

FIGURE 1. Architecture of DCBITS

The Student module facilitates a student to interact and learn through the use of DCBITS. The student can register, choose course/topic to learn, access oneself. The student can access the system through web portal where s(he) has an account. The system models the student's learning behavior. The design architecture of DCBITS is shown in figure 1. Figure 1, shows the core of the system is the Student modeling concept. The tutor is the main part of the system architecture. It is the system coordinator, dispatcher and monitor at the same time. The pedagogical strategies are implemented in the tutor module. It analyzes the data of the student model (model of particular student) and uses its teacher knowledge to acquire the proper learning contents. The expert module maintains the references of domain knowledge and rule base. The reasoning machine processes the request of the tutor and composes the learning content. This content can include the text, the picture or some other multimedia. In the test phase the content is represented by the test sets or by the problems that students have to solve. These contents the tutor sends back to the system.

The student module generally helps students to master the knowledge. The system enables a student to declare his/ her interests, favorites, predisposition and real skills. These data help the system to initiate a student model and determine a student stereotype. While the student uses the system, different tools provide her/him navigation through the learning space, marks for important things, contextual help and skills measurement. The student can also collaborate with other students, teachers and experts. This is a way that DCBITS provides high cohesion and synergy of efforts from all the subjects in the learning process. The system knowledge is transparent and distributed on the Web. It becomes possible to use concepts of the Semantic web integration process in the adaptive composing of learning materials. Different specialized pedagogical knowledge becomes accessible for all interested subject (users of the system) over the semantic web.

If the performance of the student is not satisfactory then *ITS* presents the appropriate remedial course content to the student. Only after achieving the learning objectives the student is allowed to move to a new topic. If any topic (case) is not available at the local node then the system searches it from other nodes within the network through course agent and with the help of course index. If the case is not found on the entire network then the *ITS* interacts with the human tutors for "plug-in topics". These new plug-in topics are stored on the node in use and at least one more node for fault tolerance. The course index is also updated.

Network Architecture of DCBITS

DCBITS is a web based distributed system that can be hosted on both the Internet and Intranet. This means the learning content is distributed to the students via the Web server. The user is on any learning node and (s)he accesses the learning resources using the web browser. The node sends the request through standard web page. The web server forwards this request to the application server. The application server processes the request and returns the results usually in the form of dynamically generated standard web page. The web server dispatches this page to the appropriate client/ student. A typical network architecture of DCBITS is shown in figure 2 where every node has the architecture shown in figure 1:

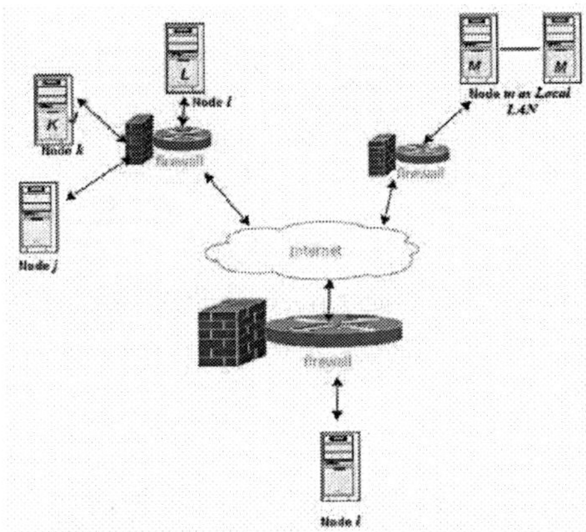

FIGURE 2. Network architecture of DCBITS

Any student can register as a user at any node over the entire network. The student can select any topic for learning. This can be done through navigation of topics provided by the ITS. The ITS presents the right content to the students. The management and communication with students and tutor is done with the help of appropriate agents. DCBITS uses three agents' namely *personal agent*, *course agent* and *tutor agent*. These agents are described in the next section.

Agent Architecture of DCBITS

DCBITS presents an environment for developing an agent based ITS which works in a fully distributed mode with the help of a number of specialized agents with different expertise. It uses three agents namely *personal agent*, *course agent* and *tutor agent*. All the communication between the tutor and system, or, learner and system is maintained by these agents (figure 3).

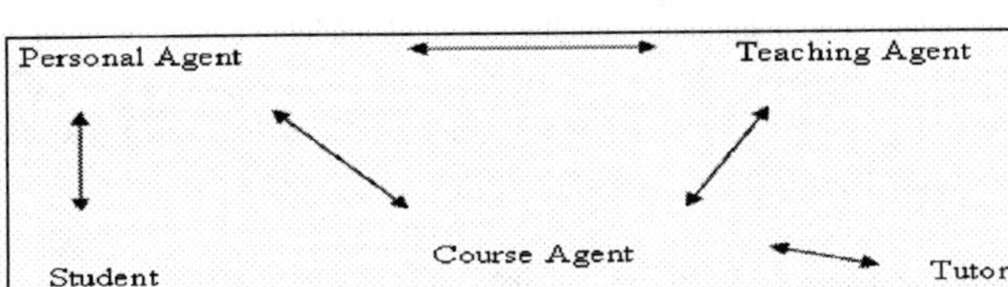

FIGURE 3. Communication model among Agents

The *personal agent* manages the student's personal profile, including knowledge background, learning style, interests, courses enrolled in, etc. The *personal agent* talks to other two agents in the system namely *teaching agent* and *course agent* through different communication channels situated in distributed environment

In the distributed environment of DCBITS the student module, the tutor module and case based student modeling reside on every node of the network. *The teaching agent* indexes all the cases globally and *personal agent* is responsible for indexing all the registered or aspirant students (learner) with their personal profiles which include the static background profile (such as GPA or percentage, majors, interests and course taken) and the students' dynamic profile globally (based on their interaction on the agent). The *course agent* manages the course material as per student's requirement and *teaching agent* decides the topics or lessons, to be supplied to the student according to student performance. The student modeling keeps the record in case base for future reference and manages the student performance and course material accordingly.

Thus the whole system is managed in the distributed environment with just three agents:

- Personal Agent,
- Teaching Agent, and
- Course Agent

Associated with these agents there are also three global indices: Student Index, Case Index, and Course Index. The Student Index maintains the record of system log file (IP addresses of the entire network) and student log file (User ID) registered in the entire network managed by the personal profiler or *personal agent.*

The *personal agent* maintains the students' profile master, students' learning profile, and usage profile. In profile master user ID, registered node, user personal profile, course ID, and preliminary or past performance, level of student (poor, average, excellent), category of the student(slow learner, normal learner, fast learner) are stored. Similarly, learning profile maintains the record of subject ID, topic ID and topic name, pre-leaning performance, and post learning performance, performance of individual topic and subject, type of errors performed by the student. The usage profile keeps the record of system ID, time in, time out, topic name, number of attempts, questionnaires etc.

The Case Index maintains the record of cases managed by the *teaching agent.* In Case Index, case type, content of each case, case adaptation condition, case indexing strategy, case retrieval method, case recall method, case updation (addition, deletion, modification), error rate, teaching strategy, location of the case where it is stored, and the remedial strategy used by the system, are stored and maintained.

The *teaching agent* decides the teaching strategy by using learning profile of the student and thereafter presents the right content to the student with the help of *course agent* using Case Index. The Course Index managed by the course agent maintains the record of available course content like ordering or sequencing of course material, set of related examples, assignments, questionnaires, set of exercise problems to access the student's understanding of the top and remedial content.

For the purpose of fault tolerance all storage in the distributed environment are maintained in duplicate. It means if a new case is diagnosed and stored at node *i*, a copy of it is also stored on node *j*, *j* being the next nearest node available. Later, one can research on optimum redundancy parameters for it will be better that the node where it is duplicated is chosen in such a manner so as to make network diameter minimum. Thus the complete learning process is managed with the help of these three agents; *personal agent*, *teaching agent* and course *agent*, and three global indices; student index, case index and course index.

In DCBITS web based course is supported jointly by a *teaching agent* and a *course agent* that manage course material and course-specific teaching techniques and student modeling strategy. Multiple *course agents* may exist on distributed sites to provide greater efficiency, flexibility, and availability. The *teaching agent* can be personalized i.e. one for each student curentely loged in. The *teaching agents* can talk to any *course agent*, and often choose one nearby for better performance. The *course agents* also act as mediators for communication among students and tutors. A *teaching agent* interacts with a student and serves as an intelligent tutor of a topic or course. From a *course agent*, each *teaching agent* obtains course material and course-specific teaching techniques and then tries to teach the material in the most appropriate form and pace based on the background and learning style of the student [10]. Lecture note, presentations, multiple examples with different difficulty level are used to make difficult concepts and operations easy to understand.

CBR Based student modeling system can be divided into three parts. A Case Based ITS (Intelligent Tutoring System), a student interface module, and, the instruction processing engine or agent communication system. The ITS is divided into a student module, a tutor module with expert module (human tutor interaction), and an interface module. The student module consists of a Case Base, which is based on an overlay or bug part library system.

Any student can register as a user at any node over the entire network. The student can select any topic for learning. This can be done through navigation of topics provided by the ITS. However, before allowing for navigation, the *personal agent* of the system checks the past record of the performance of the student to present the right content

If the performance of the student is not satisfactory then *teaching agent* presents the appropriate remedial course content to the student with the help of *course agent*. Only after achieving the learning objectives the student is allowed to move to a new topic. If any topic (case) is not available at the local node then the system searches it from other nodes within the network with the help of course index. If the case is not found on the entire network then the *teaching agent* interacts with the human tutors through *course agent* for "plug-in topics". These new plug-in topics are stored by the *course agent* on the node in use and at least one more node for fault tolerance. The *Course Agent* also updates the course index. This is the responsibility of the *course agent* to decide as to where should the duplication be made. Thus *course agent* is responsible for the course management.

Within the distributed student modeling, the *teaching agent* and *course agent* work for the tutor module of the ITS. The *teaching agent* does not interact directly with a student but through *personal agent* and serves as an intelligent tutor of a course. The basic components of a *teaching agent* are domain expert module and pedagogy. The architecture of *teaching agent* is shown in the figure 4.

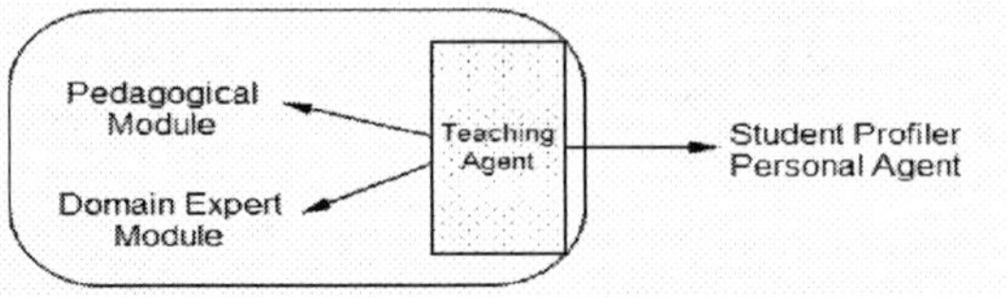

FIGURE 4. The Teaching Agent

The domain expert module is responsible for storing and managing the teaching strategy about learning content, exercises, assignments, and questions with the help of learning profile and *course agent.* It contains a problem generator, a problem solver, an explanation generator, and a domain knowledge base. The pedagogy module determines the timing, style, and content of the teaching agent's interventions. It is a rule based production system that uses the student model and pedagogical knowledge to determine the appropriate actions. The student profiler provides a model of a student based on learning style, knowledge background, and interests. It may also incorporate the information about student's learning profile gathered through dialogue with the student such as the actions the student performed and the explanations which student made

Let us think of a scenario of a student interacting with the ITS:

1) The student logs in either as a new user or a registered user from any node *i* of the network. In the first case the *personal agent* records the student's personal profile at node *i* and updates the global student index. In the second case there could be two possibilities : node *i* is the node which contains this student's profile, or , the student's profile is present at any other node *j* of the network (*j=1, n; j<>i*). In any of these two cases the student's personal profile shall be available to the system with the help of the *personal agent.*

2) Student starts session with the ITS. If new user, then the ITS first assesses the student's level and creates his/her learning profile whereas for a registered user it obtains this information with the help of *personal agent*. The *teaching agent* then decides the pedagogy and proceeds with the session. It obtains the desired courseware with the help of *course agent*.

3) The learning behavior of the student is continuously, monitored by the *teaching agent,* and compared with the previous cases, stored in the case base. If a new pattern is observed the *teaching agent* records it as new case and flashes the message to the *course agent* for obtaining the right courseware for the case. This is done through tutor module of ITS which is not dealt with in the present work. Thus the system employs machine learning.

4) Student goes through the learning content within finite time and goes through the self assessment session. The p*ersonal agent* records the performance (learning extent, learning rate and pattern) and takes further action based on Student Modeling output which uses CBR.

Access Architecture of DCBITS

DCBITS is a Distributed Intelligent Tutoring System for learning bridge bidding. A set of generic tasks is distributed over the network among four specialists (a tutor, a problem solver, an explainer and a problem generator), who perform their tasks in parallel with the help of three agents. These agents are heterogeneous. They own a private working memory and communicate by asynchronous message passing. Increasing flexibility is a key point in improving the pedagogical capacities of ITS. The strategic level of flexibility results from the determination of the curriculum, the choice of the pedagogical strategy and the degree of expertise used in the problem solver. The choices made at this level must be based mainly upon the student model. The tactical level of flexibility results from the choice of exercise, the advising, the corrections and the explanations. In DCBITS the pedagogical actions are gathered and sequenced by means of plans, which are chosen by meta rules and dynamically assembled from pieces that are memorized in library.

In DCBITS the Tutoring System provide a complete platform in the areas of logging, assessing, planning, delivering contents, managing records and reporting. It improves both the self-paced and the instructor-led learning processes. All these activities are represented to the end user (or organization) as a group of Web services.

DCBITS has been implemented as a web based system having access to three types of users the administrator, the tutor and the learner. The access architecture is chosen in such a way so as to provide authorized and controlled access to the courseware whereas public access to the content. Roughly we could identify three parts in the system: one part to handle the Interfaces or presentation of content/data to the users, second part which would hold the actual data and third part as the processing logic, and act as an interface between the presentation of data and the actual data. Now the next step was to find the best way to organize these three parts so as to result in an efficient overall design. At first thought, we considered an option of following a two tiered architecture, where presentation part and processing logic would be clubbed together in the first tier and the data residing on the second tier. This option was certainly better than designing the system as a monolithic application, as the data was in a separate tier. But the disadvantage of this design strategy was that the presentation and business parts were still clubbed together, so any change required in either of them would require changes in the entire tier. Organizing presentation and business logic together would also be clumsy. Taking into account the disadvantages of the 2 tier architecture, we decided to follow the 3-tiered architecture instead. A three tiered architecture enforces general separation between all the three parts as above (See figure 5).

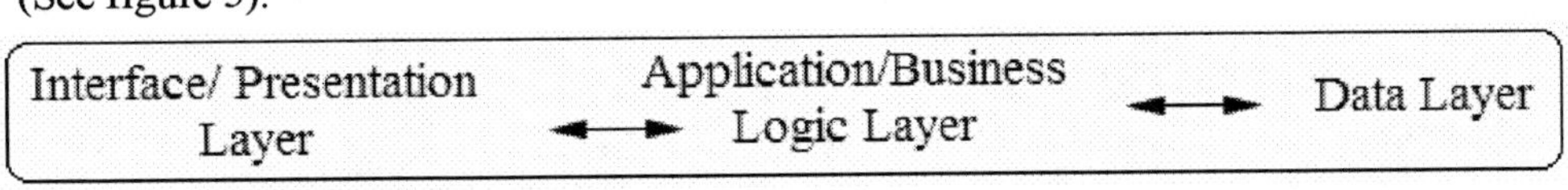

FIGURE 5. Three layered access Architecture of DCBITS

The Tutoring System architecture as three layered organization is shown in figure 5.

- Client tier: known as the presentation tier, handles the user interface or presentation logic.
- Middle tier: known as the application / business logic tier, handles the processing logic adhering to policies of the system
- Case Base Storage & Data storage tier

Overall Design & Features of DCBITS

DCBITS apart from being an ITS is also an authoring system and adapting the learning strategy that a topic is learnt by sufficiently practicing the right model of the concept in the student's mind. In case of wrong mental model, DCBITS diagnoses it and tries to correct it by presenting right remedial material to the student.

The principal features of methodology in concept learning adapted in DCBITS are:

- the modeling of all learning by the practicing teachers.
- the learning of structure through the practice of patterns of order and form rather than by explanation.
- the summarizing of the main principles of structures for the student's use when the structure are already familiar.
- the shortening of the time span between a performance and the pronouncement of its rightness or wrongness, without interrupting the response. This enhances the factor of reinforcement in learning.
- the study of basic concept of the course content in context sustained practice in the use of the language.

The major characteristics of the system architecture are:

- The system is fully distributed (not bounded with any network topology) i.e. both the domain knowledge and strategic knowledge are distributed.
- All indices are fully distributed i.e. Student Index, Case Index, and Course Index are stored on all the nodes.
- Redundancy (duplicacy) is maintained in the storage of case base for fault tolerance. Load balancing of the cases is achieved at each node. Later one can research on optimum redundancy parameters.
- If the node is a LAN, the domain knowledge can be localized and strategic knowledge (case base) can be global.

The tutoring system is made intelligent by incorporating two types of knowledge. One is the knowledge of subject domain, and, second of the tutoring principles and strategies [11]. In DCBITS these are presented in two separate but interdependent modules viz. the domain knowledge base and the strategic knowledge base.

The domain knowledge contains the declarative knowledge of the course or topic appropriate for the students as per their level of knowledge and skills. The strategies about how and what to teach (procedural knowledge) are dealt by this knowledge base. It uses the conceptual network of the domain knowledge base and facilitates to the system the reasoning about the student's behavior. It is the component responsible for making the ITS develop in the student avoidance and overcoming of errors.

DCBITS uses a structured curriculum. A student is moved to the next lesson in the curriculum only when he/she mastered the current topic (advancement). The lessons are linearly structured but the topics are componentially structured. Thus a student exercises several, topics/ skills through one lesson. For problem generation and explanations the tutoring module consults the student module. The tutoring module integrates textual knowledge processing, presentation methods and the content to be presented. It communicates with the student, selects problems, monitors student's performance and on error provides remedial material. The problem handled by this module stem from the fact that teaching is a skill requiring much more knowledge than domain knowledge only.

The teacher expert conveys the strategic knowledge i.e. lessons, topics covered by the lessons, examples, exercises and solutions to the context dependent exercises by inputting all these through various interface screens, especially designed for each of the above. During the knowledge acquisition phase, the interface also provides learning in the system. It provides the teacher expert a guided path to follow the domain knowledge conceptual network in designing the curriculum.

Keeping in mind that the student is the learner, not the expert of the domain, the student interface is designed carefully in a user friendly and attractive manner. Various dialogues form screens are chosen for presenting the tutoring material to the student.

STUDENT MODELING IN DISTRIBUTED ENVIRONMENT

The process of student modeling [7] is shown in figure 6.

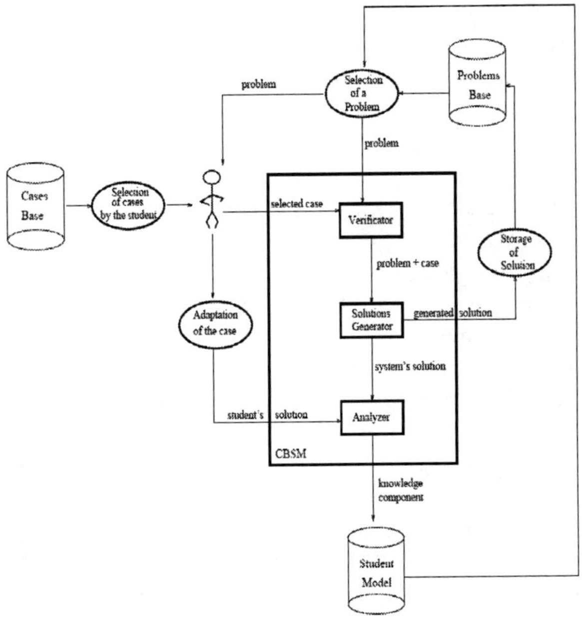

FIGURE 6. The process of Student Modeling

The following activities take place in the system during the student modeling when the student interacts with the system.

1. Selection of right course material for the student and getting student's response.
2. Analyzing the student's behavior
3. Selection of appropriate case from the Case Base
4. Adaptation of the case
5. Achieving the knowledge component of the student model through case base retrieval
6. Solution generation
7. Presenting the next problem / content to the student

In a distributed environment the complete system works as follows:

Any student can register as a user at any node over the entire network. The student can select any topic for learning. This can be done through navigation of topics provided by the ITS. However, before allowing for navigation, the *personal agent* of the system checks the past record of the performance of the student to present the right content, thus the system is adaptable.

If the performance of the student is not satisfactory then *teaching agent* presents the appropriate remedial course content to the student with the help of *course agent.* Only after achieving the learning objectives the student is allowed to move to a new topic. If any topic (case) is not available at the local node then the system searches it from other nodes within the network with the help of course index. If the case is not found on the entire network then the *teaching agent* interacts with the human tutors through *course agent* for "plug-in topics". These new plug-in topics are stored by the *course agent* on the node in use and at least one more node for fault tolerance. The *Course Agent* also updates the course index. This is the responsibility of the *course agent* to decide as to where should the duplication be made. The *course agent* is responsible for course management.

The design of such case based student modeling requires algorithms for the following operations:

- Case creation in active case base.
- Case indexing in active case base
- Update case in case base
- Adaptation of cases in the case base
- Case storage Algorithm
- Pattern matching of case in case base or Retrieval of cases from case base
- Algorithm to check duplicate case from case base
- Case Deletion from case base and store deleted case in to archive for future reference.
- Student modeling

Within the distributed student modeling, the *teaching agent* and *course agent* work for the tutor module of the ITS. The *teaching agent* does not interact directly with a student but through personal agent and serves as an intelligent tutor of a course. The basic components of a teaching agent are a domain expert module and pedagogy. The architecture of teaching agent is shown in the figure 4.

The domain expert module is responsible for storing and managing the teaching strategy about learning content, exercises, assignments, and questions with the help of learning profile and *course agent.* It contains a problem generator, a problem solver, an explanation generator, and a domain knowledge base. The pedagogy module determines the timing, style, and content of the teaching agent's interventions. It is a rule-based production system that uses the student model and pedagogical knowledge to determine the appropriate actions. The student profiler provides a model of a student based on learning style, knowledge background, and interests. It may also incorporate the information about student's learning profile gathered through dialogue with the student such as the actions the student performed and the explanations which student made.

CONCLUSION

In the present work we have proposed distributed case base reasoning for intelligent tutoring system: an agent based student modeling paradigm. It uses the three agents' personal agent, course agent, and teaching agent. All three agents timely interact with the system and manage the student personal profile along with learning activities, tutoring activities, and course content management. Also it uses three indexes i.e. student index, course index, case index, which are fully distributed, to keep the record of every student, course content and stored cases by the system with in entire network.

A prototype of this distributed case based reasoning architecture for Intelligent Tutoring System is currently under development, where a 'C' Language Tutor is being designed, developed and implemented for the topic on "Operators in C". The result will be published subsequently.

ACKNOWLEDGMENTS

The authors are grateful to the anonymous referees for their constructive and helpful comments.

REFERENCES

1. Yi Shang, Hongchi Shi, and Su-Shing Chen Deptt of computer engineering and science, University of Missouri-Columbia;An Intelligent Distributed Environment for active Learning, may 1-5,2001 hong Kong. ACM 1-5113-348-0/01/0005.
2. Ang Yang, Ray Kemp, Kinshuk; 2001, Web based intelligent tutoring system.
3. Sinha M. & Govil R,; 1995, PRABODH: A Distributed online Hindi Grammar Teaching Learning System. IJMS vol 11, No.3.
4. Brusilovsky, P., Schwarz, E. Weber, G. ELM-ART: An intelligent tutoring system on World Wide Web. Intelligent Tutoring Systems Lecture Notes in Computer Science, Vol. 1086. Berlin: Springer Verlag. (1996) 261-269.
5. Aamodt, A.; 2005, Knowledge Intesive Case-based reasoning and Intelligent tutoring system.
6. Leen-kiat Soh; 2005, Incorporating Intelligent tutoring system in to CS1,SIGCSE06.

7. Shiri M. E. , Aimeur A. E. and Frasson C.; 1998, Student modelling by Case Based Reasoning, Lecture Notes in Computer Science, vol 1452
8. C. Burger and K. Rothermel, A Framework to Support Teaching in Distributed Systems, ACM Journal of Educational Resources in Computing, Vol. 1, No. 1, Spring 2001,
9. Amruth N. Kumar, Generation of Problems, Answers, Grade, and Feedback – Case Study of a Fully Automated Tutor, ACM Journal of Educational Resources in Computing, Vol. 5, No. 3, September 2005
10. Aamodt, A., Knowledge-intensive case-based reasoning in Creek. In Proceedings of the Seventh European Conference on Case-Based Reasoning. Berlin: Springer, (2004), pp. 1-15.
11. Anderson, J.R., Prirolli, P., and Farrell, R., Learning to program recursive functions. In The Nature of Expertise. Chi, M.T., Glaser, R., and Farr, M.J. (eds.). Hillsdale, NJ: Lawrence Erlbaum Associates, (1988), 151-183.
12. VanLehn, K., Student Modeling. Foundations of Intelligent Tutoring Systems, Polson M.C. and Richardson J.J. (Eds), (1988).

Quantifying Learning and Teaching Performance of an Engineering Class

Tang-Hung Nguyen and Tariq Shehab

Department of Civil Engineering & Construction Engineering Management, California State University at Long Beach, Long Beach, CA 90840 U.S.A

Abstract. The assessment/evaluation of a course refers to the process of selecting and implementing the methods that will be used to determine whether and how well the learning objectives have been achieved and interpreting the results. Researchers/educators have extensively discussed various methods to assess learning performance of students. The primary purpose of this research is to develop a computer tool for quantifying the learning and teaching performance of an engineering course with respect to the selected learning objectives as well as identifying necessary actions to be taken for improving the course quality. For this particular project, the ABET (Accreditation Board of Engineering and Technology) Engineering Criteria 2000 (also known as eleven ABET Engineering Outcomes 3A-K) were used as the learning objectives.

Keywords: ABET Engineering Criteria, Engineering Education, Learning Assessment, Learning Performance, Teaching Performance.
PACS: 01.40.Di

INTRODUCTION

The ABET (Accreditation Board of Engineering and Technology) Engineering Criteria 2000 (also known as eleven ABET Engineering Outcomes 3A-K) were introduced in the middle 1990s and have been used to evaluate all American engineering programs since then [1]. Since the new ABET accreditation system was first introduced, researchers/educators have extensively discussed various methods to assess Outcomes 3a-3k, as shown in Table 1. Obviously, the potential of the new system to improve instruction depends strongly on how well engineering faculty understand it and achieve its specified outcomes. When developing a course to meet the pre-determined educational criteria (i.e. ABET), usually faculty goes through three major activities: planning (identifying course content and defining measurable learning objectives for it); instruction (selecting and implementing the methods that will be used to deliver the specified content and facilitate student achievement of the objectives); and assessment/evaluation (selecting and implementing the methods that will be used to determine whether and how well the objectives have been achieved and interpreting the results) [1]. Among these activities, assessment usually plays an important role in enhancing the course quality; as it determines the level of achievement of the course objectives, which in turn identifies necessary actions to be taken for improving the learning and teaching performance. The primary purpose of

CP1007, *Current Themes in Engineering Technologies*, edited by S.-I. Ao, M. A. Amouzegar, and S.-S. Chen

this paper is to describe an electronic tool that was developed to measure the level of achievement of the ABET engineering outcomes of a civil engineering and construction course at California State University at Long Beach. The paper is organized as follows: (i) the overall assessment process for an engineering class is briefly described; (ii) the basic concept of quality function deployment (QFD) is explained; and (iii) the development of the proposed assessment tool using QFD method is presented. Findings and results are discussed, followed by conclusions.

TABLE 1. ABET Criterion 3 Outcomes

Criterion	Description
A	an ability to apply the knowledge of mathematics, science, and engineering.
B	an ability to design and conduct experiments, as well as to analyze and interpret data.
C	an ability to design a system, component, or process to meet desired needs.
D	an ability to function on multi-disciplinary teams.
E	an ability to identify, formulate, and solve engineering problems.
F	an understanding of professional and ethical responsibility.
G	an ability to communicate effectively.
H	the broad education necessary to understand the impact of engineering solutions in a global and societal context.
I	a recognition of the need for, and an ability to engage in lifelong learning.
J	a demonstrated knowledge of contemporary issues.
K	an ability to use the techniques, skills, and modern engineering tools necessary for engineering practice.

ASSESSMENT OF STUDENT LEARNING

The basic starting point for assessing student learning begins with asking two basic questions: 1.) what should students know and be able to do at the end of the semester?, and 2.) what evidence will indicate that they have reached these goals? The basic tools that are often used for assessing knowledge in a particular course are the student works such as homework, quizzes, tests/exams, and projects/labs reports. These basic tools may help to answer the two basic questions above, however, they are hardly used to determine how well the course achievement addresses each of the ABET Criterion 3 outcomes. In addition, the teaching performance of instructor usually impacts the success of a class and should be taken into consideration when assessing the class achievement. This leads to a need for a tool that is designed to assess both teaching and learning performance with respect to predefined learning objectives. The assessment tool will help the instructor to collect and analyze data containing information linking student expectations (i.e. learning objectives or ABET criteria) to a set of teaching performance metrics that the instructor can then measure and control.

In this particular study, the student expectations or learning objectives include the eleven ABET Engineering Criteria 3A-K (see Table 1), whereas the teaching performance metrics represent the critical factors that most affect the instructor's performance. The teaching performance factors were obtained from the results of a previous research found in [2] and are presented in the following table:

TABLE 2. Teaching Performance Factors

Factors	Description
Communication skill	The ability of the instructor to communicate effectively
Student focus	The degree of importance that an instructor places on student relationships and student satisfaction and the degree of knowledge about student needs or learning objectives. Examples: displaying a personal interest in students and their learning, inspiring students to set and achieve goals which really challenge them, and involving students in 'hands-on' projects such as research, case studies, or 'real life' activities.
Knowledge/Expertise	The instructor's knowledge or expertise about the subject being taught. Examples: Capability of making it clear how each topic fit in the course, explaining course material clearly and concisely, and relating course material to real life situations
Attitude about enhancing student learning	The degree of encouragement by the instructor so as to maximize learning of all students with varying learning styles and abilities. Examples: finding ways to help students answer their own questions, forming teams or discussion groups to facilitate learning, encouraging students to use multiple resources (e.g. data banks, library holdings, outside experts) to improve understanding, and providing timely and frequent feedback on tests, reports, projects, etc. to help students improve.
Interaction with students	The degree of effectiveness of the instructor to interact with students in class or outside of class (office visits, phone calls, emails, etc.). Examples: Asking students to share ideas and experiences with others whose backgrounds and viewpoints differ from their own, introducing stimulating ideas about the subject, and stimulating students to intellectual effort beyond that required by most courses.
Teaching performance improvement	The attitude of the instructor about continuously improving teaching performance. Examples: identifying and evaluating student satisfaction and class performance results to take appropriate actions for improving teaching performance.

These teaching performance factors greatly impact the level of achievement of the class with respect to the learning objectives (i.e. ABET outcomes). Additionally, one teaching performance factor may have a greater impact than others and the different degrees of impact among the factors, in this study, are referred to as strengths of relationship between the teaching performance factors and the learning objectives. These strengths of relationship should be taken into consideration as measuring the learning and teaching performance of an engineering class. An appropriate assessment method for such a measuring task was selected for this particular study, i.e. Quality Function Deployment (QFD), which is briefly presented in the following section.

QUALITY FUNCTION DEPLOYMENT (QFD)

QFD, also known as the House of Quality, is defined as a structured methodology and mathematical tool used to identify and quantify customers' requirements and translate them into key critical parameters that in turn help a company to prioritize actions to improve their product or service to meet customers' expectations [3]. In an education context, the 'customers' represent students, the 'service' refers to 'teaching' or 'learning performance', and the 'company' represents the instructor teaching the course. In a business context, QFD is used for translating the 'voice of customer' through the various phases of project or service planning, designing, and manufacturing into a final product. In this education study, the QFD can be used for translating the 'voice of the student' through the three stages of the course planning, teaching, and assessment/evaluation into the learning objectives. In a typical QFD application, a company or instructor creates and analyzes a data matrix (refer to Figure 1) linking customer or student needs (i.e. learning objectives) to a set of service or teaching metrics that the company or instructor can then measure and control. For further details of a QFD application, readers are referred to [4] and [5]. In this study, the basic QFD process has been extracted and modified to develop the process of measuring the level of learning performance of an engineering class. In general, the process of evaluating the learning performance of a class involves six major steps as discussed below.

Step 1: Collecting the data

The data to be collected is organized in a data matrix, as shown in Figure 1. It contains information about student needs or learning objectives (the WHATs), the instructional characteristics of the course (the HOWs), and the strength of their interrelationships.

		HOW_1	HOW_2	HOW_3	.....	HOW_m
		H_1	H_2	H_3	....	H_m
$WHAT_1$	W_1	IR_{11}	IR_{12}	IR_{13}	....	IR_{1m}
$WHAT_2$	W_2	IR_{21}	IR_{22}	IR_{23}	....	IR_{2m}
$WHAT_3$	W_3	IR_{31}	IR_{32}	IR_{33}	....	IR_{3m}
....	.....	.....	.....	.....	....	......
$WHAT_n$	W_n	IR_{n1}	IR_{n2}	IR_{n3}	....	IR_{nm}

Instructional characteristics (the HOWs) and their importance weights (H_j) obtained from the instructor who teaches the course

The interrelationships (IR_{ij}) between the WHATs and the HOWs obtained from independent assessors

The student needs (the WHATs) and their importance weights (W_i) obtained from students

FIGURE 1. Data Matrix

Step 2: Processing of the data in the data matrix

The information contained in the data matrix will be used to create a process matrix (refer to Figure 2), in which the data in column 2 and row 2 represents the normalized importance weights for WHATs (W_i) and HOWs (H_j). The sum of the normalized important weights equals 1.

$$\sum_{i=1}^{n} W_i = 1 \qquad \text{for } 1 \leq i \leq n$$

$$\sum_{j=1}^{m} H_j = 1 \qquad \text{for } 1 \leq j \leq m$$

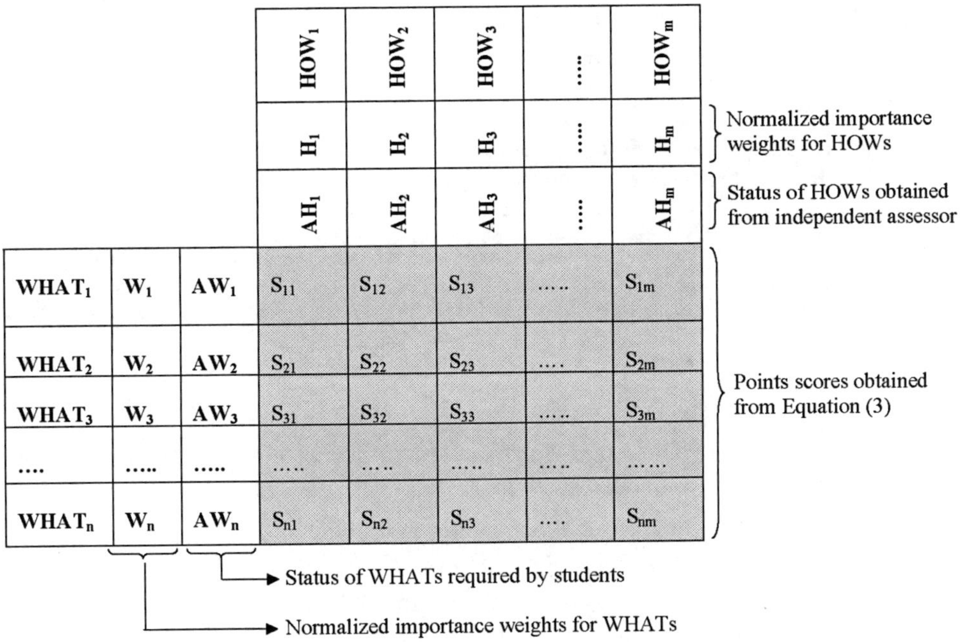

FIGURE 2. Process Matrix

The information in column 3 and row 3 represents the existing status (AW_i and AH_j for actual condition or MW_i and MH_j for maximum/expected condition) of the WHATs and HOWs in a class. The values of AW_i and AH_j (or MW_i and MH_j) are evaluated and specified by an independent assessor on a scale of 1 to 5, where 1 is 'poor' and 5 'excellent'. The information contained in the shaded matrix represents point scores (S_{ij}) for each intersection between WHATs and HOWs. These point scores are calculated by multiplying the mean of the relative importance of a HOW and that of a WHAT by the strength of their interrelationship (IR_{ij}) given from the data matrix, as shown in the following equation:

$$\text{Actual } S_{ij} = \frac{(W_i \text{ x } AW_i) + (H_j \text{ x } AH_j)}{2} \text{ x } (IR_{ij})$$

Where:

i = 1, 2, ..., n (number of learning objectives)

j = 1, 2, ...,m (number of teaching performance factors)

Actual S_{ij} = point scores in actual condition for the teaching performance factor j with respect to the ABET outcome i

AW_i = the actual status of ABET outcome i

AH_j = the actual status of teaching performance factor j

W_i = the normalized weight of importance of ABET outcome i

H_j = the normalized weight of importance of teaching performance factor j

IR_{ij} = the strength of the relationships between teaching performance factor j and learning objective or ABET outcome i.

Step 3: Calculating the actual level of learning performance

The actual level of learning performance of the class with respect to ABET outcome i is given by

$$\text{Actual } LP_i = \sum_{j=1}^{m} \text{Actual } S_{ij} \qquad \text{for } 1 \le j \le m$$

Similarly, the actual level of learning performance of the class with respect to teaching performance factor j is computed as follows:

$$\text{Actual } LP_j = \sum_{i=1}^{n} \text{Actual } S_{ij} \qquad \text{for } 1 \le i \le n$$

The actual level of learning performance (Actual LP) for the class is calculated as follows:

$$\text{Actual LP} = \sum_{i=1}^{n} \text{Actual } LP_i \quad = \sum_{j=1}^{m} \text{Actual } LP_j \qquad (1)$$

Step 4: Calculating the maximum level of learning performance

In addition, point scores in maximum (expected) condition for each teaching performance factor with respect to a particular ABET criterion or learning objective are calculated by using the following formula.

$$\text{Max } S_{ij} = \frac{(W_i \text{ x } MW_i) + (H_j \text{ x } MH_j)}{2} \text{ x } (IR_{ij})$$

Where:

$i = 1, 2, \ldots, n$ (number of learning objectives)

$j = 1, 2, \ldots, m$ (number of teaching performance factors)

Max S_{ij} = point scores in maximum condition for the teaching performance factor j with respect to the ABET outcome i

MW_i = the max status of ABET outcome i (i.e. all MW_i score maximum 5)

MH_j = the max status of teaching performance factor j (i.e. all MH_j score maximum 5)

W_i = the normalized weight of importance of ABET outcome i

H_j = the normalized weight of importance of teaching performance factor j

IR_{ij} = the strength of the relationships between teaching performance factor j and learning objective or ABET outcome i.

Therefore, the maximum level of learning performance of the class with respect to ABET outcome i is given by

$$\text{Max } LP_i = \sum_{j=1}^{m} \text{Max } S_{ij} \qquad \text{for } 1 \le j \le m$$

Similarly, the maximum level of learning performance of the class with respect to teaching performance factor j is computed as follows:

$$\text{Max } LP_j = \sum_{i=1}^{n} \text{Max } S_{ij} \qquad \text{for } 1 \le i \le n$$

The maximum level of learning/teaching performance (Max LP) for the class is calculated as follows:

$$\text{Max LP} = \sum_{i=1}^{n} \text{Max } LP_i \quad = \sum_{j=1}^{m} \text{Max } LP_j \qquad (2)$$

Step 5: Calculating the learning performance index

The overall performance index (PIndex) of the class is computed using equations (1) and (2) as follows:

$$\text{PIndex} = (\text{Actual LP}/\text{Max LP})\text{x}100\% \qquad (3)$$

In addition, the performance index for a particular learning objective i ($PIndex_i$) or teaching performance factor j ($PIndex_j$) can be computed as follows:

$$PIndex_i = (\text{Actual } LP_i/\text{Max } LP_i)\text{x}100\% \qquad \text{for } 1 \le i \le n \qquad (4)$$

$$PIndex_j = (\text{Actual } LP_j/\text{Max } LP_j)\text{x}100\% \qquad \text{for } 1 \le j \le m \qquad (5)$$

Step 6: Identifying the weakness of the class

Finally, the learning objectives that have low performance indexes, i.e. the weakness of the class, can be identified. It is obvious that the desirable learning performance index should be as close to 100% as possible. Consequently, the learning or teaching performance factors having lower performance indexes should be well taken care of for improvement. In addition, the same process to calculate the level of learning performance can be applied to different classes or the same class in different semesters or academic years. Different classes may be ranked according to their relative performance. The ranking of classes relative to learning performance can be invaluable to the engineering faculty who seeks the suitable actions to improve the course quality and to meet the course learning objectives.

DEVELOPMENT OF THE PROPOSED TOOL

The proposed assessment tool was developed through three steps. First, based upon the QFD concept on how to determine the service quality performance of an organization, the reasoning of logics (or algorithms) for calculating the learning performance index were established. Second, the data structure of the QFD model was used to develop the user interface for the proposed computer assessment tool. Basically, the user will be prompted to enter the information necessary for calculating the expected level and the actual level of learning/teaching performance with respect to the ABET criteria. Finally, the assessing algorithms and the user interface for the proposed assessment tool were implemented into a computer platform. Microsoft Visual Basic 6.0 was used for the research implementation since it is widely used software programs and the development of the prototype can be easily beta tested and demonstrated to users. The proposed assessment tool was designed to measure the overall performance index of an engineering class and was named PIndex (Performance Index). PIndex was used to evaluate learning/teaching performance of several classes at the Department of Civil Engineering and Construction Engineering Management at California State University, Long Beach (CSULB). Below is an example input and output of PIndex for an engineering class: CE 130 – Surveying and Mapping at CSULB. The assessment process in PIndex can be summarized as follows:

First, the instructor is prompted to define the learning objectives of the class. In this example, ABET Engineering Criteria A, B, C, D, E, and K were selected (Figure 3).

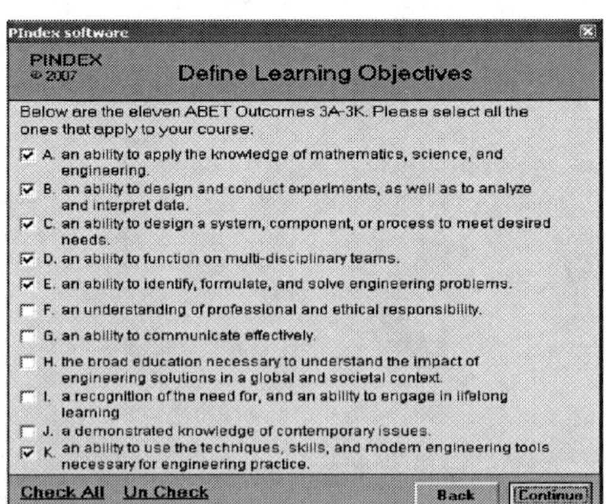

FIGURE 3. Define Learning Objectives

Second, the data about the importance weights of the learning objectives as well as the teaching performance factors are input (see Figures 4 and 5). These importance weights were reported from students and senior professors. Also, the strengths of relationship between the teaching performance factors and the learning objectives which were given by senior professors are entered.

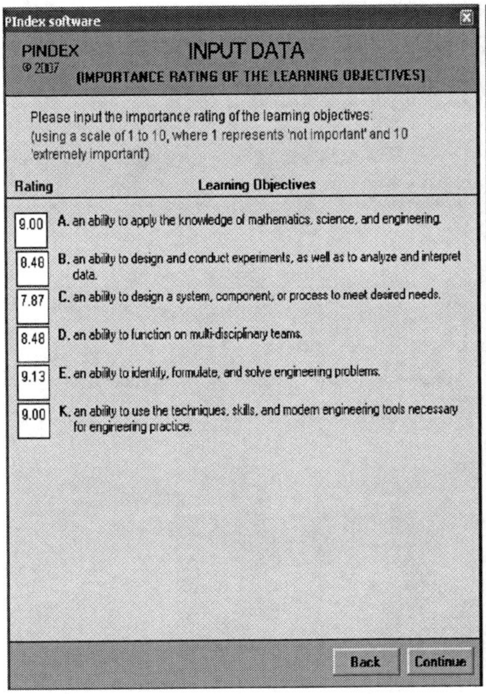

FIGURE 4. Importance Weights of Learning Objectives

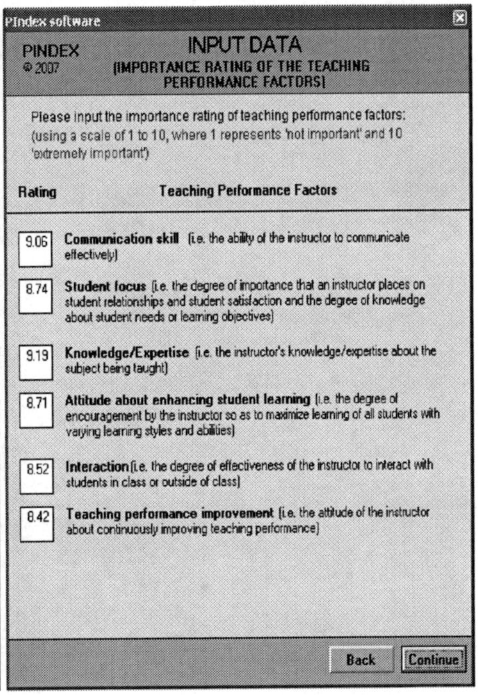

FIGURE 5. Importance Weights of Teaching Performance Factors

Third, the course evaluation with respect to the teaching performance factors as well as grades of student works (e.g. homework, quizzes, tests/exams, lab reports, projects) addressing different learning objectives were entered as averaged percentages. These data were used to determine the level of learning performance under the actual condition of the class (Actual LP).

Finally, the overall learning performance index of the class can be determined using equations (1), (2), and (3) and is presented in the assessment report (Figure 8). In the final assessment report, the learning objectives or teaching performance factors that have low performance indexes are displayed so that the instructor is aware of the weakness and should take appropriate actions for improving quality of future classes. In addition, the performance index of the class with respect to each learning objectives (i.e. ABET criteria) as well as each teaching performance factor can also be computed by using the equations (4) and (5), respectively, as shown in Figures 6 and 7.

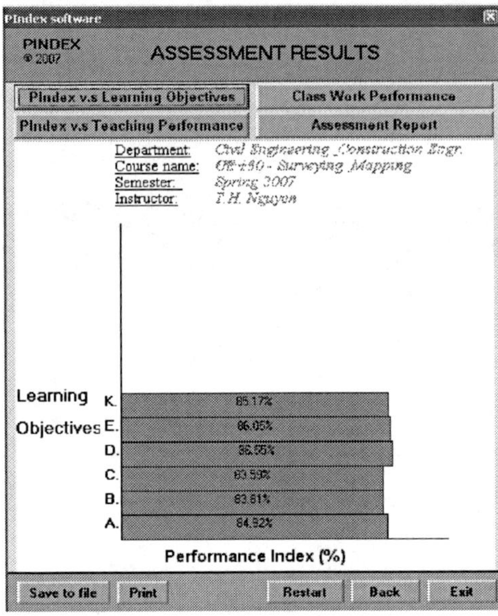

FIGURE 6. PIndex v.s. Learning Objectives

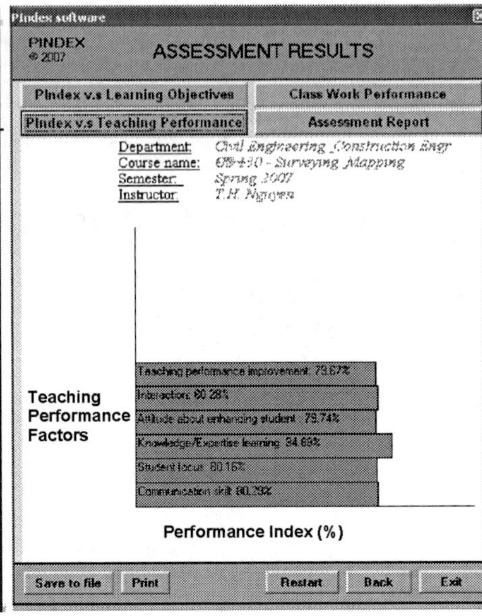

FIGURE 7. PIndex v.s. Teaching Performance Factors

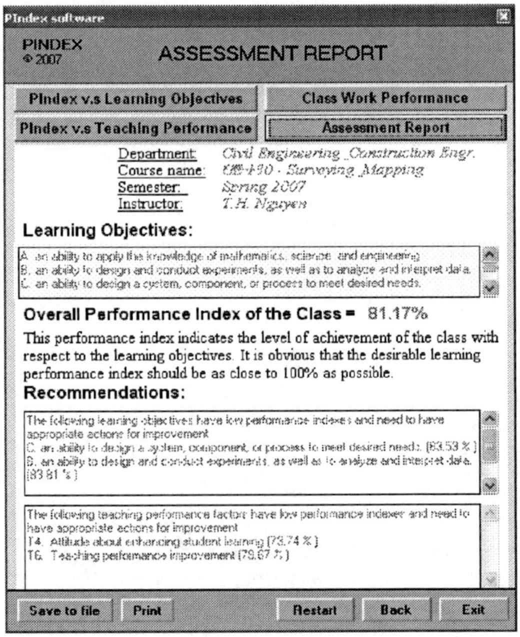

FIGURE 8. Final Assessment Report

CONCLUSION

The assessment of an engineering course with respect to student expectations or learning objectives is a necessary activity for enhancing the course quality. While traditional assessment tools such as homework, quizzes, tests, exams, lab reports, oral presentations, and projects may be used as evidences of the overall student achievements with respect to the learning objectives, they do not address the level of performance of each individual learning objective as well as the instructor's teaching performance. This study made use of ABET Engineering 3A-K Outcomes as learning objectives of an engineering class to be measured. Also, the most critical factors that impact teaching performance of an instructor were evaluated together with the learning objectives. These factors include communication skill, student focus, knowledge/expertise, attitude about enhancing student learning, interaction with students, and teaching performance improvement. In this paper, an electronic assessment tool was developed in the computer platform of Microsoft Visual Basic 6.0. The tool can be used to measure the learning performance index as the level of the achievement with respect to the learning objectives of an engineering class, which indicate the strength as well as the weakness of the class so that appropriate actions will be taken for enhancing quality of future classes. The assessment method employs the concept of quality function deployment (QFD), a technique to measure the customer service quality of an organization, and applies it to academia by linking students expectations (i.e. learning objectives) and teaching performance of an instructor. The values of the proposed assessment tool can be summarized as follows:

- A comprehensive and weighted learning performance index can be determined for a particular course. This performance index can be used as a reference point to compare learning performance of the same course in different semesters. The performance index can also serve as a base of comparison between different courses with respect to teaching performance and learning objectives.
- Student needs/expectations and teaching performance factors can be individually assessed to determine areas of strength or weakness in order to provide continuous course improvement as well as enhanced student learning.
- The assessment method can be used as a template for developing similar tools to measure learning and teaching performance of any course other than engineering courses.

ACKNOWLEDGMENTS

The authors gratefully acknowledge the support of the *Scholarly and Creative Activity Committee program* at California State University Long Beach for funding this project.

REFERENCES

1. Felder, M.R. and Brent, R., *Designing and Teaching Courses to Satisfy the ABET Engineering Criteria,* ASEE Journal of Engineering Education, Vol. 92, No. 1, 2003, pp. 7-25.
2. Nguyen, T.H., McIntyre, C., and Diab, M., *Measuring Learning Perofrmance of an Engineering Capstone Course with respect to the ABET Engineering Criteria,* Proceedings of the 2005 ASEE North Midwest Regional Conference, Brookings, SD, Oct. 13-14, 2005.
3. Hoyle, D., *ISO 9000 Quality System Handbook,* 3rd Edition, Butterworth-Heinemann Ltd., Boston, MA, 1998.
4. Akao, Y., *Quality Function Deployment: Integrating Customer Requirements into Product Design,* Productivity Press, Cambridge, MA.
5. Akao, Y., Mizuno, S., and Ishihara, K., *QFD: the Customer-Driven Approach to Quality Planning and Deployment,* Asian Productivity Organization, Tokyo, 1994.

Collaborate to Learn: A Case Study of Technology-enhanced, Collaborative Language Learning

YanJun Wang

School of Asian Languages and Studies, University of Tasmania, Launceston, Locked Bag 1340, Tasmania, 7250, Australia

Abstract: Technology Enhanced Language Learning (TELL) TELL refers to any technology used in the classroom such as video, tape recorders or even entire language labs. In collaborative learning, students work together as members of a learning community, working on problem-solving tasks by questioning each other, and discussing and sharing information. This paper examines the notion of collaborative learning, the relationship between technology and collaboration, and the integration of technology into collaborative language learning. A case study was conducted to examine how the use of technology-enhanced, collaborative language learning enables students to work collaboratively in a university context. Participants from a range of students in a variety of majors at the University of Tasmania were invited to participate in a questionnaire to present their views on this style of learning and their perception of technology-enhanced language learning.

Key words: collaborative learning, technology-enhanced language learning

INTRODUCTION

The advent of innovative information and communication technologies has induced certain changes in the present educational system. These interactive technologies have affected the very nature of teaching and learning. From an educational perspective, the Internet has the potential to radically change the way that learners assimilate information. Students who learn via the Internet have to develop cognitive abilities which enable them to research, to identify, to analyze and to synthesize new information in order to construct their knowledge base. The interactive media of the Internet, such as e-mail and

CP1007, *Current Themes in Engineering Technologies*, edited by S.-I. Ao, M. A. Amouzegar, and S.-S. Chen

hypertext navigation, gives learners the chance to participate actively in the learning process and to communicate easily with other learners. Therefore, the role of computers in language instruction has now become an important issue confronting large numbers of language teachers throughout the world.

Additionally, there has been a significant expansion in distance education technologies with the potential to improve the quality of teaching and learning. Furthermore, the new computer-based technologies provide opportunities for enhancing the quality of teaching and learning. Language is closely associated with economic modernization and technological development. Technological advances may have an impact on language teaching.

In a review of existing evidence of technology's impact on learning, [10] found strong evidence that educational technology "complements what a great teacher does naturally," extending their reach and broadening their students' experience beyond the classroom. "With ever-expanding content and technology choices, from video to multimedia to the Internet," [10] suggests that "there's an unprecedented need to understand the recipe for success, which involves the learner, the teacher, the content, and the environment in which technology is used." Teachers are expected to be competent in utilizing technology for their own productivity and also to improve their students' learning outcome. In addition, computer hardware and software are developing fast and this requires teachers to adapt quickly to changes. Ongoing and quality professional development is seen as necessary to help teachers cope with pace and the use of current change [20], [4].

BACKGROUND

In the 1980s, the application of technology in language classrooms included the use of film, radio, television, language labs with audio and videotapes, computers, and interactive video [3]. Computer technology became more accessible to both individuals and schools. Today, the use of multimedia, the Internet (especially the World Wide Web), and various forms of distance learning are widespread. Interest in using computers as tools to support language learning is growing, both from the perspective of a language educator and that of a language learner.

Why Technology?

Technology stimulates learning motivation through collaborative learning and it also improves learning efficiency by integrating classroom learning. Classroom research suggests that students do not get much speaking time during regular classroom interactions. Technology can provide a way to give speaking practice, and an improvement is suggested over the traditional use of audio cassettes. Audio Portfolios is a web-based tool that collects and manages students' audio recordings in an online environment.

Technology as a powerful and convenient tool which can provide learners with a rich resource, a visual environment as well as an instructional platform, plays a vital role in language learning. Each technology is likely to play a different role in students' learning. Rather than trying to describe the impact of all technologies as if they were the same, researchers need to think about what kind of technologies are being used in the classroom and for what purposes. Students can learn "from" computers—where technology is used essentially and serves to increase students basic skills and knowledge; and students can also learn "with" computers—where technology is used as a tool that can be applied to a variety of goals in the learning process and can serve as a resource to help develop critical thinking, creativity and research skills [13].

Currently, there are various computer technologies used in education in order to improve teachers' productivity and students' learning outcomes. Three of the common technologies being used are highlighted in this paper. Computer Assisted Instruction (CAI) is the term used to describe computer settings that is specifically design for drill and practice, tutorials, simulations and problem solving [5], [15]. CAI is also used as a supplement for traditional teaching and learning which is more teacher-directed [5]. It can be seen that CAI is used when discrete lessons are presented to achieve specific but limited educational objectives. Computer-based multimedia is the most current and most attractive computing technology used in education today [2], [8]. In a computer-based multimedia learning environment, various types of multimedia elements may be used at the same time [2], [8]. The elements of multimedia include pictures, animations, sound and video which are also integrated with other computer technologies such as the Internet [8]. The combination of the technologies used is hoped to be able to enrich the learning experience for the learners including adults [15]. Computer mediated communication (CMC) is mainly about computer applications that facilitate communication [1], [11]. Students and teachers are able to communicate with one another through emails, video-conferencing and newsgroups. CMC provides means of better and more flexible communication in which students and teachers can keep in touch even at long distances and at anytime.

Why Collaborative Learning?

According to [6], the concept of collaborative learning, the grouping and pairing of students for the purpose of achieving an academic goal, has been widely researched and advocated throughout the professional literature. The term "collaborative learning" refers to an instruction method in which students at various performance levels work together in small groups toward a common goal. The students are responsible for one another's learning as well as their own. Thus, the success of one student helps other students to be successful.

Some supporters of collaborative learning [6], [9], [12] argue that it enhances both social and cognitive skills. For example, according to [5], there is strong evidence that cooperative teams achieve at higher levels of thought and retain information longer than students who work quietly as individuals. The shared learning gives students an opportunity to engage in discussion, take responsibility for their own learning, and thus become critical thinkers [18]. In collaborative learning classrooms where students are engaged in a thinking curriculum, each student is a member of the learning community, and no student is deprived of the opportunity of making contributions and appreciating the contributions of others. Of course, collaborative learning does not always work. There are many factors to take into consideration and it is important to examine those conditions under which collaborative learning is found to be either efficient or not [12], [16], [17]. Collaborative learning appears to be crucial to the effectiveness of online learning environments.

It should be noted that collaborative learning is not confined to an educational institution. It can apply to different working environments. Advances in technology and changes in the organizational infrastructure put an increased emphasis on teamwork within the workforce. Workers need to be able to think creatively, solve problems, and make decisions as a team. Teamwork is a common feature of current institutions. Therefore, the development and enhancement of critical-thinking skills through collaborative learning is one of the primary goals of technology education.

A CASE STUDY

A case study was conducted to investigate students' perspectives towards technology-enhanced, collaborative learning in language study.

Aim and Methodology

The aim of my study is to explore some technological aspects of language education, to consider these aspects in terms of language studies and to enable each student to develop their ability to improve this style of learning. This ability helps students to learn another language effectively.

A questionnaire was the tool used in this study. It was designed to collect two broad types of information:

- Background information from each respondent about their personal and demographic characteristics.
- Information relating to respondents' knowledge and experiences of technology-enhanced, collaborative language learning.

Data Analysis and Results

Data gathering involved the distribution of 45 questionnaires. Demographic information was obtained from the questionnaire which was completed by 45 respondents. The data collected from the questionnaire in this study was entered into the SPSS (Sphinx Survey or Ethnography respectively) for analysis, focusing on any relationship between independent variables such as male/female, and dependent variables such as views and attitudes on technology-enhanced, collaborative language learning in a university context. The items in the questionnaire provided an opportunity to gather information on the various participants' perceptions of technology-enhanced, collaborative language learning. The data provided information on two major aspects:

- Participants' Demographic Background
- Attitudes towards and experience of technology-enhanced, collaborative language learning

As mentioned above, a total of 45 students participated in this study. A fifteen item questionnaire was developed to collect descriptive data on the participants. Results of the questionnaire revealed that the majority of all respondents were aged between 19-30 years (80%), and female students (60%) were 20% more than male (40%). All of them are doing undergraduate study, more than half are first year students (51.2%) and the second largest group is the second year (30.2%). The educational specialization of these students is mainly Arts (52.3%), and the second major area of study is "others", such as commerce and education (25.0% respectively). And also 15.9% of the participants are undertaking study for combined degrees, such as computing/arts, arts/law.

Students generally saw benefits in the technology-enhanced, collaborative approach, as shown by the summary of results in Table 1, 2, 3 and 4. For questions 8 and 9, more than half the students agreed that the approach was beneficial. It is interesting that a large percentage of students have never chosen any online course, and that approximately half the students have broadband at home. Agreement on the value of the method was very high on some questions. For questions 1, 5, 7 and 10, more than 80 percent of respondents indicated agreement. Results also indicate that students were also interested in the Internet and the various audio-visual aids and software available.

TABLE1. Result of Student Survey 45 Responses

Question	Percentages				
	Always	Frequently	Sometimes	Rarely	Never
1. Use technology and/or collaboration in your learning	24.4	37.8	31.1	6.7	0
2. Use language lab to improve your learning	2.3	9.3	20.9	14.0	53.5
3. Use online dictionaries to help with your learning	11.1	20.0	24.4	24.4	20.0
4. Use any other online tools (e.g. generic software or task-based web activities) to enhance your learning	6.7	26.7	22.2	26.7	17.8
5. Use audio-visual aid/materials	9.1	15.9	45.5	18.2	11.4
6. Encounter technical difficulties	2.3	6.8	29.5	52.3	9.1
7. Use WebCT or email	18.2	34.1	20.5	15.9	11.4
8. Participate in a study group or group work	4.4	22.2	31.1	20.0	22.2

TABLE2. Result of Student Survey 45 Responses

Question	Percentages			
	Very much	Slightly more	A little	Little
9. Learn more as part of a group than individually	23.3	48.8	20.9	7.0

TABLE3. Result of Student Survey 45 Responses

Question	Percentages			
	In the library	At home	In the language	PC
10. Where do you use this style of learning	31.0	57.1	2.4	9.5

TABLE4. Result of Student Survey 45 Responses

Question	Percentages	
	Yes	No
11. Have you ever chosen any online course	20.0	80.0
12. Do you have broadband at home	57.8	42.2

*Notes: Questions are abbreviated from those actually given in the survey.

From the above, it is clearly seen that the technology-enhanced, collaborative learning method was generally well received by students, with a clear majority seeing social, learning and skill development advantages. Some student groups were more positive than others. For example, female students tended to be more positive than males, especially in agreeing that the method enhanced communication skills, encouraged them and promoted understanding.

However, the collaborative learning did not always work in an efficient manner. Some students complained of technical difficulties which greatly hampered communication. These technical difficulties created a high level of frustration amongst the learners. Lack of computer literacy may place novice computer users at a disadvantage. These students had little or no knowledge of word-processing software, the Internet and its communication tools. Since communication was problematic, the collaborative process was not able to function at an optimal level.

What Do the Data Tell Us?

Firstly, concerning problems and concerns regarding technology-enhanced, collaborative learning, this research study shows mixed results. Respondents had divergent views on this matter. Students made common remarks about the effectiveness of the Video conferencing such as:

Remark 1

Video-link can prove to have difficulties with gaining a clear understanding of lessons.

Remark 2

Slight time lag makes it difficult for oral tasks; also lack of transmission of body language makes it hard to communicate.

Remark3

In video linkups it is hard to interact with other students, but the new technology is a big improvement.

Moreover, there were also some negative findings with regard to the use of technological approaches in language learning process. For instance,

Remark 1

Online dictionaries and translations are not usually very reliable.

Remark 2

Malfunctions occur too often, much time is spent trying to fix technological issues, maybe because of the complexity in operating the software. Software should be made easier to operate.

Remark 3

Different computers run and display files in different ways. A program that runs correctly on a home computer may malfunction on a school computer.

Secondly, considering the experience of working in a group or with the help of technology, and the resultant academic and social benefits, the responses indicated that a clear majority of students saw social, learning and skill development benefits in this style of learning. Because of flexibility, learners can choose to make their own learning paths, choose when and where to learn, and even with whom they wish to learn, no matter where they are in the world. As a result, the majority were in favor of working in a group or with the help of technology, primarily because they enjoyed the process of working with partners and sharing their learning experiences with their peers. The majority of these students revealed that they would have liked more contact with other students. Here are some of the remarks:

Remark 1

Bonding makes it more enjoyable and encourages participation and attendance.

Remark 2

Getting a real understanding of other people's point of view and working with groups reflects the real work place where working in groups is highly important.

Remark 3

Working in a group helps me remember rather than monologing at home. Working together in class, I think, makes people more comfortable and therefore more confident.

Remark 4

Technology takes away embarrassment and groups instill confidence.

Remark 5

Working in a group one is less likely to make mistakes and one gains understanding of different points of view; while technology is more essential, than beneficial, to learning.

However, interestingly there was one participant holding a completely opposite opinion, "I don't benefit from group work as we end up talking." As [17] point out, the perspectives, experiences, and backgrounds of all students are important for enriching learning in the classroom. As learning beyond the classroom increasingly requires understanding diverse perspectives, it is essential to provide students opportunities to do this in multiple contexts in schools. The fact is that not everyone makes an equal contribution [16], [18], [19]. A critical characteristic of collaborative classrooms is that students are not segregated according to supposed ability, achievement, interests, or any other characteristic. Segregation seriously weakens collaboration and impoverishes the classroom by depriving all students of opportunities to learn from and with each other. Contribution to collaborative learning should be meaningful and purposeful.

Lastly, concerning any changes about the behavior or approach in future technology-enhanced, collaborative learning situations, and some students expected to engage in and have greater access to technology-enhanced, collaborative learning. One statement was that technology is helpful for multimedia (especially audio); drill (character recognition) at higher levels of language study, while "blogs" and extended writing pieces can be helpful, too. In the current context of teaching and learning in a tertiary discourse, a lecture in its traditional sense does not necessarily meet the demand of learners as a lecture can only function in a very limited context. However, the internet provides an educational discourse in which learners can interact widely with other members of a learning community, and at the same time learners are in control of their own learning. Their interaction for learning can be immediate, prompt, widely shared and resource-supported and this may not be possible in a traditional mode of teaching in which teachers and students are heavily constrained by the physical conditions of a classroom.

In summary, since the data pool was fairly limited, the results of this study pertain only to certain undergraduate courses at the University of Tasmania. Some further studies need to be done in this field to explore a very effective way to attract and take learners to new language learning environments.

CONCLUSION

As stated previously, without a doubt, technology has revolutionized society in many places around the globe, including how language instruction is taught and delivered. Technology has evolved so quickly that what was known as new technology a few years ago is now already viewed as old and obsolete. It should be noted that if we try to integrate technology in our teaching as presented in this article, staff development and teacher training will need to be continuous and ongoing. [7] reinforce this point in their claim that traditional staff development models of workshops and conference presentations will not meet the need for continuous on-going technology. Current and future educators will need to be exposed and trained to use the new technologies available so that they can utilize technology to deliver a more effective and valuable learning experience to meet learners' needs.

In a transmission model of learning, knowledge is given from teachers to learners and collaborative learning does not play an important role. Currently there has been a shift from teaching to learning, from transmitting to interacting. Teachers and learners are members of a learning community in which collaboration is one of the key learning experiences.

Future developments in networked communication, multimedia, and artificial intelligence will likely converge, creating a potentially more central role for the computer as a tool for authentic language exploration and use in the second language classroom. As our focus of attention gradually shifts from the computer itself to the natural integration of computers into the language learning process, we will know that we can make technology and the Internet a more rewarding partner in the language teaching and learning process.

REFERENCES

1. Z. Berge and M. Collins, "Computer-mediated communication and the online classroom in distance learning.," *Computer-Mediated Communication Magazine*, 2(4), (1995), http://www.ibiblio.org/cmc/mag/1995/apr/berge.html, accessed June 13, 2005
2. B. Bilan, "Applying computer-based multimedia in learning and instruction," (1995), http://www.ucalgary.ca/pubs/Newsletters/Currents/Vol2.4/06_Bilan.html, accessed June 5, 2005
3. C. J. Bonk and D. J. Cunningham, "Searching for learner-centered, constructivist and sociocultural components of collaborative educational learning tools," in C. J. Bonk, and K. King, Ed. *Electronic collaborators: Learner-centered technologies for literacy, apprenticeship, and discourse*. Mahwah, NJ: Lawrence Erlbaum Associates, 1998.
4. D. C. Caverly and L. MacDonald, "Techtalk: Keeping up with technology," *Journal of Developmental Education*, 28(2), 2004, pp.38-39.
5. K. Cotton, "Computer-assisted instruction," (1991), http://www.nwrel.org/scpd/sirs/5/cu10.html, accessed June 5, 2005
6. A. A. Gokhale, "Collaborative learning enhances critical thinking," *Journal of Technology Education*, 7, 1995, pp. 89-93.
7. C. E. Hopey and L. Ginsburg, "Distance learning and new technologies: You can't predict the future but you can plan for it," *Adult Learning*, 8(1), 1996, pp.22-23.
8. R. S. Houghton, "Rationale for multimedia use and instruction in education," (2004), http://ceap.wcu.edu/houghton/MM/rationale/rationaleMMframes.html, accessed June 12, 2005
9. R. T. Johnson and D. W. Johnson, "Action research: Cooperative learning in the science classroom," *Science and Children*, 24, 1986, pp. 31-32.
10. J. M. Marshall, *Learning with technology: evidence that technology can, and does, support learning*, CA: Cable in the Classroom, San Diego, 2002.
11. J. Palme, "Why do people use Internet forums (Computer-Mediated Communication, CMC)," (2000), http://dsv.su.se/jpalme/why-people-use-cmc.html, accessed June 13, 2005
12. W. Rau and B. S. Heyl, "Humanizing the college classroom: Collaborative learning and social organization among students," *Teaching Sociology*, 18, 1990, pp. 141-155.
13. T. C. Reeves, "The impact of media and technology in schools." A research report prepared for the Bertelsmann Foundation, 1998.
14. G. B. Shelly, T. J. Cashman, R. E. Gunter and G. A. Gunter, *Teachers discovering computers: Integrating technology in the classroom* (2nd ed.), MA: Course Technology Thomson Learning, Boston, 2002
15. A. Shepard, "A case for computer-based multimedia in adult literacy classrooms," (2003), http://coe.sdsu.edu/eet/articles/adultliteracy/index.htm, accessed June 12, 2005
16. R. E. Slavin, "Research on cooperative learning: an international perspective," *Scandinavian Journal of Educational Research*, 33, 1989, pp. 231-243.
17. M. B. Tinzmann, B.F. Jones, T.F. Fennimore, J. Bakker, C. Fine, and J. Pierce, *What is the collaborative classroom?* Oak Brook, New York, 1990.
18. S. Totten, T. Sills, A. Digby, and P. Russ, *Cooperative learning: a guide to research.* Garland, New York, 1991.
19. N. Webb, "Student interaction and learning in small groups: a research summary," *Learning to cooperate, cooperating to learn.* Garland, New York, 1985.
20. A. A. Zucker., "The growing need for professional development," (2001), http://www.nctm.org/dialogues/2001-05/20010505.htm, accessed May 23, 2005.

EMC Corporation Provides Colleges with a Course in Storage Technologies

Ed Van Sickle

EMC Corporation
55 Constitution Boulevard
Franklin, MA 02038 USA

Abstract. EMC Corporation, the world leader in data storage, created the EMC Academic Alliance Program to educate students on storage and close the education gap that exists. EMC developed a Storage Technology course to teach students about the design of storage technologies and the "big picture" of an information infrastructure. The course is "open" and focused on storage technologies, not products. College and universities use the course to teach students about a very important topic in IT: Storage. EMC collaborates with colleges and universities by providing the course, knowledge transfer sessions to faculty and program support. There is no cost to join and no cost to obtain the courses. EMC requires partners to sign an agreement for course use. Several colleges are using the course as an upper level elective and the course is taught by faculty. The alliance program has reduced faculty time to develop a storage course and time to learn the topic. Faculty is responsible for credentialing students and they supplement the course with additional materials. Students are being recruited for jobs by EMC and others, including internships. The Alliance program provides academic institutions with a way to differentiate. This paper will explain the program and the Storage Technology course.

Keywords: Computer Science Education, Industry - Education Relationships, Information Science Education, Storage Courses and Curriculum, Storage Technologies
PACS: 01.40.G- Curricula and evaluation

INTRODUCTION

Corporate CIO's and IT Managers understand that the single most critical asset of any organization is their data. They understand that more data is being created, in various formats (audio, video, mpeg files) and they have regulatory requirements which require data to be available on line, for longer periods of time. IT managers face the task of creating an information infrastructure which can store, protect, manage, optimize and leverage this information. IT departments are implementing SAN (Storage Area Networks), NAS (Network Attached Storage), and Data Replication technologies to solve the problems of storage capacity, data availability, and data mobility. Spending data reveals that storage is an important topic on the minds of CIO's and increased spending on storage makes it one of the fastest growing segments of IT.

According to the *CIO Magazine* Tech Poll for 2007 [1], over 50% of respondents ranked storage and servers as the top items for spending increases over the next 12 months. Comparatively, many expected spending to be flat for networking and

CP1007, *Current Themes in Engineering Technologies*, edited by S.-I. Ao, M. A. Amouzegar, and S.-S. Chen

telecommunications equipment, infrastructure software and eBusiness. According to a recent Gartner CIO survey, storage technologies were ranked fifth of the top ten technology priorities. CIO's will need to exploit new approaches and technologies to support the business needs of their organizations [2]. In an analysis of U.S. Department of Labor job data and a survey conducted by EMC, it is estimated that over 1 million new storage jobs will be created worldwide by 2012 [3].

Despite all the news and spending data, for most IT professionals, it seems that storage infrastructures are still relatively unknown. This is interesting when considering that storage is not new and has been in existence for some time. Why then, is so much emphasis being placed on information infrastructure technologies now?

A major reason is that system downtime is expensive and not having access to data can cost a company millions of dollars (Figure 1). Information must be continuously available to support the business. An IT infrastructure that can support this primary requirement is highly desired. Today's IT departments are implementing Information Management Infrastructures to meet those needs.

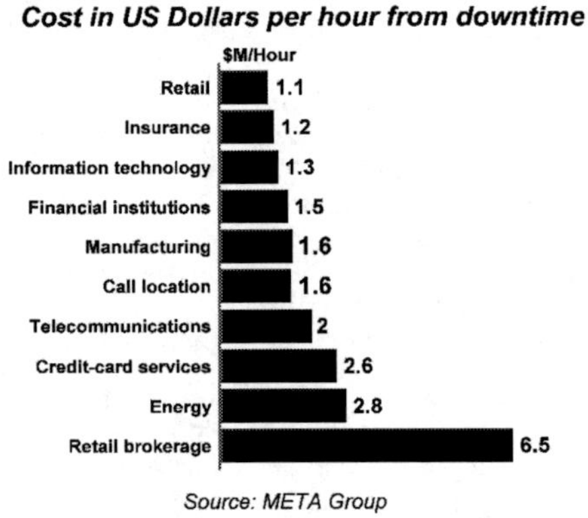

FIGURE 1 - Cost of Downtime

Typically, at the core of these Information Infrastructure solutions are intelligent storage disk arrays. Intelligent arrays provide organizations with the ability to store, protect, manage, optimize and leverage their data.

Depending on the business needs and solution requirements, different storage technologies can be used to solve the needs for managing data. These technologies include SAN (Storage Area Networks), NAS (Network Attached Storage), DAS (Direct Attached Storage) and CAS (Content Addressable Storage). These technologies and the advancements in hard disk drives and data delivery through high speed networks (fibre channel and IP networks) have enabled organizations to keep data continuously available. Additionally, for each type of storage technology, there is array-based data replication software functionality to copy and move data for backups, business continuance and disaster recovery, migration, and testing efforts.

So, if these storage technologies have become so prevalent, who provides the education on them? Most vendors usually provide training on their product offerings. Typically, the focus of vendor courses is on training, not education. Most Computer Science and Information Technology degree programs at colleges and universities aren't either. Their focus for technical course offerings are on computer architecture, operating systems, databases, networking and software application development – but not storage technologies. If we look at today's IT infrastructure, it's important to note that OS, RDBMS, Networks, Applications **and Storage** are integrated together to form the Five Pillars of IT (Figure 2) [4]. For example, we can see all these technologies in use when an end user using a software application makes a read or write request for data over the network. The operating system processes the request, the database has organized the data into tables and the disk array sends or receives the data, protects it and secures it.

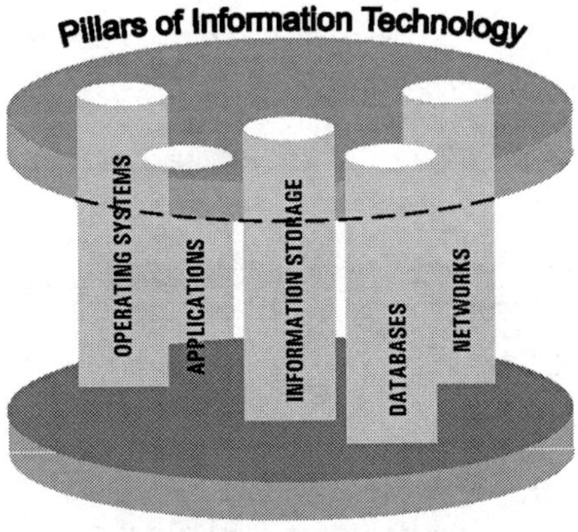

FIGURE 2- Five Pillars of IT

Most colleges and universities educate students on four of the five Pillars of IT. EMC conducted an extensive survey of IT professionals to acquire data about the storage skills gap. The results of the study drove EMC to develop solutions to meet the needs of industry and academia.

EMC STORAGE STUDY AND FINDINGS

Storage and information management infrastructures are being sold rapidly – it's a $60 billion (USD) market in 2007 and growing, which is creating high demand for people with these skills and knowledge (labor). But the supply of available people is low. Because the 5th Pillar is not being addressed in industry and academia, a skills shortage exists in the market. IT departments have the need and hiring demand as they create and implement these information infrastructures with storage arrays as the backbone layer.

EMC conducted a study (global survey of over 1,200 IT Professionals) and identified the key challenges for managing increasingly complex information infrastructures [5]. Figure 3 identifies the key challenges faced by IT / Storage Managers [5].

Challenges Identified by IT and Storage Managers and Professionals
• Managing storage growth
• Designing, deploying, and managing backup and recovery
• Designing, deploying, and managing disaster recovery solutions
• Making informed strategic/big-picture decisions
• Designing and deploying multi-site environments
• Designing and deploying emerging storage technologies (such as storage virtualization, IP SAN, GRID, etc.)
• Lack of skilled storage professionals
• Managing data availability/data retention compliance

FIGURE 3- Key Challenges for Managing Storage

An activity such as Back Up has been in practice for decades, yet professionals believe they are not doing it to the desired level. Why? EMC believes that this is due to the complexity of the environment and partially due to the fact that individuals managing the environment have not been provided a comprehensive education on storage technologies. Figure 4 further substantiates the complexity issue and identifies key pain points, by priority [5].

IT and Storage Managers	
80%	Managing storage growth
61%	Designing, deploying, and managing backup and recovery
58%	Designing, deploying, and managing disaster recovery solutions
48%	Making informed strategic/big-picture decisions
40%	Designing and deploying multi-site environments
36%	Designing and deploying emerging storage technologies (such as storage virtualization, IP SAN, GRID, etc.)
30%	Lack of skilled storage professionals
27%	Compliance regulations
Storage Professionals	
73%	Managing storage growth
62%	Designing, deploying, and managing backup and recovery
61%	Designing, deploying, and managing disaster recovery solutions
42%	Making informed strategic/big-picture decisions
42%	Designing and deploying multi-site environments
36%	Designing and deploying emerging storage technologies (such as storage virtualization, IP SAN, GRID, etc.)
31%	Lack of skilled storage professionals
23%	Compliance regulations

FIGURE 4- Key Pain Points for Managing Storage in order of priority by Managers and Professionals

With an "educated" professional, EMC contends that many of these issues would not be pain points, but tasks performed to desired levels. Figure 5 reveals hiring plans and requirements for storage personnel over the next 12 months [5]. The lack of skilled professionals becomes a serious bottleneck. A major factor for this skills shortage includes the lack of storage technology education in academia and industry. This skills gap needs to be closed.

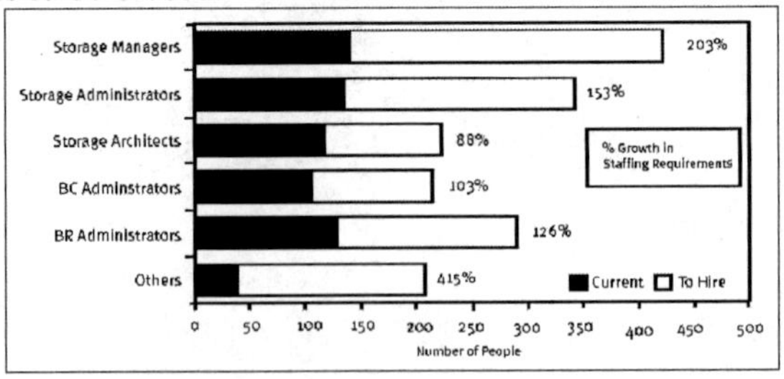

FIGURE 5- Hiring Requirements of next 12 months

Colleges and universities can close this gap and provide students with a needed education on Storage Technologies through the EMC Academic Alliance Program. The following section describes the program.

EMC ACADEMIC ALLIANCE PROGRAM ORIGIN AND CONCEPT

To address the Storage Skills Gap issue, the EMC Education Services team took action with the following steps:

Research with storage customers revealed that the most common pain point was finding individuals who understood "the big picture" of an information infrastructure. IT organizations needed people that understood, at a high level, the Five Pillars of IT.

Research revealed that IT organizations expect to provide an individual with hands-on, product training on the tools that they use. The storage industry has not standardized on a common set of tools and there are many product offerings to perform the same task. Therefore, a focus on equipment based, task-based learning would be a mistake and posed a risk to students.

If a course was created for college students on storage technologies, it would benefit EMC, EMC partners, EMC customers and the storage industry in general (non-EMC customers who have the same needs).

EMC Education had employees with experience in academia and decided to create a set of courses focused on storage technologies, not products. It was decided to develop course materials for use by academia that focused on Storage Theory, Storage Design and Information Management skills, not on using storage equipment.

This "open" curriculum was needed if it were to be accepted by most academic institutions worldwide. Additionally, the course needed to use case study and classroom discussion for the applied learning aspect of the program.

Students preparing for roles in areas like Database or Networking would benefit because they will encounter storage infrastructures in their career and they will need to know about these technologies.

EMC also realized that they could educate far many more people through a partnership program than they could alone. By creating an Academic Alliance program, they could infuse storage education to tens of thousands of college students worldwide.

Executive level support at EMC was provided. Joe Tucci, Chairman, President and Chief Executive Officer of EMC, who serves on the President's Council of Advisors on Science and Technology (PCAST) and as Chairman of the Business Roundtable Task Force on Education and the Workforce, views the EMC Academic Alliance Program as serving the mission of PCAST, as well as a vehicle to improve technology education worldwide. With executive support and funding from the highest levels, the EMC Academic Alliance Program was initiated in late 2005 and launched in July 2006.

EMC ACADEMIC ALLIANCE PROGRAM STRUCTURE AND GOALS

The characteristics of the EMC Academic Alliance Program include a.) a teaching focus that provides CS/IT students with an education on storage b.) a course focused on storage design and management that explains theory and concepts, not products c.) a program that is provided at no cost to the institutes d.) supports the need for academic freedom, and where warranted, the need to supplement the materials d.) EMC support for knowledge transfer, student enrollments, guest lectures, and site visits e.) the exploration of potential research opportunities with EMC CTO Office f.) recruitment and hiring of students by EMC, partners and customers.

The structure of the program includes:

- There is no cost to join the program. EMC courses are also provided at no cost.
- EMC and University/College complete an Agreement.
- EMC remains as the owner of the course.
- Course has to be offered in an undergraduate or graduate degree program, for credit.
- Use of the materials in an adult education or other for-profit program is restricted and must be done through EMC's Learning Partner Channel.
- Course becomes part of the University/College degree program offerings. University/College schedules course delivery and lists the course in their course catalogue and Web site.
- Universities/Colleges determine how they implement the course, as a special topic elective to Jr. / Sr. students, or as a core offering or permanent elective.
- University/College provides student instruction and the faculty member to teach the course. Credentialing is done by University/College.
- University/College uses the full materials to teach the course and can supplement the materials. Sections of the course can be used in other courses, provided a complete course offering is given. Sections of the course can not be used in other courses unless the complete course is being taught.
- EMC provides program support; training to faculty on material (no cost for attending a session), trademark/logo for use in collateral materials, guest lectures/ site visits, recruiting, research.
- EMC has included the course as part of the EMC Proven Professional Program so that if students are inclined, they can become EMC Certified. This is not a mandatory requirement of the program, but a supplemental offering and option.

Assessing the program will be done by soliciting feedback from partners and students. EMC will adopt changes that benefit the alliance. The next sections of this paper will describe the Storage Technology Foundations course and its merits.

STORAGE TECHNOLOGY COURSE FROM EMC

The Storage Technology course provides a comprehensive introduction to Data Storage technology fundamentals. The selection of these topics for the course is based on the wide use, acceptance and adoption of these technologies by industry. Also, EMC's experience as the world leader in storage in providing storage solutions with these technologies (topics) and ability to provide knowledge transfer. Course participants will gain knowledge of the core logical and physical components that make up a Storage Systems Infrastructure. Throughout the course, students will be exposed to the following themes:

i.) The explosion in demand from businesses for highly available and secure access to data.

ii.) The Storage systems and infrastructure architectures and solutions available to support business needs.

iii.) The complexities and challenges in managing storage infrastructures.

Upon successful completion of the course, students should be able to:

a.) Describe storage technology solutions such as Storage Area Networks (SAN), Network Attached Storage (NAS), Content Addressed Storage (CAS).

b.) Understand and articulate the technologies and solutions available to support an IT Infrastructure including Business Continuity, Information Availability, Local and Remote Replication, Backup and Recovery and Disaster Recovery needs of businesses.

c.) Understand the key tasks in successfully managing and monitoring a data storage infrastructure

The course consists of the following sections and modules.

Section 1 - The Complexity of Information Management

Module 1.1 - Meeting Today's Data Storage Needs

- Define the challenges in data storage and the importance of data.

Module 1.2 – Data Center Infrastructure

- List the core elements of a data center including, applications, databases, servers/operating systems, networks (LAN/SAN), storage arrays.
- Describe how these core elements are integrated and show an application example.

Section 2 - Storage Systems Architecture

Module 2.1 – The Host Environment

- List the hardware and software components of the host environment such as CPU, Memory, bus, I/O devices, Registers, L1/L2 cache, file systems, volume management, Host Bus Adapters.

Module 2.2 – Connectivity

- Describe the physical and logical components of a connectivity environment.
- Define the key protocols and concepts including; Bus technology, PCI, IDE/ATA, SCSI, and Fibre Channel.

Module 2.3 – Physical Disks

- Describe the major physical components of a disk drive and functionality including platters, spindle, motor, actuator arm, R/W heads, assembly, drive controller.
- Define the physical structure of disk including sectors, tracks and cylinders.
- Define the access characteristics for disk drives and performance implications for factors such as rotation, positioning and transfer rates.

- Define the partitioning of physical drives using zoned bit recording.

Module 2.4 – RAID Arrays

- Define the concept of RAID for data protection.
- Review and understand the common RAID levels, including RAID 0, RAID 1, RAID 3, RAID 4, RAID 5, RAID 0+1, RAID 1+0, Hot Spares.

AN APPLIED LEARNING ACTIVITY- STUDENT COMPLETES A RAID CASE STUDY

Module 2.5 – Disk Storage Systems

- List the benefits of and components of an intelligent storage system.
- Compare and contrast integrated and modular storage systems.
- Explain how a storage system handles I/O flow.
- Describe the logical elements of an intelligent storage system.
- Define the Cache Structure and data flow through cache.

AN APPLIED LEARNING ACTIVITY– STUDENT COMPLETES A DATA FLOW EXERCISE

Section 3 - Introduction to Networked Storage

Module 3.1 – Direct Attached Storage

- Describe the benefits of a DAS based storage strategy and the physical elements in DAS.
- Define the connectivity options for DAS and distinguish between IDE, ATA and SCSI protocols.
- Describe the I/O flow in a DAS environment.

Module 3.2 - Network Attached Storage

- Define NAS and difference between NAS devices and general purpose file servers.
- Describe the NAS device components and file services protocols used (NFS and CIFS)connectivity options for NAS.
- Describe the I/O flow in a NAS environment.
- List NAS management considerations for performance, storage capacity, back up.
- Provide NAS examples and solutions to technology problems.

Module 3.3 – Storage Area Networks

- Provide an overview of the physical and logical elements of a SAN.
- Define SAN components including, Host Bus Adapter, Fiber Optic cabling, FC Switch, Disk Array, Management SW.
- Define Fibre Channel and connectivity layer including FC ports, World Wide Names, FC Addressing, FC log-in process and Fabric topologies.
- List SAN management considerations such as zoning and LUN masking, security, capacity and performance management.

AN APPLIED LEARNING ACTIVITY– STUDENT COMPLETES SAN CASE STUDY

Module 3.4 – IP SAN

- Define IP SAN topologies including iSCSI, FC/IP, iFCP.

- Describe the differences between FC/IP and iFCP.

Module 3.5 – Content Addressable Storage (CAS)

- Describe the benefits of a CAS based storage strategy.
- Provide an overview of the physical and logical elements of CAS.
- Define the connectivity options for CAS.
- Define the I/O flow in a CAS environment.

Section 4 – Information Availability

Module 4.1 – Business Continuity Overview

- List reasons for planned and unplanned outages and describe the impact of downtime. Differentiate between Business Continuity (BC) and Disaster Recovery (DR).
- Define Information Availability and its importance to the business.
- Define Recovery Time Objective, Recovery Point Objective.

Module 4.2 – Back Up and Recovery

- Define planning requirements for Back Up and Recovery.
- Define Back Up and Recovery Strategies.
- Define how a backup works, plus the business and data decisions for back ups.
- Describe LAN and SAN backup methods and Back Up for databases.

AN APPLIED LEARNING ACTIVITY – STUDENT COMPLETES A BACKUP & RECOVERY CASE STUDY

Module 4.3 – Business Continuity Local

- Describe potential areas of information vulnerability within a data center.
- List the local information availability technologies within the data center.
- Identify the appropriate local information availability technology based on criteria.

AN APPLIED LEARNING ACTIVITY – STUDENT COMPLETES REPLICATION CASE STUDIES 1 & 2

Module 4.4 – Business Continuity Remote

- Describe potential areas of information vulnerability between local and remote data centers.
- List the remote information availability technologies between local and remote data centers.
- Identify the appropriate remote information availability technology based on criteria.

AN APPLIED LEARNING ACTIVITY – STUDENT COMPLETES REMOTE REPLICATION CASE STUDY

Section 5 – Managing and Monitoring

Module 5.1 – Monitoring In the Data Center

- Define the areas to monitor for an information management infrastructure.
- List the issues for monitoring capacity, performance, security, servers, SAN, storage arrays and networks.

Section 6 – Security and Virtualization

Module 6.1 – Securing the Storage Infrastructure

- Define storage security.
- List the critical security attributes for information systems.
- Describe the elements of a shared storage model and security extensions.
- List and analyze the common threats in security domain.

Module 6.1 – Securing the Storage Infrastructure

- Identify different virtualization technologies.
- Describe block-level and file level virtualization technologies and processes.

EMC ACADEMIC ALLIANCE PROGRAM MEMBERS

As of January 2008, EMC has signed agreements with the following institutions in the United States.

- Pennsylvania State University, University Park, PA. College of Information Science and Technology
- Ball State University, Muncie, IN. Center for Information and Communication Sciences
- University of Massachusetts at Dartmouth, Dartmouth, MA. Charlton College of Business, MIS Program
- North Carolina A&T State University, Greensboro, NC. School of Technology, Computers, Electronics and Information Technology
- North Carolina State University, Raleigh, NC. School of Engineering, Computer Science
- Northeastern University, Boston, MA. School of Engineering Technology, Computer Engineering Technology
- Indiana University of Pennsylvania, Indiana, PA. Eberly College of Business and Information Technology, MIS
- Salisbury University, Salisbury, MD. Perdue School of Business, Information Systems
- Springfield College, Springfield, MA. Department of Math, Physics and Computer Science
- Quinnipiac University, Hamden, CT. College of Liberal Arts, Department of Computer Science
- Southern New Hampshire University, Manchester, NH. Information Systems
- Coleman College, San Diego, CA. Computer Networks
- Kennesaw State University, Kennesaw, GA. Computer Science & Information Systems
- Howard University, Washington, DC. School of Engineering & Computer Science, Department of Systems and Computer Science
- St. Edward's University, Austin TX. School of Natural Sciences, Department of Computer Science.

- Quinsigamond Community College & Worcester Consortium (13 colleges in Worcester, MA area)
- Framingham State College, Framingham, MA. Department of Computer Science.
- Worcester State College, Worcester, MA. Department of Computer Science.
- Northwest Missouri State University, Maryville, MO. Computer Science.
- University of New Orleans, New Orleans, LA. Computer Science.
- LeTourneau University, Longview, TX. Computer Science.
- Georgia Southern University, Statesboro, GA. College of Information Technology.

The alliance program is a global program with many institutions participating throughout the world. The program has been successfully launched in the following countries with a number of signed institutes;
India – 110 institutes, China – 26 institutes, Russia – 7 institutes, Europe – 3 institutes, Mexico and Brazil – 5 institutes.
At this time, EMC is expanding the program and is willing to discuss the alliance program opportunity with other institutions.

COMPARISON OF EMC STORAGE TECHNOLOGY COURSE TO TEXTBOOKS AND OTHER STORAGE COURSES

There are a plethora of storage textbooks that have been written and are currently available (including an entry from the "For Dummies" series). A quick search on Amazon.com using storage networks or storage networking as the search criteria will yield an ample number of choices. In this author's opinion, the text books were developed for working IT professionals, not students. Authors such as Marc Farley, Tom Clark, Meeta Gupta and Daniel Pollack were reviewed and compared to the EMC Storage Technology course.
The Farley book (Storage Networking Fundamentals, Cisco Press) is a very good fundamentals book. It covers Fibre Channel, SCSI, ATA, and SATA and their use in network storage subsystems. The book also covers volume management, storage virtualization, data snapshots, mirroring, RAID, backup, and multipathing. It does not cover specific storage technologies such as CAS and does not cover intelligent disk arrays using caching algorithms and cache processes to the depth that the EMC course does.
The Gupta and Clark books are heavily focused on SAN technology. A large portion of these books are devoted to Fibre Channel, SCSI and iSCSI network protocols. The EMC course is comprehensive and covers other storage technologies such as NAS, CAS and replication.
There are relatively few other storage course offerings to compare against the EMC course. An offering from the Storage Networking Industry Association (SNIA) was reviewed. SNIA has developed a certification program (SNCP) for IT professionals who are interested in becoming storage certified. SNIA do not provide courses, just exam standards. SNIA has outlined the topics for which certification exams were written and are used to measure skills and knowledge. Training courses are developed

by training firms using the standards developed by SNIA. A three day, lecture only, Storage Networking Concepts Foundation class is offered by Knowledge Transfer. The class is designed for working professionals. Like the textbooks, the course covers Fibre Channel, SCSI, iSCSI, SAN and NAS. This course appears to cover a breadth of topics but has no, or limited, applied learning.

CONCLUSION

EMC's Storage Technology Foundations course covers all the storage technologies used by today's IT departments. The course covers "the big picture" including SAN, NAS, CAS, Back Up and Recovery and array based replication. The course has been through an instructional design process and meets the requirements of pedagogical learning.

This course will provide college and universities an opportunity to meet the needs of industry by providing students with an education in the fifth pillar of IT; Storage. IT is changing and the demand to store, protect, manage, optimize and leverage the massive amounts of data being created is driving corporations to adopt information management infrastructures that utilize storage disk arrays as the backbone. Colleges and universities are urged to join the EMC Academic Alliance Program and use the Storage Technology Foundations course to teach students and prepare them for the emerging challenges of data and information management. The goal of this partnership program is to prepare tomorrow's Storage leaders today.

REFERENCES

1. CIO *Magazine*, December 2006. http://peoplepolls.com/results/CIO/120706.asp?user=CIO and http://www.cio.com/info/releases/122906_techpoll.pdf
2. "Gartner EXP Survey of More than 1,400 CIOs Shows CIOs Must Create Leverage to Remain Relevant to the Business" http://www.gartner.com/it/page.jsp?id=501189
3. Hecker, Daniel, "Occupational employment projections to 2012," *Monthly Labor Review*, February 2004, pp. 80-105.
4. Clancy, Tom (EMC), "EMC, IIT-B to Bridge Storage Knowledge Gap," *CyberMedia News*, March 14, 2007.
5. Shrivastava, Alok (EMC), Managing Storage: Trends, Challenges and Options 2007-2008. EMC White Paper. May 2007. http://education.emc.com/main/common/documents/H2159_ManagingStorage_EP.pdf

Intelligent E-Learning Systems: Automatic Construction of Ontologies

Jesús del Peso[a] and Fernando de Arriaga[b]

[a]*J. del Peso, a Ph.D. candidate, is with the Department of Applied Mathematics, Universidad Politécnica de Madrid, 28040, SPAIN (e-mail: jdelpeso@mat.upm.es).*
[b]*F. de Arriaga is with the Department of Applied Mathematics, Universidad Politécnica de Madrid, 28040 SPAIN (phone: 34-91-336-7286; fax: 34-91-336-7289; e-mail: farriaga@mat.upm.es).*

Abstract. During the last years a new generation of Intelligent E-Learning Systems (ILS) has emerged with enhanced functionality due, mainly, to influences from Distributed Artificial Intelligence, to the use of cognitive modelling, to the extensive use of the Internet, and to new educational ideas such as the student-centered education and Knowledge Management. The automatic construction of ontologies provides means of automatically updating the knowledge bases of their respective ILS, and of increasing their interoperability and communication among them, sharing the same ontology. The paper presents a new approach, able to produce ontologies from a small number of documents such as those obtained from the Internet, without the assistance of large corpora, by using simple syntactic rules and some semantic information. The method is independent of the natural language used. The use of a multi-agent system increases the flexibility and capability of the method. Although the method can be easily improved, the results so far obtained, are promising.

Keywords: Ontology, ontology construction, multi-agent system, intelligent e-learning systems.
PACS: 07.05.Mh, 89.20.Ff

INTRODUCTION

Ontologies are becoming more important for Knowledge-Based systems. Among them Intelligent E-Learning Systems (ILS) are an extended area, suitable to use ontologies not only for communication purposes among several ILS, sharing different ontologies, but also to enlarge their corresponding Knowledge-Bases.

The new learning revolution, known as E-learning (EL), raised tremendous expectations during the last thirty years. The world of Internet and the web technologies have spread over even in the construction of internal applications running on an intranet. However, once again the actual situation does not match forecasts [1] because in many cases EL has only been a coined name used to describe old applications and ancient optics, with little concern for the learning problem which still lies at the core of the educational process.

To overcome those drawbacks we have followed an approach [2] that aims at the integration of all the elements present in learning, which are: psychological ingredients of learning [3], [4], educational and tutorial aids, and advanced processing techniques to implement all those constituents and manage the whole process. There

CP1007, *Current Themes in Engineering Technologies*, edited by S.-I. Ao, M. A. Amouzegar, and S.-S. Chen

has been, so far, several attempts and concrete work on this line [5], [6], but there are just a few of them and unfortunately not very well known.

Although some of those systems have already been built by means of agents, a new ILS generation has emerged, characterized by its enhanced functionality and complexity, advanced tutorial functions and cognitive modeling [7], [8], [9]. The introduction of several machine learning techniques, fuzzy logic not only for student's evaluation but also to represent and model the student's and expert's behaviors [10], [11], [12], and affective computing [13], make possible to design robust and powerful capabilities, and to implement them in a more distributed way, with a better and more efficient integration. The consequences are all related to a closer or deeper insight of the learning process: better and more detailed representation of the domain knowledge, a more realistic design of the student behavior model, personalization of the tutoring aids and advices, and an efficient data processing implementation with learning capabilities. The evaluations so far carried out [14], [15], proof our statements.

The paper deals with the automatic construction of ontologies, a feature that could increase in great manner the capabilities of ILS. The fact that most of today's ILS rely on multi-agent systems, emphasizes the importance of the automatic or semi-automatic construction of such ontologies, task that can be undertaken by the multi-agent system.

AUTOMATIC CONSTRUCTION OF ONTOLOGIES

Several steps can be followed in the automatic construction of ontologies: a) identification and retrieval of concepts (terminology); b) identification and obtainment of taxonomic relationships (relationships of hyponyms [16]; c) identification and extraction of non-taxonomic relationships as, specific relations of interest or relationships of meronyms [17].

Among the main techniques used for extracting information from textual sources we can quote: statistical procedures for natural language processing, by frequently using a *corpus* of documents that includes at least a representative set of environment related documents, and another set of general documents, not limited to the environment of the ontology we want to build; techniques for natural language analysis including morphologic, syntactic and lexical-semantic "shallow" analysis; and finally data mining techniques, even with schemes for automatic learning.

Statistical techniques are usually followed to identify the ontology terms [18]--[21]. First of all the frequencies of appearance of the concepts within the environment related documents are obtained. Then the same frequencies of appearance are obtained, but using now the set of general documents, not related to the domain. The selection of concepts is done as a function of all these frequencies. Standard measures, such as TF (*term frequency*) or TFIDF (*term frequency-inverted document frequency*) can be used [22] or even more elaborate measures [23].

For the construction of the conceptual hierarchy lexical-syntactic patterns could be used. They are based on heuristic methods using regular expressions for information extraction. The purpose is to define regular expressions to capture recurrent expressions, allowing so to map them into semantic structures [16]. Hierarchical clustering, other statistical method, is also used [24].

For identification of non-taxonomic relationships several methods are currently used: a) lexical-syntactic patterns, usually dependant on the domain [16]; b) semantic patterns consisting on verbal patterns [18] also dependant on the domain; c) data mining techniques for the obtainment of general rules of association [25], [26]. The procedures are based on a statistical analysis that does not allow to know the nature of those relationships; d) other statistical techniques, to obtain pairs of concepts and possible relationships between them, according to the domain based documents [19], [21].

So far, the methods most currently used for the automatic construction of ontologies are statistical, based on large *corpora*, or domain dependant.

Instead, our approach deals with the automatic construction of ontologies, based only on the knowledge of natural language. This knowledge will be fundamentally syntactic. Our purpose is to avoid statistical methods and the construction of large corpora. Besides, we attempt to obtain a general method for extracting ontologic information, independent of the discourse domain. The use of multi-agent systems allows the method to be distributed and scalable, and can be integrated into an information retrieval system.

ALTERNATIVE APPROACH

The starting point is the collection of a set of documents that will be analyzed by several agents to construct the ontology corresponding to the domain of those documents. These collections could be different for each one of the agents but all of them integrated by documents that are candidates to belong to the chosen domain. But it is also possible that not all of the documents pertain to the same domain or that the documents content could not be homogeneous.

The method is based on the concept of *ontologic network*, that will be defined later on. An ontologic network is, actually, a formal representation of an ontology, able to be constructed and processed by autonomous agents in a distributed way. This representation allows several operations, as aggregation of ontologic networks to obtain a more general network. It allows also the evaluation of the relevance of the components of an ontologic network as well as the application of different functions, as *pruning*, to select elements with real interest from the point of view of their information content.

Principles of the System

The basic principles of the system are the following:

a) several agents are dedicated to the construction of the ontology which corresponds to a certain domain;
b) each agent will carry out the analysis of a set of documents, candidates to belong to the domain;

c) each agent will combine the different ontologic networks produced for each of the documents in order to obtain an enhanced network with information relative to the collection of documents analyzed by the agent. This enhanced network will be called *final A_n-agent network*;

d) agents finally interchange their own final networks to elaborate a *global aggregated network*, that includes by aggregation, the ontologic information obtained by all the agents;

e) finally, the global aggregated network will be evaluated and pruned to select the elements with relevance for the searched ontology.

Agents coordination can be realized in a centralized or decentralized way. The method can be used in combination with any kind of agents coordination.

In the case of a centralize coordination there is a *control agent A_c* in charge of: a) controlling the activities of the remainder agents; b) assigning candidate documents to be processed by each agent; c) construct the global final network, as a result of the aggregation of the final networks elaborated by each agent.

That way it will be necessary a first phase for agents initialization. In that phase the control agent assigns domain candidate documents to each agent. Once they have finished up their work the control agent gets the final networks for each agent, and proceeds to construct the global final network. Finally, it selects the relevant elements of the global network to obtain the definite ontologic network.

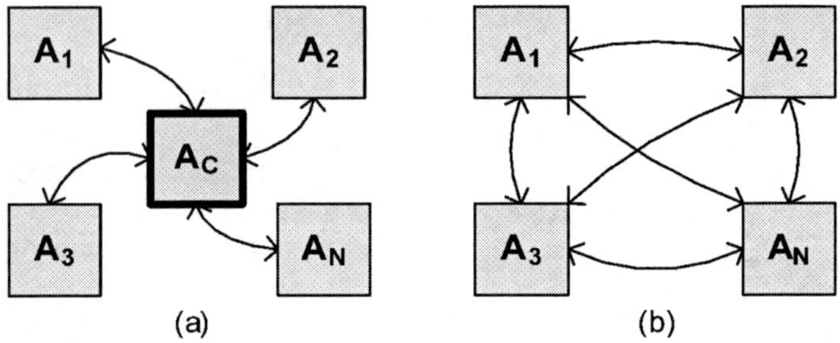

FIGURE 1. Centralized agents control (a) and distributed agents control (b).

In a distributed control scenario each of the system agents share its responsibility for the construction of ontologic networks. This is the case of an heterogeneous environment, composed of multiple agents which may be part of very different systems with diverse final objectives, but which are able to cooperate. Each agent is idependent from each other and autonomous to build its own ontologic networks.

So, the main steps in the processing carried out by each of the agents are: a) selecting candidate documents to be analyzed; b) find out which of these documents have been already processed by other agents; this could be done by means of dialogues between agents (whose format has not yet been defined); c) analyzing those documents not yet processed by any other agent; d) building up the ontologic networks corresponding to the newly analyzed documents; e) retrieving the ontologic networks already built by other agents; f) building up the global ontologic network through the aggregation of the networks obtained in phases d) and e).

The formal model explained here has been specially designed to accommodate this multi-agent distributed processing environment.

FORMAL MODEL

The method so far exposed is based on a formal approach that lies outside the scope of this paper due to space limitation. But at least we want to outline some definitions and explain the meaning of several operations and functions used by the method.

Definition 1: Simple ontologic network

A simple ontologic network on a natural language L, can be defined as an ordered pair

$$O_s = (T_s, R_s) \tag{1}$$

where T_s is the set of the network terms, being $T_s \subseteq N_T^L$, N_T^L is the set of terms defined over the natural language *L*. In general both sets of terms are different; in fact the last one can have an infinite cardinality, however we will only consider finite ontologic networks.

Similarly R_s is the set of relationships among the network terms or equivalently:

$$R_s = \{ r_s = (t_1, t_2, n_{rs}) / t_1, t_2 \in T_s , n_{rs} \in N_R^L \} \tag{2}$$

where N_R^L is the set of verbs in the language L that are tagging or giving name to the relations of the network.

An ontologic network may be described as a directed graph, where nodes corresponds to terms and edges to their semantic relations. Each of the edges of the graph is labelled by the verb that explains its associated semantic relation.

Starting out with this definition it is easy to define the concept of extended network.

Definition 2: Extended ontologic network:

An extended ontologic network on a natural language L can be defined as an ordered triple

$$O = (T, R, M) \tag{3}$$

where *T* is the set of the terms of the network, being

$$T = \{(t_s, k_t) / t_s \in T_s, k_t \in N\} \tag{4}$$

with $T_s \subseteq N_T^L$. The integer numbers k_t indicates the number of ontologic networks, precedent of this network, containing the term t_s. *R* is the set of relations established on the network terms, where

$$R = \{r = (t_1, t_2, n_r, k_r) \mid t_1, t_2 \in T, n_r \in N_R^L, k_r \in \mathbb{N}\} \tag{5}$$

and the integer number k_r indicates respectively the number of ontologic networks containing the relation r.

M is the number of extended ontologic networks from which O has been obtained. $M = 1$ if O does not proceed from any other, and $M = 0$ for the case of empty ontologic networks.

It is rather obvious to check that every simple ontologic network can be expressed as an extended ontologic network. Similarly, the concepts of sub-networks and equality of extended ontologic networks can be defined.

Now certain operations can be defined on extended ontologic networks, such as reduction, enlargement and aggregation.

Definition 3: Aggregation of (extended) ontologic networks:

Given two (extended) ontologic networks $O_1, O_2 \in \mathcal{O}$, $O_1 = (T_1, R_1, M_1)$, $O_2 = (T_2, R_2, M_2)$ their aggregation $\oplus$ is defined as the following function:

$$\oplus : \mathcal{O} \times \mathcal{O} \to \mathcal{O} \tag{5}$$

where $O_3 = O_1 \oplus O_2 = (T_3, R_3, M_3)$, being

$$\begin{aligned} T_3 = \{t_3 = (t, k_{t_3}) \mid \exists k_{t_1}, k_{t_2} \in \mathbb{N},\ (t, k_{t_1}) \in T_1 \wedge (t, k_{t_2}) \in T_2 \wedge k_{t_3} = k_{t_1} + k_{t_2}\} \cup \\ \cup \{t_3 = (t, k_{t_3}) \mid \exists k_{t_1} \in \mathbb{N},\ (t, k_{t_1}) \in T_1 \wedge \neg \exists k_{t_2} \in \mathbb{N},\ (t, k_{t_2}) \in T_2 \wedge k_{t_3} = k_{t_1}\} \cup \\ \cup \{t_3 = (t, k_{t_3}) \mid \neg \exists k_{t_1} \in \mathbb{N},\ (t, k_{t_1}) \in T_1 \wedge \exists k_{t_2} \in \mathbb{N},\ (t, k_{t_2}) \in T_2 \wedge k_{t_3} = k_{t_2}\} \end{aligned} \tag{6}$$

$$\begin{aligned} R_3 = \{r_3 = (r', k_{r_3}) \mid \exists k_{r_1}, k_{r_2} \in \mathbb{N},\ (r', k_{r_1}) \in R_1 \wedge (r', k_{r_2}) \in R_2 \wedge k_{r_3} = k_{r_1} + k_{r_2}\} \cup \\ \cup \{r_3 = (r', k_{r_3}) \mid \exists k_{r_1} \in \mathbb{N},\ (r', k_{r_1}) \in R_1 \wedge \neg \exists k_{r_2} \in \mathbb{N},\ (r', k_{r_2}) \in \\ \in R_2 (r', k_{r_2}) \wedge k_{r_3} = k_{r_1}\} \cup \\ \cup \{r_3 = (r', k_{r_3}) \mid \neg \exists k_{r_1} \in \mathbb{N},\ (r', k_{r_1}) \in R_1 \wedge \exists k_{r_2} \in \mathbb{N},\ (r', k_{r_2}) \in \\ \in R_2 (r', k_{r_2}) \wedge k_{r_3} = k_{r_2}\} \end{aligned} \tag{7}$$

where $r' = (t_a, t_b, n_r), r' \in R'_3 \wedge r' \in R'_2 \wedge r' \in R'_1$, and

$$M_3 = M_1 + M_2 \tag{8}$$

The aggregation of ontologic networks is an associative and commutative operation, which makes it appropiate to be used for the building up of ontologic networks in a distributed processing environment.

It is also possible to define concepts such as: degree of a term relationship and degree of a generalized relationship. With those elements, the concepts of network connectivity and partition of an ontologic network can be introduced, by using the ideas of ontologic network generated by a term and difference of ontologic networks;

it is shown that the partition so defined is unique and therefore a partition function can be obtained.

Several functions have been introduced on the domain of ontologic networks, such as: size of a simple and an extended network, and also the concept of grading the network. For this last purpose several possible grading functions have been defined in order to reach the concept of pruning the network according to a certain grading function. Finally, several functions for measuring the size of an ontologic network and the similarity of two of them are used.

PRACTICAL ASPECTS OF THE METHOD

Construction of Ontologic Networks from Documents

This is the first step of the method. Although it is not necessary, we will assume that for each text an ontologic network will be obtained. Text analysis is mainly syntactic: identification of the most important elements of the ontology. For this purpose the following syntactic elements have been considered: nouns (simple and compound) and noun syntagmas.

In fact only the following principal noun syntagmas have been taken into account: subject and predicate (attribute, direct object, indirect object). It has been assumed that relationships appearing in the network are specified or tagged by the verb they include in the sentence. In consequence this verb will identify the relationship type and its meaning.

As possible interesting relationships, within each phrase, those integrated by terms which are part of the subject, the verb and the predicate terms have been considered. That way each relationship is made of a first term (subject term), a second term (predicate term) and the verb. In case of complex verbs only the principal verb is considered, omitting auxiliary or modal terms. As a first approximation tenses have also been ignored. Ontologies often contain permanent information (non-time-dependent). For this reason it could be necessary to rule out relationships with non-present tenses. In all cases analyzed elements, nouns and verbs, are lemmatized by means of the Porter algorithm [27].

As far as semantic information, the meaning of special verbs, that develop important relationships within ontologies, are considered. So, for example, *to be* (that gives birth to generalization relationships integrating the ontology taxonomy), and *to have* (producing relations of meronyms). Verbs denoting spatial relations (and their synonyms) can also be easily introduced.

At this step of the method only terms that take part in relevant or “special” relationships are considered in the analysis. By special relations we mean those relations tagged by a verb whose frequency within the text is greater than a minimum threshold. That way we take into account the effect of real syntactic analyzers and their frequent errors.

Construction of the final and global ontologic network

Once all initial ontologic networks nave been obtained, they are combined to obtain the final one according to the following procedure.

First each agent obtains its own final network by using the operation of aggregation on the set of initial networks it has produced. Then the global network is also obtained by aggregating the final agent networks. Once this global network is produced it is necessary to select the elements of real interest by means of the following procedure: a) grading the global network by means of functions that take into account the terms and relations extensions; b) pruning of the global network after setting up a threshold for the terms grades; c) partition of the pruned network into sub-networks as a result of pruning, but including on each sub-network the elements related with any element of it; d) selection of the most significant sub-network. For this purpose the concept of network size and the functions introduced above for measuring it, are interesting as measure of the information content of a network.

SOME RESULTS AND ANALYSIS

Several tests have been and are been carrying out. Among them we want to briefly describe some results obtained in the automatic construction of an ontology for "geology".

According to our objectives, a small number of documents was selected (only 70) from the Internet, because most of our Intelligent E-Learning Systems Knowledge Bases are updated with that sort of information. No associated corpus was used to aid this task. The documents were articles related to aspects of the domain of interest. Some other spurious texts were introduced in the set of documents to test the power and capacity of the construction method for selecting the appropriate information. Besides, the domain related documents are not fully homogeneous, in the sense that some of them include out-of-domain elements able to cause errors to ontology construction methods based on pure statistical analysis.

For the syntactic analysis the Stanford Parser [28], provided by the Stanford Natural Language Processing Group, was used. It has to be pointed out that our approach is independent of the natural language used, but it depends on existing parsers if we do not want to get involved in the construction of such a tool.

The rules for terms identification are simple, basically the same introduced in the previous section, able to identify simple, compound nouns and noun syntagmas acting as subject or object kernels within the analyzed sentences. Anaphoric resolution has not been introduced, although its consideration will improve the efficiency of the described method.

As far as relations identification is concerned, all relations of generalization and of meronyms have been included. For the remainder relations, in order to avoid spurious ones, due to limitations or errors of the parser, a threshold has been used for the appearance frequency of related terms.

As the terms grading function *P(t)*, in this example, we have used the following one:

$$P(t,O) = P_1(t,O) \cdot P_2(t,O) \cdot P_3(t,O) \tag{9}$$

where

$$P_1(t,O) = \alpha \cdot e_T(t,O)^2,\ P_2(t,O) = \beta \cdot \sum_{t' \in \Phi_R(t,O)} e_R(t')^2,\ P_3(t,O) = \gamma \cdot \sum_{t' \in \Phi_T(t,O)} P(t') \tag{10}$$

being $\alpha = \beta = 1$, and $\gamma = 0.1$ and $e_T : N_T^L \times \mathcal{O} \to \mathbb{N}$ is a function that returns the extension value of a term of an (extended) ontologic network; $e_R : \mathcal{R} \times \mathcal{O} \to \mathbb{N}$ returns the extension value of a relation of an (extended) ontologic network; $\Phi_R : \mathcal{T} \times \mathcal{O} \to \mathcal{R}^n$ returns the relations in wich a term of a network takes part; $\Phi_T : \mathcal{T} \times \mathcal{O} \to (N_T^L)^n$ returns the terms directly related to a network term.

This function takes into account for each term not only the square of its own grading value, $e_T(t)^2$, but also the sum of the square of the grading values of all directly related terms, $\Sigma e_R(t')^2$, $\Sigma P(t')$. This formula gives more importance to terms close to other terms that have high grades.

As size function for ontologic networks the following has been used:

$$S(O) = \begin{cases} 0 \text{ if } \#T = 0 \vee \#R = 0 \\ \sum_{t=(t',k_i) \in T} P(t',O) = \sum_{O_i \in \Pi(O)} \sum_{t=(t',k_i) \in T_i} P(t',O_i) \end{cases} \tag{11}$$

where P is the grading function defined previously and $\Pi(O)$ is the partition of network O.

Many different sets of tests have been carried out looking for distinct objectives. One of them, which is briefly described here, tries to obtain a general ontology for *geology*. Several proofs have been carried out with the same set of documents, but changing some parameters. So far the results obtained depend on the threshold value for pruning the aggregate ontologic network. In fact a threshold too low will produce an ontologic network with too many terms and relations, which should not be included.

But in spite of this, and considering the simplicity of the methods here used for information retrieval, the results are promising, and the same conclusion has been obtained in the remainder domains of the tests so far carried out. Choosing an appropriate threshold for pruning, the selection of the terms that integrate the last ontologic network is good. The same thing happens to the set of chosen relations, although here some less relevant relations appear in some more cases. The reason for that could be some deficiencies in the parser or in the grading function that does not consider relations.

The results shown here correspond to the processing of an small collection of brief documents, extracted directly from the WWW, for *geology*. Small sections of the ontological networks built for this subject is shown. These graphs are composed of some of the concepts extracted, as well as the taxonomic and non taxonomic relations identified. Precisely this one is one of the main advantages of the model: not only it is possible to determine the existence of relations between terms, but it also makes possible to interpret their meaning.

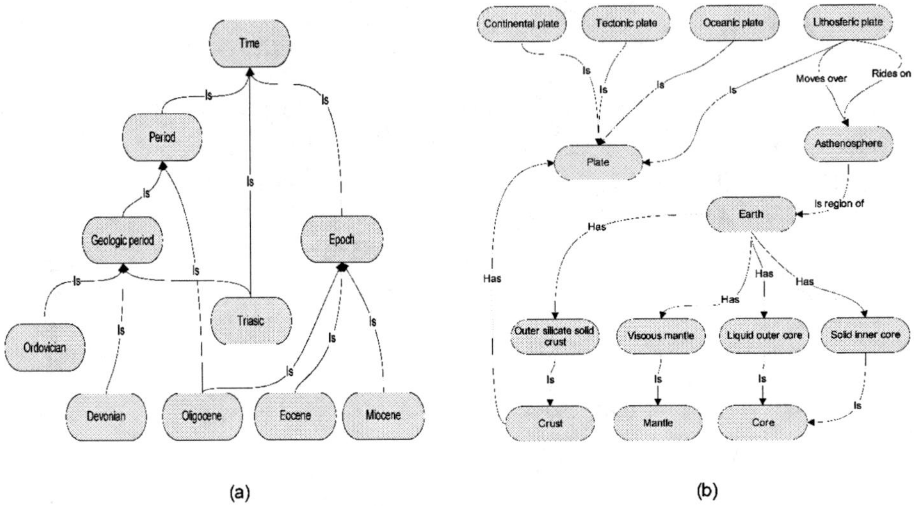

FIGURE 2. Small parts of an automatically generated ontology obtained for "geology", with taxonomic (a) and no taxonomic (b) relations.

Some improvements could easily be introduced in the method; for example, by making more perfect the used parser, or by improving or increasing the semantic information, mostly in relationship to the meaning of verbs conducting the relations identified in the texts. Also the functions and parameters used can be easily refined according to the formal method already developed.

Ontologic networks are a useful model for the representation and processing of semantic information. However, the final construction of an ontology must result in a standard representation, universally comprehensible and usable. Thus, the resulting ontologic network, once built, has to be translated into its final representation. To do this a mapping from ontologic network model to OWL Lite [29] has been used.

CONCLUSION

So far, an alternative has been presented to the existing methods for automatic construction of ontologies. Our approach is based on a formal framework that covers not only the basic definitions of an ontologic network and sub-networks but also the definition of concepts such as aggregation, connectivity, partition, grading and pruning of an ontologic network. The concepts of size and similarity of ontologic networks is also introduced.

This formal framework constitutes the knowledge that has been introduced into our agents architecture, previously developed, allowing them to carry out in parallel the different tasks for the automatic construction of the ontology.

The method uses syntactic and possibly semantic information that can be extracted from the texts by means of standard tools. But, probably one of its advantages relies on the use of non-statistical methods, allowing in consequence the construction of the ontology based only on a small set of documents, without the assistance of large corpora. Another advantage of the method is to be independent of the natural language used.

The proofs carried out show the practicality and efficiency of the method even with basic syntactic rules, provided the threshold value for pruning elements in the global ontologic network is not very low. Besides, the multi-agent system, working in parallel, can deal simultaneously with a lot of similar problems, increasing that way the system throughput.

The method can be used for many different purposes, such as the automated updating of knowledge bases from the Internet. We are now using this procedure to update knowledge bases of different Neocampus2 spin-off ILS, such as MEDIC2, CentMed2 and Finance.

The formal model and the construction method described in this paper can also be used for semantic annotation of documents. They can also be used for automatic information summarization. The elements for the summary could be the homogeneous text sections corresponding to the sub-networks of the partition of the global network obtained after pruning it.

Another possible application of the method could be the categorization of texts or documents. If we have previously obtained a set of different ontologic networks for all possible text categories, it is possible to compare those networks with the one corresponding to the text to categorize. The similarity function developed in our formal frame work would yield the most similar network within the set.

REFERENCES

1 F. de Arriaga, E-Knowledge Management, E-Learning and E-Commerce: An Evaluation of Their Situation and Tendencies, International Computer Science Institute, Technical Report, Berkeley, 2003, pp. 1-56.

2 F. de Arriaga, M. El Alami, A. Arriaga, F. Arriaga, J. Arriaga, NEOCAMPUS: Multi-agent Software Environment for On-Line Learning, Proceedings International Conference on Technology and Education, ICTE'2002, Badajoz, 2002, pp. 1355-1360.

3 A.L. Laureano, F. de Arriaga, M. García-Alegre, Cognitive Task Analysis: a Proposal to Model Reactive Behaviors, Journal of Experimental & Theoretical Artificial Intelligence 13 (3), 2001, pp. 227-239.

4 F. de Arriaga, M. El Alami, A. Ugena, Acceleration of the Transfer of Novices into Experts: The Problem of Decision Making, Proceedings International Conference BITE'2001, Eindhoven, 2001, pp. 157-178.

5 D. Carmel, S. Markovitch. Learning models of Intelligent Agents. Proceedings 13th National Conference AAAI'96, Vol. 2, 1996, pp.137-144.

6 S. K. Das, J. Fox, D. Elsdon, P. Hammond, Decision making and plan management by autonomous agents. Proceedings of the International Conference on Autonomous Agents 97. Marina del Rey, 1997, pp. 276-283.

7 F. de Arriaga, M. El Alami, A. L. Laureano, "Multi-Agent Simulation for Natural Systems", Proceedings IASTED International Conference on Modeling and Simulation, Philadelphia, 1999, pp. 257-268.

8 A. L. Laureano, F. de Arriaga, "Multi-Agent for Intelligent Tutoring Systems", Interactive Learning Environments, Vol. 6, nº 3, 1998, pp. 23-48.
9 A. Arriaga, M. El Alami, F. de Arriaga, Web-Based Tutors for Collaborative E-Learning, Advances in Technology-Based Education, ed: A. Mendez , Badajoz, 2003, pp. 2009-2013.
10 F. de Arriaga, A. Arriaga, M. El Alami, A.L. Laureano, J. Ramírez, "Fuzzy Logic Applications to Students' Evaluation in Intelligent Learning Systems", Proceedings II ANIEI International Congress on Informatics and Computing, Vol. 1, La Paz, 2003, pp. 161-167.
11 F. de Arriaga, A.L. Laureano, M. El Alami, A. Arriaga, Some Applications of Fuzzy Logic to Intelligent Tutoring Systems, Proceedings International Conference on Technology and Education, ICTE'2002, Badajoz, 2002, pp. 1222-1227.
12 F. de Arriaga, A. Arriaga, M. El Alami, "Fuzzy Intelligent E-Learning Systems: Assessment", Journal of Advanced Technology on Education, Vol. 1 (12), 2005, pp. 228-233.
13 F. de Arriaga, M. El Alami, "Affective Computing and Intelligent E-Learning Systems", Proceedings IADAT International Conference on Education e-2006, 2006, pp. 115-120.
14 F. de Arriaga, M. El Alami, A. Arriaga, "Evaluation of Fuzzy Intelligent Learning Systems", in Recent Research Developments in Learning Technologies, Vol. I, ed: A. Méndez, J. Mesa, Formatex, 2005, pp. 109-114.
15 F. de Arriaga, A. Arriaga, M. El Alami, "Guidelines for the Evaluation of Intelligent E-Learning Systems", in Technological Advances applied to Theoretical and Practical Teaching, Iadat, 2005, pp. 142-147.
16 M. A. Hearst, "Automatic acquisition of hyponyms from large corpora" Proceedings of he 14th International Conference on Computational Linguistics, Nantes, 1992, pp. 539-545.
17 H. Sundblad, Automatic acquisition of hyponyms and meronyms from question corpora, Technical Report SE-58183. Department of Computer and Information Science, Linköping University, Sweden, July 2002.
18 C. Roux, D. Proux, F. Rechenman, L. Julliard, An Ontology Enrichment method for a pragmatic information extraction system gathering data on genetic interactions, Proceedings of the ECAI2000 Workshop on Ontology Learning(OL2000), Berlin, Germany. August 2000, pp.245-252.
19 A. Maedche, S. Staab, "Mining ontologies from text". Proceedings of EKAW-2000, Lecture Notes in Artificial Intelligence (LNAI-1937) France, Springer, 2000, pp. 59-67.
20 A. Maedche, S. Staab, "Ontology learning for the semantic web" IEEE Intelligent Systems, vol. 16 n. 2, 2001, pp. 72-79.
21 A. Maedche, R. Volz, "The Ontology extraction and maintenance framework text-to-onto", Proceedings ICDM Workshop on Integrating Data Mining and Knowledge Management, San José, 2001, pp. 189-198.
22 C. D. Manning, H. Schuetze, Foundations of Statistical Natural Language Processing, M.I.T. Press, Cambridge, 1999.
23 R. Navigli, P. Velardi, A. Gangemi, "Ontology learning and its application to automated terminology translation", IEEE Intelligent Systems, vol. 18, n. 1, 2003, pp. 22-31.
24 A. Maedche, V. Pekar, S. Staab, "Ontology learning, Part 1: on discovering taxonomic relations from the web" in Web Intelligence, Springer, 2002, Chapter 1.
25 R. Srikant, R. Agrawal, "Mining generalized association rules" , Proceedings of VLDB '95, 1995, pp. 407-419.
26 A. Maedche, S. Staab, "Discovering conceptual relations from text", Proceedings of the 14th European Conference on Artificial Intelligence, Berlin, 2000, pp. 21-25.
27 M. F. Porter, "An Algorithm for suffix stripping", Program, vol. 14, n. 3, 1980, pp. 130-137.
28 StPar, Java implementation of natural language parsers. Available: http://nlp.stanford.edu/software/lex-parser.shtml
29 OWL Web Ontology Language. Available: http://www.w3.org/TR/owl-ref

Computational Evolutionary Methodology for Knowledge Discovery and Forecasting in Epidemiology and Medicine

Dhananjai M. Rao*, Alexander Chernyakhovsky† and Victoria Rao**

*CSA Department, Miami University, Oxford, OH 45056, USA
†William Mason High School, Mason, OH 45040, USA
**Cybernetic Evolution Inc., Mason, OH 45040, USA

Abstract. Humanity is facing an increasing number of highly virulent and communicable diseases such as avian influenza. Researchers believe that avian influenza has potential to evolve into one of the deadliest pandemics. Combating these diseases requires in-depth knowledge of their epidemiology. An effective methodology for discovering epidemiological knowledge is to utilize a descriptive, evolutionary, ecological model and use bio-simulations to study and analyze it. These types of bio-simulations fall under the category of computational evolutionary methods because the individual entities participating in the simulation are permitted to evolve in a natural manner by reacting to changes in the simulated ecosystem. This work describes the application of the aforementioned methodology to discover epidemiological knowledge about avian influenza using a novel eco-modeling and bio-simulation environment called SEARUMS. The mathematical principles underlying SEARUMS, its design, and the procedure for using SEARUMS are discussed. The bio-simulations and multi-faceted case studies conducted using SEARUMS elucidate its ability to pinpoint timelines, epicenters, and socio-economic impacts of avian influenza. This knowledge is invaluable for proactive deployment of countermeasures in order to minimize negative socio-economic impacts, combat the disease, and avert a pandemic.

Keywords: Descriptive models, Markov process, Avian Influenza (H5N1), Epidemiology, Lyme Disease, Bio-Simulations
PACS: 02.50.Ga, 7.05.Tp, 28.41.Ak, 87.18.Vf, 87.64.Aa

MOTIVATION

Communicable and vector-borne diseases, primarily induced by viral and bacterial infections, are the most common forms of ailments amongst humans and domesticated farm animals [1]. Despite numerous advances in epidemiology and medicine many of these diseases continue to defy human efforts to treat and control them [1]. The source of such resistance to treatment and control is two fold. First, the incessant evolutionary processes such as genetic reassortment, recombination, and mutation induce changes to organisms, thereby constantly morphing their antigenic characteristics. Second, the vectors of the disease, such as birds and animals, often disperse the pathogens over wide areas. The scenario is further convoluted due to complicated symbiotic processes between the vectors and other organisms including humans. Rapid changes in phylogeny and bio-diversity pose significant challenges in vaccine manufacturing. In addition, socio-political restrictions and technological limitations further magnify the challenges involved in controlling intercontinental spread of infectious diseases.

CP1007, *Current Themes in Engineering Technologies*, edited by S.-I. Ao, M. A. Amouzegar, and S.-S. Chen

The aforementioned issues are salient characteristics of avian influenza which is caused by H5N1, a highly virulent strain of the influenza-A virus [2, 3, 4]. In Spring 2006, it was established that infected migrating waterfowl causes transcontinental dispersion of H5N1. The virus has a devastating impact on poultry causing 100% mortality within 48 hours of infection [5]. Currently, mass culling of poultry is the recommended approach to contain the disease and this aspect causes significant economic hardships [6]. Moreover, the pathogen also spreads to humans through direct contact with infected poultry and contaminated surfaces [5]. Many researchers have stated that H5N1 has the potential to cause the next pandemic [5, 7].

Proactive administration of suitable vaccination is the prevention mechanism for influenza [2, 5]. Unfortunately, a myriad of technological and socio-political issues have rendered manufacturing and distribution of H5N1 vaccine a significant challenge [2, 5]. Some of these issues include: *(i)* rapid mutations in the virus make it practically infeasible to predict antigenic characteristics for manufacturing vaccines; *(ii)* current vaccine manufacturers do not have sufficient capacity; and *(iii)* targeted distribution of limited quantities of the vaccine is a challenge [5, 8]. These issues necessitate that preventing and controlling avian influenza requires proactive international coordination and strategic deployment of countermeasures. The effectiveness of proactive strategies, including those proposed by the World Health Organization (WHO), will greatly benefit from prediction and forecasting of epicenters and timelines of H5N1 outbreaks. However, forecasting and proactive deployment of countermeasures requires global knowledge about the current and future epidemiology of the disease.

Analogous scenarios apply to current and future communicable and vector-borne diseases as well. Humanity needs to be prepared to proactively and effectively combat emergent diseases to minimize loss of human lives and impacts to global economy. Combating disease requires extensive knowledge about epidemiology, vaccinal characteristics, epicenters of disease outbreaks, timelines, economic impacts, and human mortality. Accordingly, we propose a computational evolutionary methodology to address the aforementioned need.

PROPOSED APPROACH

One of the most promising methodologies to rapidly analyze a complex, stochastic system consisting of millions of organisms is bio-simulation [9, 10]. Bio-simulations have shown to be intuitive, realistic, and provide close reflection of epidemiological characteristics [10, 11]. The proposed approach aims to leverage bio-simulations to perform epidemiological analysis through the use of evolutionary eco-models of selected organisms in a global ecosystem. In the eco-models individual organisms are modeled (typically as smart agents) to evolve in a virtual ecosystem [12]. As the ecosystem evolves, necessary attributes at different levels of resolution are recorded to infer various characteristics. Note that in this evolutionary approach the behaviors of organisms (modeled as smart agents) are not coerced. Instead they are permitted to proceed "naturally" reacting to changes in the ecosystem and reflecting their real world counterparts. Once verified, the evolutionary bio-simulations can be used to further analyze a wide variety of scenarios to discover more comprehensive knowledge.

Background on Computational Evolutionary Approaches

Computational evolutionary methodology employs computers and digital models to study and analyze evolution, phylogeny, and associated biological characteristics of organisms [12, 13, 14]. This methodology includes a wide variety of other modeling, simulation, and analysis techniques as well [12, 13, 14]. Since our proposed methodology uses bio-simulations to analyze the natural progression and evolution of avian influenza, it falls under the category of computational evolutionary methods.

The fundamental motivation for computational approaches, that enable study organisms in silico rather than in vitro, stems from many of their advantages [12]. The primary advantage is that existing knowledge can be used to create virtual organisms, rapidly and economically analyze numerous future generations, and draw inferences [12, 13, 14]. Computational evolutionary approaches have shown to be excellent candidates for analyzing phylogeny of microorganisms such as bacteria and viruses [12]. These approaches are gaining significant importance in biology and medicine.

Epidemiology of Avian Influenza

The proposed computational evolutionary methodology can be applied to a broad range of biological systems [12, 13, 14]. However, here we will focus on the issues involved in applying the proposed methodology to study and analyze the epidemiology of avian influenza along with its socio-economic impacts. The objective is to use existing information on various organisms in the ecosystem, namely migrating waterfowl, poultry, and humans, to discover additional knowledge for forecasting timelines and epicenters of disease outbreaks. The motivation is to empower disease control centers and emergency management agencies with crucial information required to contain and control outbreaks to save human lives and minimize socio-economic impacts.

SEARUMS: A COMPUTATIONAL EVOLUTIONARY ENVIRONMENT FOR AVIAN INFLUENZA

Realizing the advantages of the proposed computational, evolutionary methodology requires the use of an effective software environment for eco-modeling, bio-simulation, and epidemiological analysis [10]. Specifically, in this case, the software environment must be conducive for analyzing the epidemiology of avian influenza. The software environment must facilitate rapid modeling and simulation (M&S) to minimize analysis time frames, thereby accelerating deployment of countermeasures. Moreover, the software must be portable and accessible to enable its widespread use [10]. Accordingly, we have endeavored to design and develop such a software environment called SEARUMS.

SEARUMS is an acronym for Studying Epidemiology of Avian Influenza Rapidly Using Modeling and Simulation. It is an Java-based, integrated, graphical modeling, simulation, and analysis environment that is specialized for epidemiological study of avian influenza. SEARUMS provides an extensible, agent-based, spatially explicit modeling front-end coupled with a discrete event simulation kernel. The conceptual model of the software agents has been developed from corresponding Markov processes that provide a strong mathematical grounding. SEARUMS has been integrated with libraries for plotting graphs and charts that ease visualization of simulation results and aid in analysis.

In addition, SEARUMS includes comprehensive models that incorporate real-world statistical data on: waterfowl migration, published by Global Register Of Migratory Species (GROMS) [15]; waterfowl species that are at higher risk to carry the virus [6]; global poultry population and distribution, published by the Food and Agriculture Organization (FAO) of the United Nations [16]; and human population in metropolitan areas of the United States, obtained from the U.S. Census Bureau [17]. The model is stored as a portable XML document that can be readily reused and further extended. This aspect alleviates many procedural hurdles and promotes sharing and collaboration. SEARUMS is envisioned to serve as a global, multi-disciplinary environment that seamlessly integrates knowledge from various fields so that epidemiologists, economists, and disease control centers can collaboratively use it to combat avian influenza.

Conceptual Epidemiological Model

The epidemiology of avian influenza currently involves three primary types of entities, namely: migrating waterfowl flocks, humans, and poultry [6, 8]. Migrating waterfowl are the primary vectors of the H5N1 virus [18]. They indirectly transmit the virus to poultry in their vicinity through infected food, water, and feces [19, 20]. The disease is transmitted to humans that come in direct contact with contaminated surfaces or infected poultry [21]. In other words, the transmission of infections is determined both by space and time. Consequently, an epidemiological model needs to incorporate temporo-spatial interactions between the aforementioned three entities. Accordingly, we have developed our conceptual model using spatially explicit, discrete-time Markov processes as described further below [22]. Note that, since the goal of this section is to present factors steering the design and implementation of SEARUMS, it does not delve into mathematical details that are available in the literature [10].

The proposed conceptual model consists of a set of interacting entities. An entity may represent a flock of waterfowl, poultry, or confederation of humans. Each entity is mathematically modeled by a corresponding Markov process. A Markov process undergoes instantaneous, probabilistic state transitions to reflect the behavior of its real-world counterpart [10]. Temporal state transitions induce changes to state variables, which in this case include population counts, location, infection percentage, and migration attributes. These changes corresponding various conditions such as: migration, healing or regeneration, and death or reduction in population. In conjunction with temporal changes, spatial interactions between entities are also captured as state transition probabilities. The spatial interactions, that occur in a spatially explicit models, have been mathematically modeled using principles of Euclidean geometry [23]. The surface of the earth is represented as an Euclidean plane and each entity has a circular area of influence. Interactions between entities occur when their circle of influence overlaps. Interactions between entities and the degree of overlap are used to model the intensity and spread

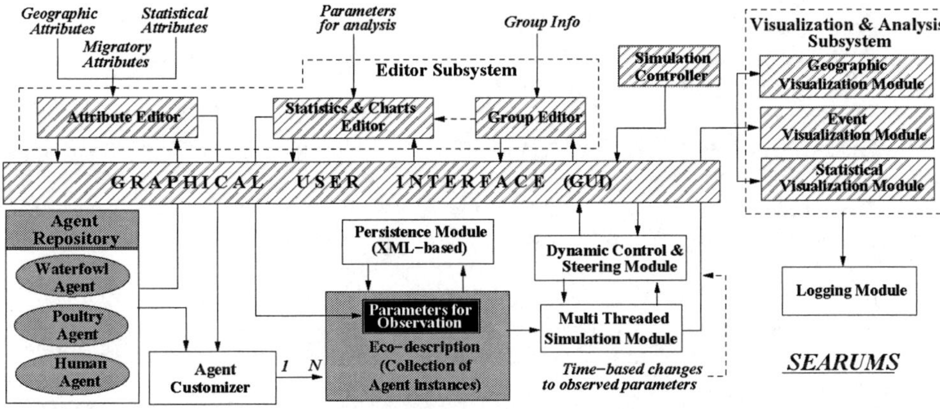

FIGURE 1. Architectural overview of SEARUMS. GUI modules are highlighted with a striped background. SEARUMS can be downloaded from `http://www.searums.org`.

of infection. Such a modeling approach is widely used in spatially explicit ecological models to reduce computational complexity [23, 24, 25].

In SEARUMS, each Markov process in the aforementioned mathematical formalism is modeled as a smart agent [24]. The simulation consists of a collection of interacting smart agents whose temporal behavior is causally coordinated using a discrete event kernel. The state has been implemented using a set of attributes that incorporate both temporal and spatial variables. Note that our methodology does not coerce behaviors or interactions between entities. Instead, each entity mimics the behavior of their real-word counterpart which significantly simplifies implementation, reduces computational complexity, enables lucid use of real-world data, and eases addition of new instances including new types of software agents.

ARCHITECTURE OF SEARUMS

SEARUMS is an eco-modeling and bio-simulation virtual environment that uses a computational evolutionary methodology for study and analysis of epidemiology of avian influenza. It has been developed in Java by capitalizing on many of its object oriented programming features [26]. SEARUMS is designed to be a user friendly, integrated, graphical modeling, simulation, visualization, and analysis environment for conducting epidemiological analysis of avian influenza using an Agent-based, Spatially Explicit (ASE) model. These design goals have been achieved by composing the system using a collection of interdependent but loosely coupled modules. Each module has a well defined functionality that can be accessed and utilized via a set of Application Program Interface (API) method calls. API of the modules are Java interface classes that are implemented by each module. Interactions between modules is achieved through interface classes to ensure loose coupling. This approach permits seamless "plug and play" of modules and the environment is composed by loading suitable modules dynamically on-demand via Java's reflection API [26]. Such an implementation approach has been adopted to ease customization and extension of SEARUMS without requiring changes to its design or impacting existing modules.

An architectural overview of the modules constituting SEARUMS is shown in Figure 1. The modules can be broadly classified as core modules and Graphical User Interface (GUI) modules. The core modules of SEARUMS are the Agent Repository, Agent Customizer, Persistence Module, Dynamic Control & Steering Module, Simulation Module, and Logging Module. These modules provide the core M&S functionality of SEARUMS. The GUI facilitates interactions with the core modules via convenient and intuitive user interfaces. The GUI modules can be further categorized into the Editor subsystem, the Simulation Controller, and the the Visualization & Analysis subsystem. SEARUMS uses the Model-View-Controller pattern to couple the core modules, the GUI models, and the Eco-description. The design permits the GUI modules to be easily replaced with a minimal command-line text interface for running SEARUMS in offline batch mode. The batch mode is useful for performing repeated runs or analyzing different scenarios on computational clusters.

The modules and subsystems constituting SEARUMS cooperatively operate on a shared, in-memory representation of the model called the Eco-description. The Eco-description is a centralized data structure that includes all the information necessary for modeling, simulation, and analysis. It is composed using a collection of Java classes and that provide efficient access to data and information required by the various modules. The primary information encapsulated by the Eco-description relates to the smart agents [24] that constitute the model. As shown in Figure 1, the agents are organized into an `Agent Repository` to facilitate instantiation and use via Java reflection API. Currently, SEARUMS includes the following three smart agents: `Waterfowl Agent` that represents a migrating waterfowl flock, `Poultry Agent` that models behavior of poultry flocks, and `Human Agent` that models humans. Each agent has its own behavior that reflects the characteristics of its real-world counterpart. The behaviors are customized to represent specific instances of an agent by specifying suitable values for the exposed attributes via the `Attribute Editor` GUI module. The attributes of an agent include:

1. *Geographic attributes* that indicate the location (latitude and longitude) and logical association with countries and continents. In addition, each agent has a circle or influence that circumscribes its neighborhood.
2. *Migratory attributes* are specified only for agents whose location changes over the lifetime of the simulation. The migratory attributes are described as a sequence of migration points. Each migration point has geographical and chronological (arrival and departure dates) attributes associated with it. In SEARUMS, only one complete migration cycle needs to be specified. The software automatically reuses the information to simulate annual migratory cycles.
3. *Statistical attributes* for agent instances include their initial population, density and distribution, initial infection percentage, infection spread parameters, incubation periods, mortality rates, and population regrowth parameters.

The agents implement the conceptual, mathematical model of the system developed using Markov processes. They are added to a model via suitable toolbar buttons or menu options provided by SEARUMS. Agent instances are created with default attributes from the Agent Repository by the `Agent Customizer` module using Java's reflection API. Once instantiated, the attributes for agents can be modified via the `attribute editor` module. The agents are implemented as a family of Java classes by extending a common base class called `Agent`. The `Agent` class provides methods for interacting with the simulation kernel, inspecting the neighborhood, scheduling events, and interfacing with the GUI modules.

The agents in a model are logically organized into hierarchical sets called groups. SEARUMS permits multiple top-level groups with an arbitrary number of hierarchies, with one or more sub-groups at each hierarchical level. An agent can be a member of multiple groups. The groups serve several different purposes in SEARUMS. A group can be used as a parameter for statistical analysis and for plotting charts. For example, a group called "United States" can be created with 50 different sub-groups, one for each state, encompassing various agents. The main "United States" group can be selected for plotting charts and SEARUMS automatically collates and plots data for each state. Note that even though graph plotting is restricted to one hierarchical level, statistics for plotting are collated in a recursive, depth-first manner and includes data from all agents in underlying hierarchies. A modeler can use a combination of groups to perform multi-faceted analysis at different scales. In addition, groups can be included or excluded from simulations for analyzing different scenarios. The GUI modules utilize groups to provide control on visibility of agents to manage details displayed on the screen. The Group Editor module provides the user interface for managing group entries and hierarchies.

Once all the agent instances and groups have been established in a model, the parameters for observation are added to the Eco-description. These parameters are selected by the user via the Statistics & Charts Editor from a list of options. The list includes the attributes of the agents and the groups in the Eco-description. Each parameter is configured to be sampled hourly, daily, or weekly in terms of simulation time. Moreover, each parameter can be subjected to statistical operations, such as sum, mean, and median. SEARUMS can dynamically (*i.e.,* during simulation) plot and save a variety of charts including: line graphs and pie charts. Multiple charts can be simultaneously used for analyzing a variety of data.

All of the aforementioned information is stored as an integral part of the Eco-description. The Eco-description can be saved for future reuse via the `Persistence Module`. The Eco-description is unmarshalled into an XML document that is compliant with a predefined XML schema. Serializing to an XML document has its advantages. First, it enables simple scripts to be developed that can modify specific values and perform multiple simulation runs in batch mode. Second, XML documents can be readily version controlled and archived using commonly available revision control systems like CVS and Subversion. Third, it eases documentation, validation, sharing, and reuse of valuable domain-specific statistical data collated by different researchers from diverse sources. Such features play an important role in facilitating large-scale, collaborative epidemiological studies.

The `simulation module` performs the task of conducting a Discrete Event Simulation (DES) using the Eco-description. This module utilizes a multi-threaded DES kernel that manages and schedules the discrete events generated by the Agents. Multi-threading enables the DES kernel to exploit the compute power of multi-processor or multi-core machines thereby reducing the wall-clock time for simulation. The number of threads spawned by the DES kernel is configurable. Each thread processes concurrent events (events with the same timestamp) in parallel without violating the causal constraints between events. The Dynamic Control and Steering Module provides the infrastructure to control the DES kernel. In addition, it permits selected agent attributes to be modified during the course of simulation. The Simulation Controller module provides the graphical interface to the Dynamic Control & Steering Module.

ECO-MODEL FOR VERIFICATION AND USE OF SEARUMS

The initial experiments conducted using SEARUMS centered around its verification and validation. For this purpose, we have developed an Eco-description using selected, high risk species of waterfowl as reported by Hagemeijer *et al* [27]. Table 1 lists the waterfowl species, including high risk species [27], used in the model. The migratory flyways of the waterfowl and their population has been collated from data published by various organizations [2, 4, 15, 16, 17, 27]. For modeling and simulation purposes the dates for migration were approximated to the middle of the months reported in the statistics. The initial positions of the flocks were set to correspond to 01/01/2006, which is the real-world time when this specific simulation is logically set to commence. The dispersion of poultry population in different continents has been approximated to circular regions with even density in the Eco-description [16, 25]. Poultry data has been collated from statistics published by national organizations and government databases [16]. Currently, we have represented human population, as reported by U.S. Census Bureau [17], in all 26 major metropolitan areas of the United States, approximated to circular regions. However, the Eco-description can be readily extended to include other parts of the world. Note that to the best of our knowledge, it is the most comprehensive model of its kind reported to date.

VERIFICATION OF SEARUMS

Having developed the Eco-description, we verified the validity of the Eco-description by performing extensive simulations with the initial source of infection set to outbreak in Indonesia [5]. We established the validity of the Eco-description and SEARUMS by confirming that the timing and chronology of several outbreaks observed in the simulations correlate with significant real-world incidents as reported by WHO [5]. The data in Table 2 presents a comparison of real-world and simulated outbreaks. The data is presented for some of the significant, initial avian influenza outbreaks and not for repeated outbreaks that occur in these regions. Note that deviations of ± 2 weeks is expected due to approximation of migration dates. In addition, deviations in dates also occur because the Eco-description does not include all types of migratory waterfowl flocks but only the high risk species. However, the sufficiently close concordance between simulated and real-world outbreaks establishes validity and effectiveness of SEARUMS. Moreover, it significantly increases confidence in inferences drawn from the simulation. Currently, a variety of case studies using SEARUMS are already underway to predict and avert a

TABLE 1. Various agent instances used to develop the Eco-description used for experiments. Entries prefixed with * indicate high risk waterfowl species [27].

Description of AgentType	Number of Agents	Total Population	#Countries in Migration Path
Bar-Tailed Godwit	4	40,000	18
Canada Goose	16	231,700	5
Common Crane	9	225,000	21
*Eurasian Widgeon	3	1,296,000	17
Great Knot	3	231,000	8
*Mallard	1	5,000	1
Razorbill	1	148,000	4
*Red-Breasted Goose	1	44,000	4
Red-crowned Crane	1	15,000	4
Siberian Crane	3	30,000	12
Yellow-Billed Duck	2	20,000	8
Total waterfowl flocks	44	2,285,700	40
Total Poultry flocks	121	12,280,000,000	32
Total Human groups	26	125,000,000	1
Total	**191**	**12,407,285,700**	**40**

TABLE 2. Comparison of chronology of significant Real-world outbreaks with simulated outbreaks

Incident	Real-world Date	Simulated Date	Error (Days)
Outbreak in Indonesia	23-Jan-06	1-Jan-06	22
Infection in Iraq/Iran	1-Mar-06	25-Mar-06	24
Infection in China	27-Apr-06	2-Apr-06	-25
Infection in Egypt	11-Oct-06	14-Sep-06	-27

pandemic.

Scalability and Performance Evaluation

One of the important objectives underlying the design of SEARUMS is to facilitate rapid analysis of epidemiology of avian influenza. Accordingly, the validated Eco-description was also used to conduct scalability and performance evaluation of SEARUMS using three different platforms. The first platform was a Sun Netra-T12 SMP workstation with eight 1.2 GHz SPARCv9 processors with 16 gigabytes of RAM running Solaris 9. The second platform was a conventional personal computer (PC) running Windows XP (32-bit). The PC had a dual-core, 64-bit Turion processor running at 2 GHz with 2 GB of memory. The third platform was the same PC but running Linux (Fedora Core). Specifically, the PC was setup to a dual-boot configuration enabling it to run either Windows XP or Linux. These simulations were conducted by running SEARUMS

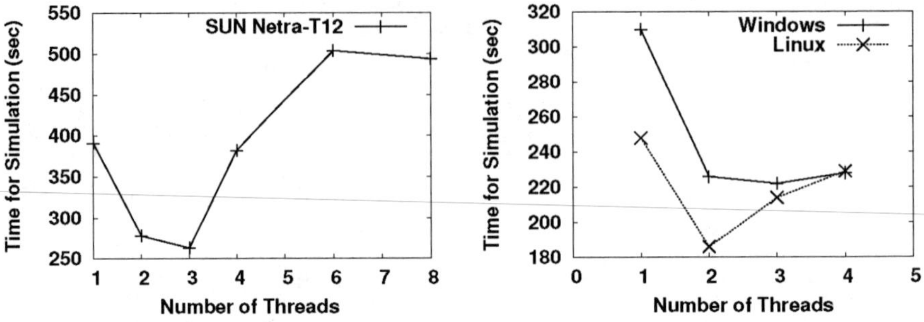

(a) Sun Netra-T12 (b) PC (Windows & Linux)

FIGURE 2. Influence of number of threads on wall-clock time for simulation on different hardware platforms and operating systems.

in batch mode, without any GUI overhead to ensure emphasis on core performance.

The graph in Figure 2 plots the wall-clock time taken to complete the simulation using a varying number of threads on the aforementioned platforms. As shown by the curves, initially the time taken for simulation reduces as the number of threads are increased. The wall-clock time decreases because the threads run in parallel on the multi-processor/multi-core machines and rapidly process concurrent events. The initial performance improvement highlights SEARUMS' scalability. However, the performance decreases as the number of threads are further increased. The deterioration occurs because the Eco-description used in this experiment does not have sufficient, inherent concurrency to leverage the available compute power. On the other hand, larger models with more concurrency will benefit from the scalable design.

These experiments highlight the portability, scalability, and performance aspects of SEARUMS design. Simulations involving more than 20 million events complete within 4 minutes in optimal cases demonstrating that SEARUMS enables rapid epidemiological analysis of avian influenza. Furthermore, the experiments provide empirical evidence that the design goals of SEARUMS have been successfully achieved.

Case Study: Spread and Impact to United States

In this case study, we analyzed the potential impact of avian influenza to poultry farming in the United States using the validated Eco-description presented earlier. The study was conducted using a number of bio-simulations with three different experimental groups of migrating waterfowl. The three flocks were chosen based on their close proximity to known primary sites of disease outbreaks. The initial infection in each experimental group was varied for analysis. In all these experiments, the initial conditions were seeded corresponding to January 1, 2006 and the simulations were run for a period of 5 years.

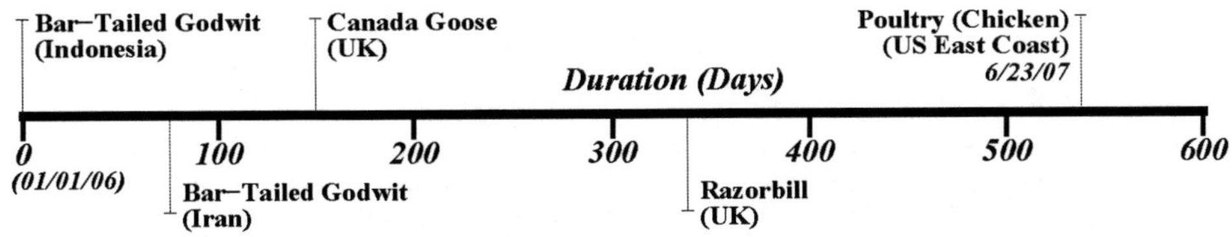

FIGURE 3. Timelines for H5N1 infection spread from experiment logically set to start on 01/01/2006.

Figure 3 illustrates one of the trans-Atlantic transmission pathways to the continental United States. We observed that the spread was determined by migratory pathways and timelines of different species of waterfowl rather than initial infection percentages. One of the notable observations is that our experiments correctly predicted an outbreak in the United Kingdom [4]. The graph in Figure 4 presents the impact of avian influenza outbreaks on poultry population in continental United States. Decrease in poultry population corresponds to H5N1 induced death and culling of birds to control the disease. Increase in poultry population reflects regeneration of poultry flocks after an outbreak. As illustrated by the graph, infections in poultry also follow a cyclic pattern that correlate with annual migration of waterfowl. The mortality figures can be translated to corresponding dollar figures for financial analysis.

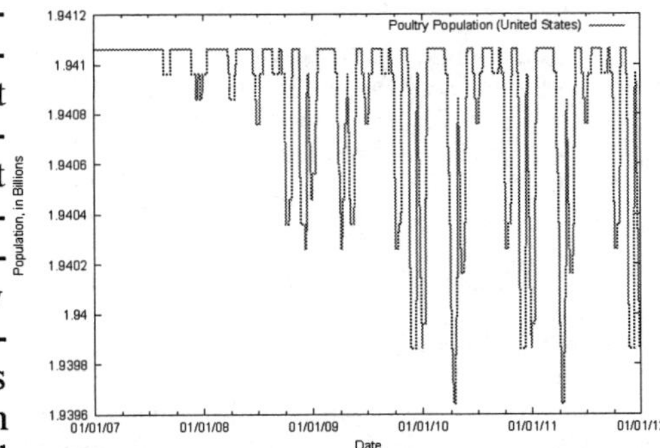

FIGURE 4. Fluctuation in Poultry Population

CONCLUSIONS

The application of computational evolutionary methodology to analyze the epidemiology of avian influenza was presented. Specifically, the practical issues involved in applying this methodology using an eco-modeling and bio-simulation environment called SEARUMS were discussed. The evolutionary epidemiological model of avian influenza was developed and empirically verified. The verification experiments highlight that the proposed methodology closely reflects real world scenarios. The bio-simulations conducted using SEARUMS provide good forecast of time lines and epicenters of outbreaks. Furthermore, it was observed that outbreaks follow a cyclical, annual pattern. This new, comprehensive knowledge gained about epidemiology of avian influenza is invaluable to combat outbreaks and prevent a pandemic. We are optimistic that the proposed software environment will enable mankind to strategically invest precious time and resources to combat diseases like avian influenza, minimize their impacts on human life and global economy.

REFERENCES

1. S. J. Flint, L. W. Enquist, V. R. Racaniello, and A. M. Skalka, *Principles of Virology*, American Society for Microbiology (ASM), Washington DC, USA, 2004, 2 edn., ISBN 1-55581-259-7.
2. Centers for Disease Control and Prevention, Avian influenza: Current situation (2006), URL `http://www.cdc.gov/flu/avian/outbreaks/current.htm`.
3. United States Department of Agriculture, Avian influenza (bird flu) (2007), URL `http://www.usda.gov/wps/portal/usdahome?navid=AVIAN_INFLUENZA`.
4. World Health Organization, Avian influenza (2006), URL `http://www.who.int/csr/disease/avian_influenza/en/`.
5. World Health Organization, Avian influenza: Situation updates (2007), URL `http://www.who.int/entity/csr/disease/influenza/H5N1-9reduit.pdf`.
6. M. Brahmbhatt, "Economic Impacts of Avian Influenza Propagation," in *First International Conference On Avian Influenza in Humans*, 2006.
7. M. Zambon, *Nature Biotechnology* **25**, 433–434 (2007).
8. D. Normile, *Science* **312**, 1855 (2006).
9. D. M. Rao, *Study of Dynamic Component Substitution*, Ph.D. thesis, University of Cincinnati (2003).
10. D. M. Rao, A. Chernyakhovsky, and V. Rao, "SEARUMS: Studying Epidemiology of Avian Influenza Rapidly Using Simulation," in *Proceedings of the 2007 International Conference in Modeling Health Advances (ICMHA'07)*, Berkeley, CA, USA, 2007, pp. 667–673, (best paper).
11. D. M. Rao, and P. A. Wilsey, "Accelerating Spatially Explicit Simulations of Spread of Lyme Disease," in *Proceedings of the 38th Annual Simulation Symposium*, San Diego, California, USA, 2005, pp. 251–258.
12. C. Ofria, and C. O. Wilke, *Artificial Life* **10**, 191–229 (2004).
13. D. N. Alstad, *Basic Populus Models of Ecology*, Prentice-Hall Inc., Upper Saddle River, NJ 07458, 2001, first edn., ISBN 0-13-021289-X.
14. S. H. Rice, *Evolutionary Theory: Mathematical and Conceptual Foundations*, Sinauer Associates, Inc, MA 01375-0407, USA, 2004, 1 edn., ISBN 0878937021.
15. GROMS: Global Register of Migratory Species, Summarising knowledge about migratory species for conservation (2006), URL `http://www.groms.de/`.
16. Animal Production and Health Division of Food and Agriculture Organization of the United Nations, Global Livestock Production and Health Atlas (GLiPHA) (2007), URL `http://www.fao.org/ag/aga/glipha/index.jsp`.
17. U.S. Census Bureau, 100 largest counties (2006), URL `http://www.census.gov/popest/counties/CO-EST2006-07.html`.
18. D. Normile, *Science* **311**, 1225 (2006).
19. H. Chen, G. Smith, S. Y. Zhang, K. Qin, J. Wang, and K. S. Li, *Nature* **436**, 191–192 (2005).
20. J. Liu, H. Xiao, F. Lei, Q. Zhu, K. Qin, X. w. Zhang, X.-I. Zhang, D. Zhao, G. Wang, Y. Feng, J. Ma, W. Liu, J. Wang, and G. F. Gao, *Science* **309**, 1206 (2005).
21. D. Normile, *Science* **311**, 1692 (2006).
22. A. R. Solow, and W. K. Smith, *Journal of Applied Ecology* **43**, 185–188 (2006).
23. G. Booth, *Artificial Life* **3**, 147–163 (1997), ISSN 1064-5462.
24. M. Hare, and P. J. Deadman, *Mathematics and Computers in Simulation* **64**, 25–40 (2004), ISSN 0378-4754.
25. R. Law, U. Dieckmann, and J. A. Metz, *The Geometry of Ecological Interactions: Simplifying Spatial Complexity*, Cambridge University Press, NY 10013-2473, USA, 2005, ISBN 9780521022095.
26. J. Bloch, *Effective Java – Programming Language Guide*, Pearson, Prentice Hall, 200 Old Tappan Road, NJ 07675, USA, 2001, first edn., ISBN 0-201-31005-8.
27. W. Hagemeijer, and T. Mundkur, "Migratory flyways in Europe, Africa, and Asia and the spread of HPAI H5N1," in *International Scientific Conference On Avian Influenza and Wild Birds*, FAO and OIE, Rome, Italy, 2006.

On HIV Prevalence and AIDS Deaths in India

B. D. Aggarwala
Department of Mathematics and Statistics,
University of Calgary,
Calgary, Alberta, T2N 1N4, Canada
e-mail:aggarwal@math.ucalgary.ca

Abstract. The National AIDS Control Organization (NACO) of India had estimated, before this year, that there were 5.134 million HIV positive people in India at the end of 2004 and that they were increasing at the rate of more than a quarter of a million people every year. In a recent publication, we estimated that, if the number of reported AIDS cases in India are only 50% efficient, i.e. if the number of actual AIDS cases in India is no more than twice the reported number, then the number of HIV positive people in India should have been no more than 2.5 million at the end of 2004. Many other people in the epidemiology community have the same point of view. Now, the government of India is also of the same view and "The latest data released by the government shows that the country has around 2 to 3 million people with HIV, much lower than last year's figure of 5.7 million". However, our assumption that the actual number of AIDS cases in India is only about twice the number reported, has been questioned, and it has been suggested that the Indian system of AIDS reporting is woefully inaccurate and the actual number of AIDS cases there could be three, four or even five times the reported number. In this paper, we consider this suggestion and show that, even if the actual number of AIDS cases was three, four, or even five, times the reported number, the number of HIV positive people in India, at the end of 2004, should still be no more than 2.5 million. This is because our previous estimate was an over estimate and had room to accommodate considerably more number of AIDS cases. We also estimate the number of AIDS deaths in India and show that it should be considerably less than those estimated by the World Health Organisation (WHO).

Keywords: HIV Prevalence, AIDS Deaths, E-M Algorithm, Gamma curves, India.

INTRODUCTION

According to the National AIDS Control Organisation (NACO) of India, the number of HIV positive people in India was estimated at 3.5, 3.7, 3.86, 3.97, 4.58, 5.106 and 5.134 million at the end of the years 1998 to 2004 respectively [1]. This gives increases of 200000, 160000, 110000, 610000, 526000, and 28000 in the years 1999, 2000, 2001, 2002, 2003, and 2004 respectively. Instead of such wildly oscillating figures of increase in HIV prevalence, it is perhaps more accurate to say that this increase was approximately 272000 per year over the last six years or 287000 per year over the last five years. We consider these figures, based on serological surveys in selected populations, to be very much on the high side but, in the absence of anything better, accept these as reliable. Also the cumulative number of AIDS cases at the end of May, 2005 was estimated at 109349 by the same organisation (NACO), so that,

CP1007, *Current Themes in Engineering Technologies*, edited by S.-I. Ao, M. A. Amouzegar, and S.-S. Chen

at the end of 2004, this number may be estimated at 100,000. We would think that the above figures for AIDS incidence are perhaps more reliable than the corresponding figures for HIV positive cases because of the difficulty of estimating HIV incidence due to the infection being asymptomatic in the beginning. In a recent paper [2], because of the difficulty of getting accurate numbers of AIDS cases from the countryside, we assumed that the actual number of AIDS cases was twice the number reported and arrived at the conclusion that cumulative HIV incidence in India at the end of 2004 was in the neighbourhood of 2.5 million as against the expert (NACO) estimate of hiv prevalence of 5.134 million. Other people in the epidemiology community have the same point of view [3]. Thus, according to one author, "there continues to be inadequate understanding of basic HIV epidemiology - especially its transmission dynamics - along with gross ignorance of how current HIV/AIDS estimates and projections are "cooked" or made up. Most of the public, policy makers, and the news media have uncritically accepted high HIV prevalence estimates and refuse even to consider the possibility that lower prevalence estimates may be more accurate" [4]. It should be pointed out that now the government of India is also of the same view and "The latest data released by the government shows that the country has around 2 to 3 million people with HIV, much lower than last year's figure of 5.7 million"[5]. However, our assumption that the actual number of AIDS cases in India is only about twice the actual number has been questioned and it has been suggested that the Indian system of AIDS reporting is woefully inadequate and the actual number of AIDS cases there could be three, four or even five times the reported number [6]. In this paper, we consider this suggestion and show that, even if the actual number of AIDS cases was three, four or even five times the reported number, the number of HIV positive people in India, at the end of 2004, should still be no more than 2.5 million. This is because our previous estimate was an over estimate and had room to accommodate considerably more number of AIDS cases. As in the previous paper [2], since the antiretroviral drugs are being administered to only a tiny fraction of HIV positive people, we shall assume that they have no impact at all on the overall HIV situation in India.

The first cases of HIV/AIDS were diagnosed in India in 1986. However, in the beginning, the number of such cases is only in the single digits and, in a vast country like India, it is practically impossible for a case to come to light if the total number of such cases in the country is, let us say, ten. We shall therefore assume that the first cases of HIV in India appeared a few years earlier, say 1981, and the first cases of AIDS appeared in 1986 and were diagnosed either in 1986 or later.

THE MODEL

As in the previous paper [2], we shall present cumulative HIV incidence estimates for India using the method of back calculation which calculates HIV incidence from the data on AIDS incidence, based on an estimate of the incubation period, the time it takes for an HIV positive person to develop AIDS. It has been estimated [4, p.117] that this incubation period in India may be as short as 8 years. Consequently, we take this period to be 7.98 years in what follows. We now write $F[t]=1-Exp[-(\lambda_1 t)^{\theta_1}]$, where $\lambda_1 = .115$; $\theta_1 = 5$. The graph of F[t] is given in Fig. 1.

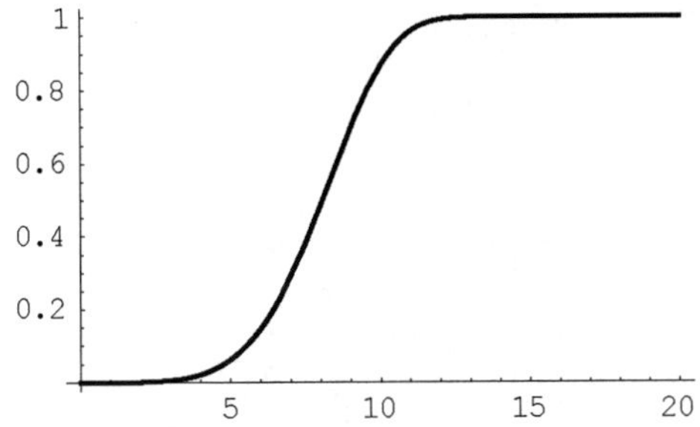

FIGURE 1. Graph of $F[t]=1-Exp[-(\lambda_1 t)^{\theta_1}]$, for λ_1=.115 and θ_1= 5.

The first few values of F[t] are given by
{ F[0], F[2], F[4], F[6], F[8], F[10], F[12], F[20]} =
{0, 0.000643427, 0.0203856, 0.144786, 0.482674, 0.866193, 0.993295, 1.}

The graph in Fig. 1 gives the cumulative probability distribution of development to AIDS. The first few values say that out of a hundred patients say, most stay asymptomatic for about four years, after which some patients start developing AIDS, and by about twelve years, most of them have done so. It follows that if g(s) is the number of people who become HIV positive *at* time s, and x(t) is the number of patients who have developed AIDS *by* the time t, then x(t), the total number of patients who are diagnosed with AIDS in the time interval (0, t) is given by the so called Back Calculation Method,

$$x(t) = \int_0^t g(s)F(t-s)ds \quad \text{.............(1)}$$

where F(t-s) is the probability that the patient has developed AIDS in t-s years or less.

The details of the back calculation method are given in our previous paper [2] and will not be repeated here. However, for the sake of completeness, we shall give a very brief outline. We shall assume that the AIDS cases started appearing in 1986 and denote the number of these cases in the subsequent years by x_1, x_2, x_3..... etc. Assuming that the total number of HIV infected people up to the end of 2004 is N, we try to maximise the expression (the likelihood function)

LF=

$$N!/(x_1!\, x_2!\, x_3! \ldots x_{19}!\, (N-xdot)!)$$

$$p_1^{x_1}\, p_2^{x_2}\, p_3^{x_3} \ldots\ldots p_{19}^{x_{19}}\, (1-pdot)^{(N-xdot)} \quad .. (2)$$

where $xdot = x_1+x_2+x_3+\ldots x_{19}$, **and**
$pdot = p_1+p_2+p_3+\ldots p_{19}$,

the year 2004 being the year number 19, and where we have assumed that the numbers x_1, x_2,....... etc. have a multinomial distribution with sample size N and cell probabilities p_1, p_2,.... etc., p_i being the probability that an individual infected sometime before the i-th year develops AIDS in the i-th year. After some algebra [2], this maximisation leads to an iterative process which successively improves upon these numbers p_i.

In this paper, we follow the procedure outlined in our previous paper [2] and give only the results. We consider three different cases, namely when the actual number of AIDS cases was three times, four times or five times the reported numbers. These reported numbers are given in our previous paper as also the appropriate exponential interpolation which gives these numbers as 1, 2, 4, 8, 14, 24, 43, 77, 137, 243, 434, 772, 1375, 2449, 4361, 7765, 13827, 24621 and 43843 in the years 1986 to 2004 respectively. Because of the excellent agreement of these numbers with the few reported numbers by NACO, we shall call these the 'reported numbers'. Further, assuming that the total number of HIV cases by 2004 was N, we calculate the yearly HIV incidence which would have produced these AIDS cases, through the back calculation method based on equation (1). The case when the actual numbers are two times the 'reported numbers' is discussed in the previous paper [2]. We now consider the case when the actual numbers are three times the reported numbers.

Case when the actual numbers are three times the reported numbers

The method of Back Calculation, as explained in [2], is an iterative process which successively produces the HIV incidence profile which will produce the desired number of AIDS cases with the help of equation (1). In the beginning, we assume this incidence profile to be flat and run the method taking N = 2,500,000. The results of the first 1000 iterations are given in Figures 2 and 3.

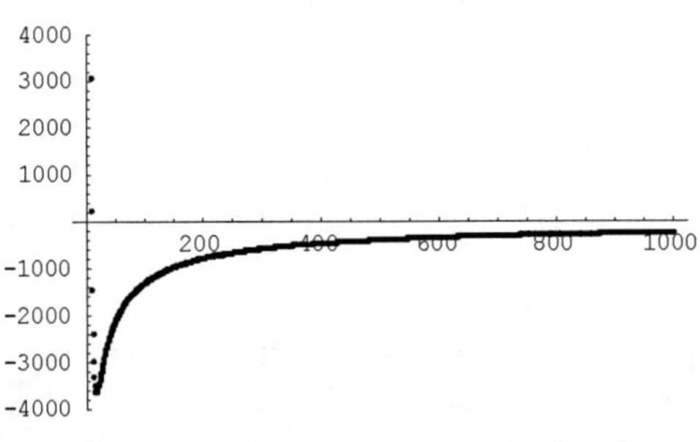

FIGURE 2. Difference between the calculated number of AIDS cases and the assumed ones at the end of 2004 for the first 1000 iterations for N = 2,500,000.

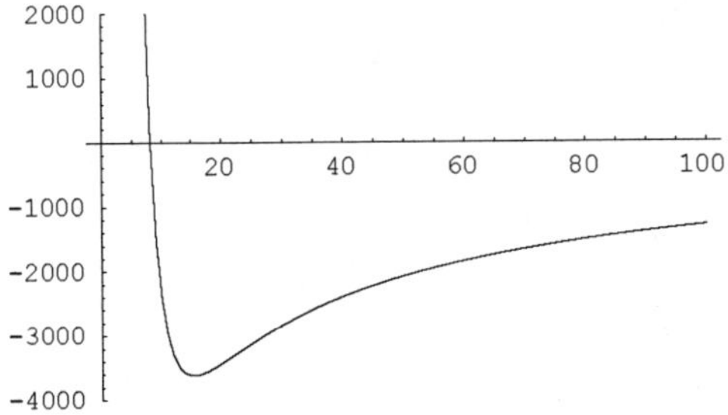

FIGURE 3. Detailed view for the first 100 iterations. The numerical results are given for the case when the curve first crosses the x-axis between eight and nine iterations. We give the results after nine iterations.

It is not clear from these figures whether the process converges. If it does, it must do so extremely slowly. Also, there does not seem to be any advantage in giving the results after 1,000 iterations as compared to only nine iterations. For this reason, we give the results for the case when the curve in Fig. 3 first crosses the x-axis, i.e. after nine iterations for N = 2,500,000 (and after 7 iterations for N = 2,300,000). The calculated yearly HIV incidence for this case is given in Fig. 4 for N = 2,300,000, and for N = 2,500,000. For N = 2,300,000 and for the last five years preceding 2004, the yearly incidence was 285568., 289352., 290223., 290343., and 290350 (in 2004). For N = 2,5000,000, as shown in Fig.4, it was considerably higher. If we therefore assume that the incidence figures given by NACO (average of 287,000 over these years) are correct, then the cumulative HIV incidence up to 2004 must have been approximately 2,300,000. Subtract from this the number of AIDS patients that died and we conclude that HIV prevalence must have been even less than that.

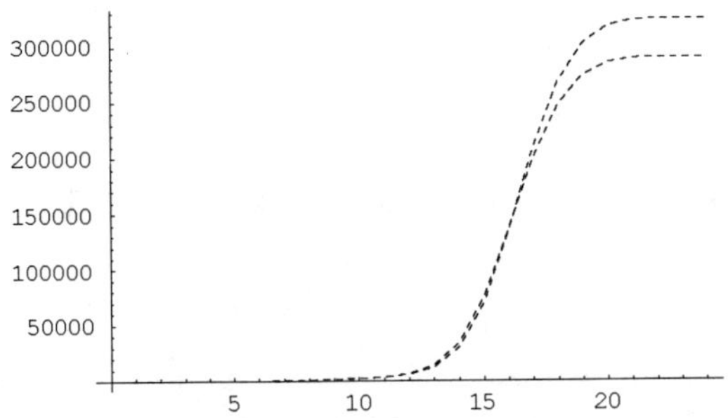

FIGURE 4. Calculated yearly HIV incidence up to 2004 when the cumulative HIV incidence is 2.3 or 2.5 million. The year 1981 is year one. The curve for 2.5 million is higher on the right

The comparison of the calculated cumulative number of AIDS cases with the 'actual' ones is given in Fig.5 for N = 2,300,000. Notice the excellent agreement.

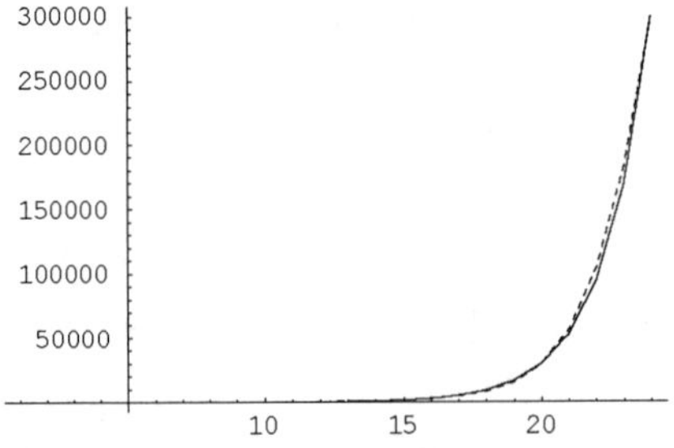

FIGURE 5. Comparison of calculated cumulative number of AIDS cases (dotted line) with the 'actual' ones in our model for N = 2,300,000 for the case when the actual number of AIDS cases is supposed to be three times the 'reported' ones.

The case when the actual number of AIDS cases was four times the 'reported' ones sheds no more light on the problem and therefore, we shall now consider the next case.

Case when the actual numbers are five times the reported numbers

The counterparts of Figs. 2 and 3 in this case were similar and we give the results for the case when the curves crossed the x-axis after four iterations for N = 2,500,000 and after five iterations for N = 3,000,000.

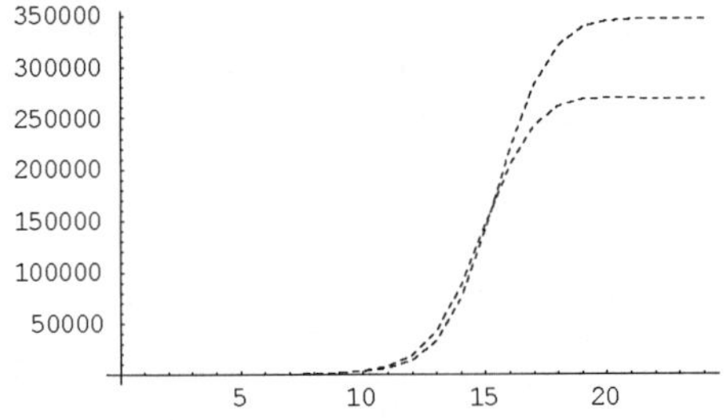

FIGURE 6. Calculated yearly HIV incidence up to 2004 when the cumulative HIV incidence is 2.5 or 3.0 million for this case. The year 1981 is year one. The curve for 3.0 million is higher on the right.

For N = 2,500,000, the yearly incidence for the last five years preceding 2004 was 269249, 268971, 268686, 268606, and 268599 (in 2004). Since the average of these numbers is only slightly lower than the corresponding average of NACO figures, we conclude that in this case as well, the cumulative HIV incidence was only slightly more than 2,500,000. However, if the NACO figures for HIV incidence are on the high side, then 2.5 million may be about right or slightly higher. Subtract from this the number of AIDS patients who died, and the HIV prevalence at the end of 2004 was even less. The yearly incidence for the last five years preceding 2004 for N=3,000,000 was 344847, 346273, 346415, 346400, 346397 (in 2004) which is considerably higher than the NACO figures. We conclude that the cumulative HIV incidence at the end of 2004 was closer to 2.5 million than to 3.0 million. The comparison of the calculated cumulative number of AIDS cases with the 'actual' ones is given Fig.7 for N = 2,500,000.

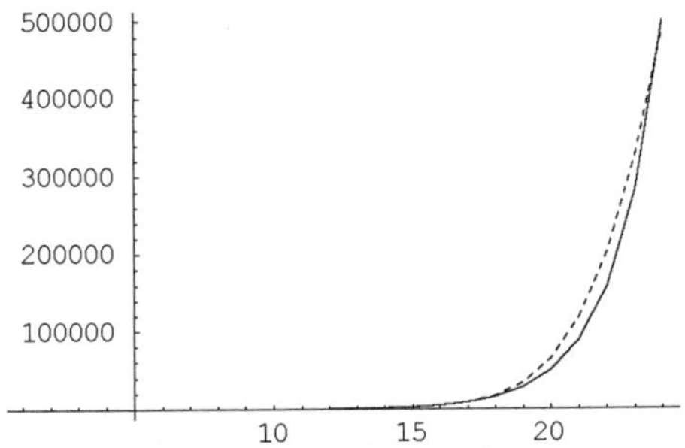

FIGURE 7. Comparison of calculated cumulative number of AIDS cases with the 'actual' ones (solid line) in our model for N = 2,500,000 for the case when the actual number of AIDS cases is supposed to be five times the 'reported' ones.

DISCUSSION

Our conclusion is that cumulative HIV incidence in India at the end of 2004 was close to 2.5 million or slightly less. With the number of HIV positive people increasing at the rate of roughly half a million per year, this number should be close to 3 million at the end of 2006. Take away from it the number of AIDS deaths, and the HIV prevalence today should be slightly less. This helps shed light on the question "Were the number of HIV positive people in India at the end of 2006, three million or six million?". This question has been asked before [4, p.192]. Our estimate would be "less than 3 million", which contrasts with the previous NACO estimate of more than 6 million, which estimate was the one reported before the new surveys in 2006. This supports the opinion that "Unrealistically high HIV prevalence estimates or projections in countries such as China and India will within a few years be exposed as, at best, naive efforts of well meaning AIDS program advocates and at worst as the work of AIDS "experts" who deny the levelling or decreasing global HIV prevalence trends to support their own political, social, economic or personal agendas"[4, p.138].

It should be pointed out that we arbitrarily took the value of θ_1 in our model to be five. This value has been estimated variously in the medical literature and some researchers have estimated it at as low as 2.3 [7]. However, with a lower value of θ_1, AIDS will progress faster in the beginning, and if we keep the same value of incubation period, this will result in an even lower estimate of HIV prevalence. Keeping the same value of incubation period, we tried a lower value of θ_1 and indeed our estimate of HIV prevalence was lower than above. Thus we tried the case of $\theta_1 =$ 2.3 and $\lambda_1 = .111$ which gives an incubation period of 7.98 as before. When the actual number of AIDS cases is five times the reported ones and for N = 2,500,000, the HIV incidence in the last five years before 2004 was 328006, 342061, 340998, 334307, 331178. These numbers are higher than the estimates of NACO implying that a lower value of N would be more appropriate for this case. The comparison of calculated AIDS cases with actual ones is given in Fig. 8.

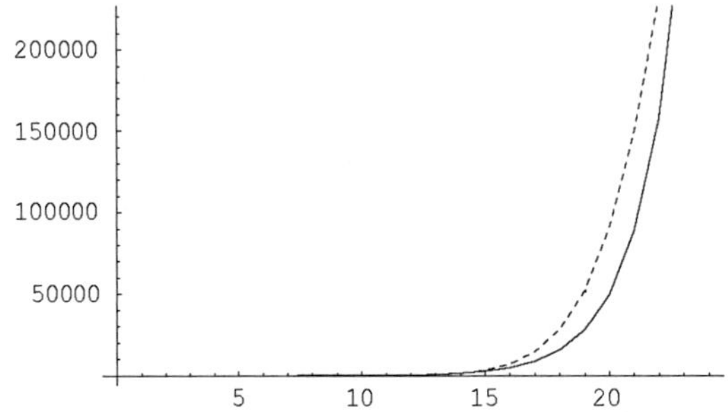

FIGURE 8. Comparison of calculated cumulative number of AIDS cases (dotted line) with the 'actual' ones for $\theta_1 = 2.3$ and $\lambda_1 = .111$ implying an incubation period of 7.98 years, for the case when the actual number of AIDS cases was five times the reported ones.

In this figure, for a given value of t, the calculated number of AIDS cases (dotted line) is much higher than the actual ones, implying that HIV positive people are contracting AIDS much too fast in the model. Comparing with the previous figures (Fig. 5, or 7), we would think that $\theta_1 = 5$ is a more reasonable value of θ_1 in this context.

We have pointed out that with cumulative HIV incidence of 2.5 million at the end of 2004, the cumulative number of AIDS cases may be 2, 3, 4, or 5 hundred thousand. It follows that if we only know the cumulative HIV incidence (or HIV prevalence) at one point in time, it is not possible to estimate historical HIV incidence without additional assumptions. Such assumptions have been made [8], and historical and future trends of HIV and AIDS estimated. Thus "assuming 5 million HIV infected adults in Sub Saharan Africa in 1990, the WHO estimated there would be 700,000 cumulative AIDS cases by the end of 1990 [9, p.270].

The infection curves of Figs. 4 and 6 may also be approximated by appropriate Weibull curves. For this purpose, we write

$H(t) = 0$ in $0 \le t < \alpha$, ... (A1)

and

$H(t) = 1 - \mathrm{Exp}[-(\lambda_1(t-\alpha))^{\theta_1}]$ in $t \ge \alpha$........ (A2).

for appropriate values of α, λ_1, and θ_1. In each case we take $\theta_1=7$ and $\lambda_1=.1$ and compare the curves for the cumulative AIDS incidence of 300,000 and 500,000 with the ones for H(t) above for $\alpha = 7$ and for $\alpha = 5.3$ respectively. The comparison is given in figures 9 and 10. The comparison for N=400,000 with the same curve for $\alpha = 6.2$ was similar and is not shown here. It can be seen that the area under the assumed infection curve for the case of AIDS incidence being 500,000 (dotted line in Fig. 9) is .4% more than the area under the calculated curve (solid line). In comparison, the approximation by any of the gamma curves (used by WHO) would be vastly more in error. This suggests that gamma curves may not be the most suitable curves for modeling HIV infection curves and that the Weibull curves, which have just one more parameter, may be more suitable. Adopting of these curves by modeling agencies like WHO may give much better estimates, particularly in countries like India and China, where because of the very large population base, yearly hiv infection rates are not likely to come down anytime soon. These curves also suggest that if the hiv infection in India started in 1986, and if we accept the NACO estimate of current yearly HIV incidence of more than quarter of a million, then the cumulative number of AIDS cases at the end of 2004 should be 500,000. If it started in 1987, then it should be 400,000 and if it started in 1988, then it should be 300,000. The first cases of HIV positivity in India were diagnosed in 1986.

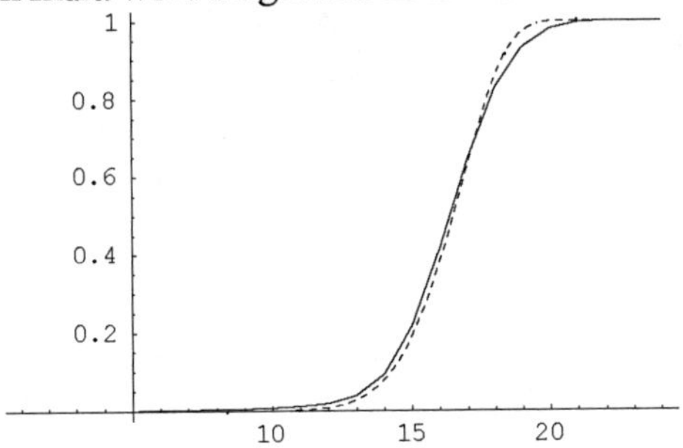

FIGURE 9. Comparison of the infection curve calculated above for the case of 300,000 AIDS incidence at the end of 2004 (solid line) with H(t) given in the text for $\alpha = 7$.

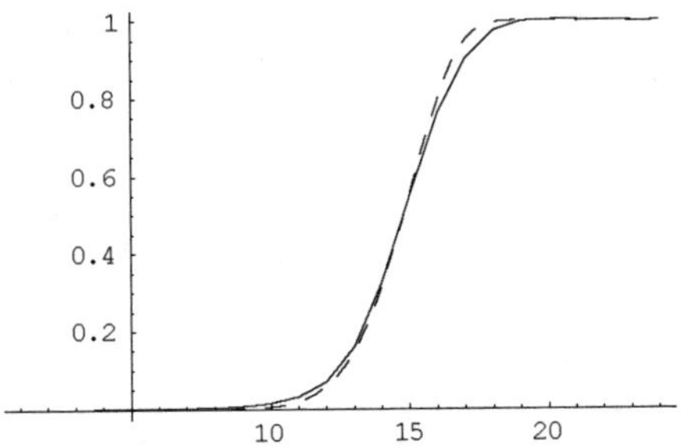

FIGURE 10. Comparison of the infection curve calculated above for the case of 500,000 AIDS incidence at the end of 2004 (solid line) with H(t) given in the text for $\alpha = 5.3$.

Once you decide upon the infection curve, it is easy to calculate the number of AIDS deaths till the end of 2004 and then HIV prevalence at the end of that year. It should be pointed out that in any serosurvey, we only estimate the HIV prevalence, so that it is important for us to estimate the number of AIDS deaths in this paper. AIDS deaths have been estimated before [10]. As in that paper, AD(t), the number of AIDS deaths at any time t, is given by

AD (t) =

$$\int_0^t g(s)k(t-s)ds \text{ ....(B1)}$$

where k (t)=

$$\int_0^1 tG[(1-w)t]F'[wt]dw \text{ .... (B2)}$$

where g(s) is the infection curve, and F[t] and G[t] define the progression from HIV infection to AIDS and from AIDS to death respectively. In our case, we have just calculated the infection curves g(s), F(t) is given in Fig.1 and G(t) is the same function as in Fig.1 but with λ_1=.9, which gives the time from development of AIDS to death as 1.02019 years, a figure which is suited to a developing country like India [4, p.117]. It turns out that for the case of AIDS incidence being 500,000 to the end of 2004, the number of AIDS deaths up to the end of 2004, would be 307060, slightly over 300,000 (the numbers AD(t) are also seen to closely follow a Weibull curve). This puts the HIV prevalence at the end of 2004 at well below 2.5 million, a figure which this author was the first to advocate [2] and which is now accepted by all people working with HIV/AIDS in India [5].

The number of AIDS deaths at the end of 2003 turns out to be 184063 implying that a total of 122997 AIDS patients died in the year 2004. This is about a quarter of the total AIDS incidence up to the end of 2004. Since we are assuming HIV incidence of over 250,000, this further implies that HIV prevalence in India is on the increase these days. Also, according to NACO, "Although data gathered by NACO in 2007 has revealed that HIV prevalence has stabilised in at least Tamil Nadu, Andhra Pradesh, Karnataka, and Maharashtra, it is increasing in at-risk populations in other states. As a result, overall HIV prevalence has continued to rise"[12].

The statement that "HIV prevalence in India is on the increase these days" is extremely important in estimating the number of AIDS deaths in India. This is because HIV prevalence curve in any country usually has a peak after which the prevalence comes down but the cumulative incidence keeps on going up. The difference between the two is the number of AIDS deaths.

In the EPI Model [4], used widely by WHO, the hiv prevalence is approximated by a gamma curve,

$$F(t)=A\, t^p \exp[-t/q] \text{........ (C)}$$

for suitable values of p and q. The curve with p = 4 has matched the prevalence in most societies where such reports are reliable [4, p.116]. The most important parameter to determine now is q because the peak occurs at t = pq, and the position of the peak is the most important parameter in these calculations. A may then be determined by knowing the value of hiv prevalence at any one point, like 2.5 million in 2006 in India. With the help of equations (B), the hiv prevalence at any time t, Pr(t), may also be written as Pr(t)=

$$\int_0^t g(s)(1-k(t-s))ds \text{ ....(D)}$$

and g(s), the infection rate, may now be determined by back calculation (because Pr(t) is known as a gamma curve). The important thing to notice here is that in equation (D) for the determination of g(s), the values of g(s) are more heavily weighted towards the end (s = t) than toward the beginning (s=0). However, in the beginning, these values are very small anyway and do not affect our results appreciably, see [11] for details. This is in contrast to the manner in which g(s) is determined from a record of AIDS cases (see eq.1), when the values of g(s) in the beginning have more weight than those at the end, so that the values towards the end, the more significant ones, are harder to estimate. Perhaps a combination of the two can give more satisfactory results.

We shall now take p = 4 in the gamma distribution (eq. C) and put the peak in 2016 and beyond in the case of India. This is because, according to the National AIDS Control Organisation of India (NACO), the number of HIV positive people in the country was estimated at 3.86, 3.97, 4.58 and 5.1 million respectively in the years 2000 to 2003 respectively. These estimates are shown in Fig.11 and clearly show that the peak is nowhere in sight.

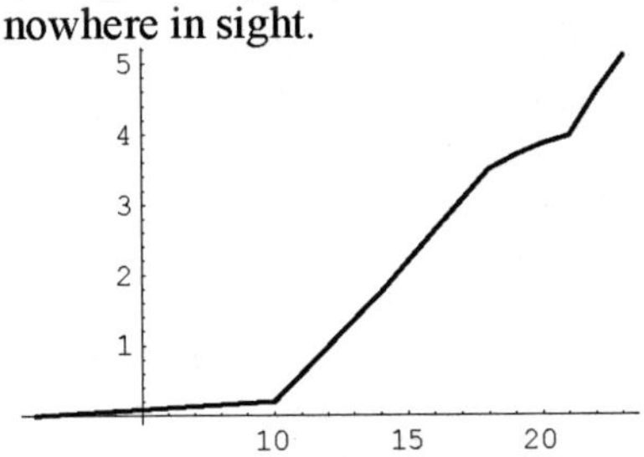

FIGURE 11. NACO estimates of hiv prevalence in India. 1981 is year one.

While it is true that these estimates were revised downwards later on, one would think that such revision will be equally valid in all the previous years as well and will not change the position of the peak, which seems to be way beyond 2003 in Fig.11. Taking the value of hiv prevalence in India to be 2,500,000 in 2006 (NACO's latest revised estimate), we come up with estimates of cumulative incidence till 2006, cumulative AIDS deaths till 2006, incidence in 2006 and AIDS deaths in 2006 as given in the table below with peaks in the years 2001 ,2004, 2006, 2008, 2011, 2016, and 2032 (in successive rows). It appears from the table that as we move the peak forward in time, the cumulative incidence (in 2006) goes down, the number of cumulative AIDS deaths goes down and so on. The comparison of these numbers with our previous estimates (above) would suggest that the peak is still farther away than 2016. Notice that the position of the peak is the most important variable in this table, and affects these numbers most significantly.

TABLE ONE:

{{5.35, 2.85, 178, 332},
{4.36, 1.86, 245, 264},
{4.10, 1.60, 272, 243},
{3.92, 1.42, 295, 227},
{3.72, 1.22, 325, 209},
{3.56, 1.06, 356, 192},
{3.26, 0.76, 434, 158}}

TABLE ONE: Values of cumulative incidence till 2006, the cumulative number of deaths till 2006, incidence in 2006 and the number of deaths in 2006 in India with peaks in 2001, 2004, 2006, 2008, 2011, 2016 and 2032 in successive rows according to our model. Notice that the cumulative number of deaths till 2006 changes by more than 30% if we put the peak 2 years before or 2 years after 2006. The times from seroconversion to AIDS and from AIDS to death are assumed to be 8 years and one year respectively in these calculations.

CONCLUSION

In a previous paper [2], we argued that the cumulative HIV incidence in India, at the end of year 2004, must not have been greater than 2.5 million in India, as against the estimates of WHO and NACO of 5.134 million at that time. This figure of 2.5 million has now been agreed upon by all people concerned [3]. In this paper, we have suggested that Weibull functions, and not gamma curves, should be used for modeling cumulative HIV infection incidence and also argued that the system of AIDS reporting in India is seriously deficient and the number of AIDS cases there, by the end of 2004, was considerably more than the 100,000 reported by NACO, provided we accept the (NACO) estimate of yearly HIV incidence of more than quarter of a million per year [1]. It is further argued that the prevalence peak in India is yet to come and the number of AIDS deaths in India should also be revised sharply downwards by WHO.

ACKNOWLEDGMENT

The contribution of ABM Scientific Research Inc. is gratefully acknowledged.

REFERENCES

1. www.nacoonline.org/facts.htm

2. Aggarwala,B.D., On estimating HIV Prevalence in India, Far East Journal Of *Mathematica*l Sciences, Vol.21, No:2 (2006). pp.197-210.

3. The Pioneer, March 23, 2007.

4.Chin, James, The AIDS Pandemic, The collision of epidemiology with political correctness, Radcliff Publishing, Oxford, U.K, (2007).

5. NDTV.com, July 6, 2007.

6. Private communication with Prof. James Chin.

7.Gruttola V. D. and Lagakos,S.W., The value of AIDS incidence data in assessing the spread of HIV infection, Statistics in Medicine, Vol.8 (1989), pp.35-43

8. Chin, J. and Lwanga, S.K., Estimation and Projection of adult AIDS cases; a simple epidemiological model, Bulletin of the World Health Organisation, Vol.69, no;4(1991).

9.Brookmeyer, R and Gail, M.H., AIDS Epidemiology, A Quantitative Approach, Oxford University Press (1994)

10. Aggarwala, B.D., On Estimating HIV Prevalence in Canada and in the United States, Far East Journal of Applied Mathematics, Vol.20, No:3 (2005) pp. 335-354.

11. Aggarwala, B.D., "On WHO and AIDS deaths in India" Submitted for publication.

12. NACO Website, "HIV Situation in India" Dec.7,2007.

Optimal inspection period and replacement policy for CBM with imperfect information using PHM

Alireza Ghasemi, Soumaya Yacout, M-Salah Ouali

Industrial Engineering and Mathematics Department, École Polytechnique of Montréal, C. P. 6079, succursale Centre-ville, Montréal, Québec, Canada, H3C 3A7,
(alireza.ghasemi, soumaya.yacout, msouali)@polymtl.ca

Abstract. This chapter introduces a model which identifies the optimal inspection period for Condition Based Maintenance (CBM) of a system, when the information obtained from the data gathered on the system does not reveal the system's exact degradation state and the data collection is costly. The proposed model uses the Proportional Hazards Model (PHM) introduced by [Cox 1972] to model the failure rate of the system. The PHM takes into consideration the system's degradation state as well as its age. Since the acquired information is imperfect, the degradation state of the system is not precisely known. Bayes' rule is used to estimate the probability of being in any of the possible states. The system's degradation process follows a Hidden Markov Model (HMM). By using dynamic programming, the system's optimal replacement policy and its total long run average maintenance cost are found. Based on the long run average cost, the optimal inspection interval and the corresponding replacement criterion are specified. A numerical example shows the behaviour of the CBM model when the inspection is costly, and finds the optimal inspection period and maintenance cost.

Keywords: Condition Based Maintenance (CBM), Imperfect Information, Proportional Hazard Model (PHM), Hidden Markov Model (HMM), Costly Inspections.
PACS: 89.20.Bb, 45.10.Db, 46.15.Cc, 87.55.de

INTRODUCTION

When a system is subjected to a Condition Based Maintenance (CBM) program, inspections are performed to obtain information (observations) about the degradation state of the system. In this work, the information acquired during the inspections is imperfect, that is, it does not reveal the exact degradation state of the system but represents some data which is stochastically related to the system's degradation state [Maillart, 2004, Ohnishi et al., 1986]. This data is used to calculate the probability of being in a certain degradation state. The hidden degradation state of the system is modeled by a Markov Chain. In CBM studies, several models have been used to take into account the system's degradation state. One of these models is the Proportional Hazards Model (PHM), introduced by [Cox 1972], which has increasingly been applied to CBM [Lin et al. 2005, Banjevic and Jardin 2001]. According to the PHM, the system's failure rate (also called hazard rate) is calculated based on its age as well as its degradation state. In this paper, the PHM is used to calculate the optimal

CP1007, *Current Themes in Engineering Technologies*, edited by S.-I. Ao, M. A. Amouzegar, and S.-S. Chen

replacement policy and long-run average maintenance cost for a system with imperfect information.

Afterwards, the unrealistic assumption of non-costly inspection is relaxed and the corresponding optimal replacement policy and long-run average maintenance cost are found. In the CBM modeling, if the inspections are done at no cost, the optimal inspection period is theoretically equal to zero, i.e. the best choice is to monitor and analyze the system continuously. This is because the higher frequency of inspections will provide more frequent information about the system's degradation state with no extra cost. Consequently, this will reduce the likelihood of making wrong decisions, such as performing unnecessary preventive replacements, hence, will result in a more cost effective maintenance system. When there is a considerable cost for collecting and analyzing observations obtained during inspection, an optimal inspection period that minimizes the *total maintenance cost*, which includes the cost for inspections, should be applied. In reality, in many cases, inspections require personnel and equipment, and sometimes it is necessary to stop or suspend the operations when performing the inspections [Lam and Yen, 1994]. Also, some destructive tests for analysis and extraction of useful information may be needed. Therefore, some costs are associated with the collection and analysis of the observations. In order to find the total optimal long-run average cost of the maintenance plan with costly inspections, comparison and selection between several possible inspection periods is necessary.

The rest of this chapter consists of a brief literature review of the principle models in replacement optimization. The next section deals with the assumptions, the details of the proposed model, and the optimal solution. A numerical example is then presented and finally, the conclusion and the areas of further research are discussed.

LITERATURE REVIEW

[Lam and Yen 1994] investigate the maintenance policy for a system whose exact degree of degradation is known through inspections. The objective is to find the optimal replacement criteria and inspection period that minimize the long-run average cost of the maintenance plan. The authors assume that there is a fixed inspection cost M, that the mean inspection time is q and that the cost rate per unit of time when the system is under inspection is m. This results in an average inspection cost equivalent to $mq+M$. Dynamic Programming is used to solve the problem.

[Christer and Wang 1995] consider a system with perfect information where inspections reveal the system's exact degradation state. The objective is to determine the next inspection schedule, based on the latest observation obtained. The next inspection point is selected in a manner that minimizes the average maintenance cost per unit time between the current and the next inspection. A constant, predetermined threshold of degradation indicating failure is considered, and there is a cost related to the inspections. In this model, the time between the inspections is not constant. The next inspection point is always determined at the current inspection point using the information obtained from the condition monitoring system. Possible actions are "inspection and replacement" or "inspection and no replacement". A closed form solution to calculate the cost per unit time is given. By minimizing the cost, the optimal inspection date is obtained.

[Hontelez *et al.* 1996] consider a system revealing perfect information with $N+1$ possible degradation states, where state N is a functional failure and state $N+1$ is a physical failure state. State N is detected through inspection but state $N+1$ is an obvious failure and is detected as soon as it happens. There is a cost associated with the inspections. By applying dynamic programming, a control limit rule such as $\Pi = [\pi; \rho_1, \rho_2, ..., \rho_{\pi-1}]$ is obtained. It means that at any inspection point, replacement takes place if and only if the system is in degradation state $i \geq \pi$, OR an inspection is performed after ρ_i periods if the degradation state is $i < \pi$. With this approach, the inspections will not take place at every inspection period unless there is an evident need to do so.

[Chelbi and Ait-Kadi 1999] consider an optimal inspection time with a hidden failure being detected through inspection. A pre-defined threshold of the system's degradation state is set to identify the failure. Associated cost of inspections, repairs and replacements are taken into account. The average long-run cost of the maintenance plan $E(AC) = E(C)/E(T)$, is minimized. $E(C)$ and $E(T)$ present the expected cost and the expected length of the renewal period, respectively. The renewal period is the time between two consecutive replacements, whether due to a failure or to a preventive replacement [Cox, 1962]. By minimizing the long-run expected cost over the renewal period, the optimal inspection period is found.

[Grall *et al.* 2002] model a CBM policy where both the replacement threshold and the inspection schedule are decision variables. Irregular inspection periods are allowed, i.e. the next inspection date is dynamically updated on the basis of the present system degradation state, which is revealed by the current inspection. $N(>1)$ Threshold values $0 < \xi_1, \ldots, \xi_N < L$ are set in the system degradation state range. L is the predefined degradation threshold assumed for the failure and ξ_N is the replacement threshold. At any inspection point t_k, where the observation value (degradation state) is X_k, if $\xi_l \leq X_k < \xi_{l+1}$ for some $0 \leq l < N$, the next inspection will be after $N-l$ period(s). Otherwise, if $\xi_N \leq X_k < L$, then a preventive replacement is performed while $X_k \geq L$ results in a failure replacement. The long-run expected cost per unit of time is minimized. The optimal value of the decision variables, which is the number of thresholds N, and the different values of the thresholds, $\xi_i, i = 1, ..., N$, are found.

[Ghasemi *et al.* 2007] consider a CBM based on imperfect information when there is no considerable cost associated with the inspections.

PROBLEM FORMULATION

This problem presents a deteriorating system subject to random failure. While the degradation process of the system can be continuous, we have discretisized the degradation state set. The degradation state of the system is illustrated by a finite set of non-negative integers, i.e. by the state space $S=\{1,2,,...,N\}$. State 1 indicates the best possible state for the system, which means that the system is new or like new. The degradation state process $\{X(t)=1,2,...,N\}$ is a discrete time homogeneous Markov chain with N unobservable states. All N degradation states are working states and do not include the failure state which is a non-working state. If a failure occurs, it is know instantaneously. Figure 1 shows the Markov transition process between degradation states, along with the transitions from each degradation state to the failure state. p_{ij} is the probability of going from degradation state i to the degradation state j during one inspection period, knowing that the system has not yet failed, while f_i is the probability of going from degradation state i to the failure state. The circles represent the states.

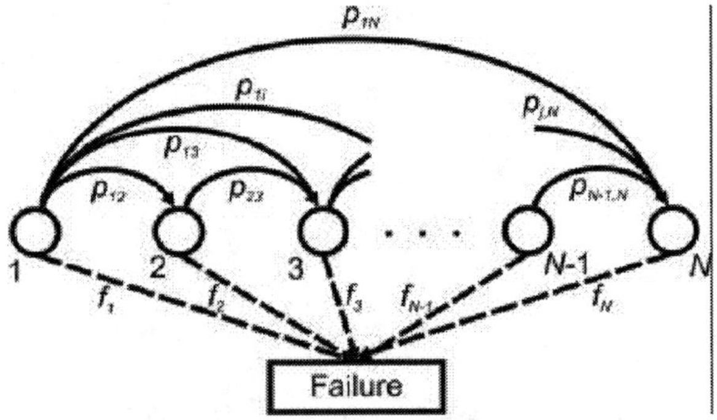

Figure 1: Markov process transition and transition to failure

The degradation states of the system are not observable except at time $t=0$ when the state of the system is certainly 1. The transition matrix P is an upper triangular matrix, i.e. $p_{ij}=\Pr(X(t+\Delta)=j\,|\,X(t)=i,T>t+\Delta), t=0,\Delta,2\Delta,...$otherwise $p_{ij}=0$ for $j<i$. $p_{ij}=0$ for $j<i$, meaning that the system degradation state does not improve spontaneously, which is true in most cases. T is a random variable representing the system's failure time. The system indicator is inspected at times $t=\Delta,2\Delta,...$. The value of the indicator (observation) is assumed to be within a finite set of M non-negative integers, i.e. $\theta\in\Theta=\{1,2,...,M\}$. It is also supposed that the value of the indicator θ is observed with a known probability $q_{j\theta}$, when the degradation state of the system is j. Q represents the stochastic matrix which specifies these probabilities, i.e. $Q=\left[q_{j\theta}\right], j\in S,\ \theta\in\Theta$.

Failure is not considered a degradation state. It is a condition that causes the system to cease functioning and is outwardly obvious. If failure occurs, it is immediately recognized and the only possible action is *Failure Replacement* (FR). Otherwise, at any inspection point, we can decide whether to perform *Preventive Replacement* (PR) or to *Do-Nothing*. The FR and the PR renew the system and return it to state 1, i.e. new or like new, and the time is reset to zero. The cost for the PR is C, while a FR costs $K+C, K, C>0$. Both actions, FR and PR, are instantaneous. Performing the inspections costs C_I per inspection, independently of the inspections' interval. More frequent inspections will cost more, but will provide information at a higher frequency that subsequently will result in a more efficient replacement policy.

The system's failure rate follows the PHM where the failure rate $h(t, X_k) = h_0(t)\psi(X_k)$ is a product of two independent functions. $h_0(.)$ is a function of the system's age only and $\psi(.)$ is a function of the system's degradation state only. $X_k = X(k\Delta)$ is the degradation state of the system at period k and Δ is the fixed inspection period. We assume that the degradation state of the system remains unchanged during each period (between two consecutive inspections). Each degradation state transfer, if any, is assumed to take place at the end of each period, just before the inspection point.

The objective is to find the optimal inspection period and corresponding replacement policy that minimizes the total long-run average cost per unit time.

Considering the failure rate of the system, the conditional reliability of the system where $X_k = X(k\Delta)$ is:

$$R(k, X_k, t) = P(T > k\Delta + t \mid T > k\Delta, X_1, X_2, ..., X_k)$$
$$= \exp\left(-\psi(X_k)\int_{k\Delta}^{k\Delta+t} h_0(s)ds\right)$$

where T is the random variable indicating the failure time of the system. The conditional reliability of the system indicates the probability of not having the failure in time t after $k\Delta$, given that the failure has not happened until time $k\Delta$ and the degradation states of the system have been $X_1, X_2, ..., X_k$ at times $\Delta, 2\Delta, \ldots k\Delta$. The conditional mean sojourn time of the system, if no preventive action is performed while the system is in degradation state X_k at period k, is: [Makis and Jardine 1992]

$$\tau(k, X_k, \Delta) = \int_0^{\Delta} R(k, X_k, t)\,dt$$

The conditional mean sojourn time of the system is defined as the expected remaining life (time to failure) of the system until the next inspection point. In what follows, Partially Observed Markov Decision Process (POMDP) is used. Since the degradation state of the system is not perfectly observable through the inspections, an alternative state space for the POMDP, indicating the "conditional probability distribution of the system's degradation state", and then an "alternative state's transition" complying with the alternative state space are introduced.

THE FORMULATION OF THE POMDP

Alternative state space

We adapt the definition of π^k as defined in [Ghasemi 2005] to indicate the *Conditional Probability Distribution of the system's degradation State* (CPDS) at period k:

$$\pi^k = \left\{ \pi_i^k;\ 0 \le \pi_i^k \le 1 \text{ for } i = 1,\dots,N, \sum_{i=1}^{N} \pi_i^k = 1 \right\},\ k = 0,1,2,\dots \tag{1}$$

π_i^k represents the probability of being at state i at the k-th inspection point. The initial value of the CPDS for a new system is defined as:

$$\pi_i^0 = \begin{cases} 1 & i = 1 \\ 0 & 1 < i \le N \end{cases} \tag{2}$$

this means that a new or renewed system is at state 1.

Alternative state transition

After each collection of observation θ at inspection point, the CPDS is updated considering the latest observation θ. Using the Bayes' formula and assuming that the observation θ has occurred at the $k+1^{st}$ inspection point, the updated CPDS $\pi_j^{k+1}(\theta)$ is calculated as [Ghasemi 2005]:

$$\pi_j^{k+1}(\theta) = \sum_{i=1}^{N} \pi_i^k p_{ij} q_{j\theta} \Big/ \Pr(\theta \mid k, \pi^k),\ j = 1,\dots,N \tag{3}$$

where $\Pr(\theta \mid k, \pi^k) = \sum_{i=1}^{N} \sum_{j=1}^{N} \pi_i^k p_{ij} q_{j\theta}$ is the probability of observing a certain value θ at $k+1^{\text{st}}$ inspection period (just before the inspection point) when the CPDS at the k-th period is π^k. This updated CPDS carries all the observations and actions history from the last replacement point. After any PR or FR, the period counter k will be reset to zero and the CPDS will be set to π^0 as introduced in Eq. (2).

Decision space

$\{0, \infty\}$ is the decision space of the POMDP, where 0 means "replace the system immediately (PR)" and ∞ means "Do-Nothing". If "Do-nothing" is selected at a decision point and a failure happens before the next inspection (decision) point, the system will be replaced immediately. The occurring cost for this event is $K+C$, $K, C > 0$. A PR costs C.

Dynamic Programming Formulation

Let $V(k,\pi^k)$ denote minimum *total cost of maintenance over the renewal period*, while the system is in the k-th inspection point and the CPDS is π^k. The *renewal period* is the time between two consecutive replacements, whether FR or PR. The *total cost* is defined as replacement cost of maintenance plus the inspections' cost.

$$V\left(k,\pi^k\right)=\min\left\{kC_I+C+V\left(0,\pi^0\right),W\left(k,\pi^k,g\right)\right\} \tag{4}$$

where $kC_I+C+V\left(0,\pi^0\right)$ is the renewal period's total cost at k-th inspection point, if one decides to perform the PR (decision=0). $W\left(k,\pi^k,g,C_I\right)$ is the renewal period's total cost, when at the kth inspection point one decides that no action takes place (decision= ∞).

$$\begin{aligned}W\left(k,\pi^k,g\right)=&\left[kC_I+K+C+V\left(0,\pi^0\right)\right]\left[1-\overline{R}\left(k,\pi^k,\Delta\right)\right]-g\overline{\tau}\left(k,\pi^k,\Delta\right)\\&+\left[\sum_{\theta=1}^{M}V\left(k+1,\pi^{k+1}(\theta)\right)\Pr\left(\theta\,|\,k,\pi^k\right)\right]\overline{R}\left(k,\pi^k,\Delta\right)\end{aligned} \tag{5}$$

$\left[kC_I+K+C+V\left(0,\pi^0\right)\right]$ represents the renewal period's total cost if the decision is "Do-nothing" (decision= ∞) and the system fails during the next inspection period. $\left[\sum_{\theta=1}^{M}V\left(k+1,\pi^{k+1}(\theta)\right)\Pr\left(\theta\,|\,k,\pi^k\right)\right]$ is the expected total future cost of the system at the $k+1$ inspection point, provided that the failure has not happened during the k-th period. $\left[1-\overline{R}\left(k,\pi^k,\Delta\right)\right]$ and $\overline{R}\left(k,\pi^k,\Delta\right)$ are the probability of the failure occurring during the k-th period and the probability that the system is still working at the beginning of the $k+1^{\text{st}}$ period consecutively, while the CPDS at period k is π^k. $\overline{\tau}\left(k,\pi^k,\Delta\right)$ and g are the mean sojourn time of the system at the $k+1^{\text{st}}$ period when the CPDS at the k-th period π^k is available, and the average cost per unit of time over infinite horizon, respectively [Ghasemi 2005]. g includes the cost of replacements only and excludes the inspections' costs.

$$\overline{R}\left(k,\pi^k,t\right)=\sum_{i=1}^{N}R\left(k,i,t\right)\,\pi_i^k \tag{6}$$

$$\overline{\tau}\left(k,\pi^k,\Delta\right)=\int_0^{\Delta}\overline{R}\left(k,\pi^k,t\right)dt \tag{7}$$

If a failure happens, the system is replaced and service begins immediately, but the next inspection will continue as scheduled as shown in Figure 2. The term $g\overline{\tau}\left(k,\pi^k,\Delta\right)$ is the expected long-run cost of the overlapped time of two consecutive replacements of the system when the system fails. For more details please refer to and replaced as shown in [Ghasemi 2005].

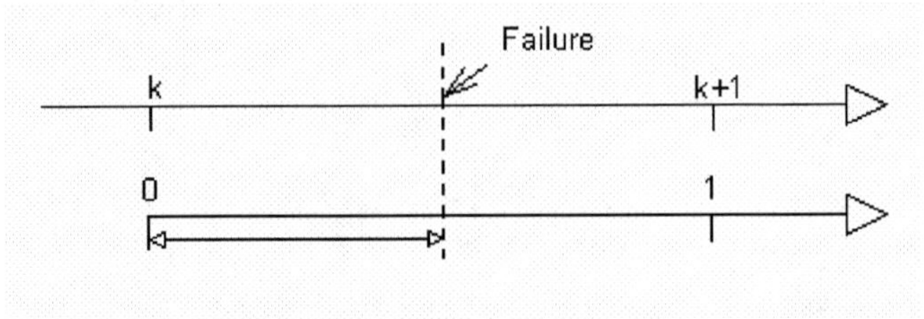

Figure 2: Observations after a failure replacement

OPTIMAL POLICY

To establish the optimal maintenance policy of the described problem, one needs an optimal decision criterion to apply at each decision (inspection) point. This criterion is a function of the observed indicator, the age of the system, the system's cost parameters and finally g, the long-run average maintenance cost of the system. In the following sections, we introduce a decision criterion depending on g. The decision criterion and the *minimum long-run average cost* of the system, g^*, together, give the *optimal* decision criterion for the introduced problem.

Decision Criterion

Considering Eq. (5) one can write:

$$W\left(k,\pi^k,g\right)=kC_I+C+V\left(0,\pi^0\right)-\left(kC_I+C+V\left(0,\pi^0\right)\right)\overline{R}\left(k,\pi^k,\Delta\right)+K\left[1-\overline{R}\left(k,\pi^k,\Delta\right)\right]$$
$$-g\overline{\tau}\left(k,\pi^k,\Delta\right)+\left[\sum_{\theta=1}^{M}V\left(k+1,\pi^{k+1}(\theta)\right)\Pr\left(\theta\mid k,\pi^k\right)\right]\overline{R}\left(k,\pi^k,\Delta\right)$$

$$\begin{aligned}W\left(k,\pi^k,g\right)-\left[kC_I+C+V\left(0,\pi^0\right)\right]=K\left[1-\overline{R}\left(k,\pi^k,\Delta\right)\right]\\-\left(kC_I+C+V\left(0,\pi^0\right)\right)\overline{R}\left(k,\pi^k,\Delta\right)\\-g\overline{\tau}\left(k,\pi^k,\Delta\right)+\left[\sum_{\theta=1}^{M}V\left(k+1,\pi^{k+1}(\theta)\right)\Pr\left(\theta\mid k,\pi^k\right)\right]\overline{R}\left(k,\pi^k,\Delta\right)\end{aligned} \tag{8}$$

Since $V\left(k+1,\pi^{k+1}\right)$ is the minimum expected renewal period cost at the $k+1^{st}$ period, then from Eq. (4):

$$V\left(k+1,\pi^{k+1}\right)\le kC_I+C+V\left(0,\pi^0\right)$$
$$V\left(k+1,\pi^{k+1}\right)\Pr\left(\theta\mid k,\pi^k\right)\le\left[kC_I+C+V\left(0,\pi^0\right)\right]\Pr\left(\theta\mid k,\pi^k\right)$$
$$\sum_{\theta=1}^{M}V\left(k+1,\pi^{k+1}\right)\Pr\left(\theta\mid k,\pi^k\right)\le\sum_{\theta=1}^{M}\left[kC_I+C+V\left(0,\pi^0\right)\right]\Pr\left(\theta\mid k,\pi^k\right)$$
$$\sum_{\theta=1}^{M}V\left(k+1,\pi^{k+1}\right)\Pr\left(\theta\mid k,\pi^k\right)\le kC_I+C+V\left(0,\pi^0\right)$$

This means that:

$$\sum_{\theta=1}^{M} V\left(k+1,\pi^{k+1}\right)\Pr\left(\theta \mid k,\pi^{k}\right)-\left[kC_I+C+V\left(0,\pi^{0}\right)\right]\le 0$$

$$\therefore\left[\sum_{\theta=1}^{M} V\left(k+1,\pi^{k+1}\right)\Pr\left(\theta \mid k,\pi^{k}\right)-kC_I-C-V\left(0,\pi^{0}\right)\right]\overline{R}\left(k,\pi^{k},\Delta\right)\le 0 \quad (9)$$

If $K\left[1-\overline{R}\left(k,\pi^{k},\Delta\right)\right]<g\overline{\tau}\left(k,\pi^{k},\Delta\right)$, then the sum of all the terms on the right hand side of Eq. (8) will be negative or zero, i.e. $W\left(k,\pi^{k},g\right)\le\left[kC_I+C+V\left(0,\pi^{0}\right)\right]$. This final equation means that the cost of leaving the system and doing no preventive action is less than the cost of the PR, so the optimal decision at k-th inspection point, i.e. optimal decision $=\infty$.

In the case that $K\left[1-\overline{R}\left(k,\pi^{k},\Delta\right)\right]\ge g\overline{\tau}\left(k,\pi^{k},\Delta\right)$, we show that the best solution is to replace the system immediately, i.e. Decision=0. To continue, we assume the contrary, i.e. $K\left[1-\overline{R}\left(k,\pi^{k},\Delta\right)\right]\ge g\overline{\tau}\left(k,\pi^{k},\Delta\right)$ and at the same time the optimal action is "Do-Nothing", from Eq. (4) based on this assumption:

$$V\left(k,\pi^{k}\right)=W\left(k,\pi^{k},g\right)<C+V\left(0,\pi^{0}\right)+C_I \quad (10)$$

Also we can write:

$$V\left(k+1,\pi^{k}\right)-V\left(k,\pi^{k}\right)=V\left(k+1,\pi^{k}\right)\overline{R}\left(k,\pi^{k},\Delta\right)+V\left(k+1,\pi^{k}\right)\left[1-\overline{R}\left(k,\pi^{k},\Delta\right)\right]-V\left(k,\pi^{k}\right)$$

$$V\left(k+1,\pi^{k}\right)-V\left(k,\pi^{k}\right)=V\left(k+1,\pi^{k}\right)\overline{R}\left(k,\pi^{k},\Delta\right)+V\left(k+1,\pi^{k}\right)\left[1-\overline{R}\left(k,\pi^{k},\Delta\right)\right]-W\left(k,\pi^{k},g\right)$$

Replacing $W\left(k,\pi^{k},g\right)$ with right hand side of (5):

$$\begin{aligned}V\left(k+1,\pi^{k}\right)-V\left(k,\pi^{k}\right)=&\left[V\left(k+1,\pi^{k}\right)-K-C-kC_I-V\left(0,\pi^{0}\right)\right]\left[1-\overline{R}\left(k,\pi^{k},\Delta\right)\right]\\&+\left[V\left(k+1,\pi^{k}\right)-\sum_{\theta=1}^{M}V\left(k+1,\pi^{k+1}(\theta)\right)\Pr\left(\theta \mid k,\pi^{k}\right)\right]\overline{R}\left(k,\pi^{k},\Delta\right)\\&+g\overline{\tau}\left(k,\pi^{k},\Delta\right)\end{aligned}$$

$$\begin{aligned}V\left(k+1,\pi^{k}\right)-V\left(k,\pi^{k}\right)=&\left[\underbrace{V\left(k+1,\pi^{k}\right)-C-kC_I-V\left(0,\pi^{0}\right)}_{1}\right]\left[1-\overline{R}\left(k,\pi^{k},\Delta\right)\right]\\&+\left[\underbrace{V\left(k+1,\pi^{k}\right)-\sum_{\theta=1}^{M}V\left(k+1,\pi^{k+1}(\theta)\right)\Pr\left(\theta \mid k,\pi^{k}\right)}_{2}\right]\overline{R}\left(k,\pi^{k},\Delta\right)\\&+g\overline{\tau}\left(k,\pi^{k},\Delta\right)-K\left[1-\overline{R}\left(k,\pi^{k},\Delta\right)\right]\end{aligned}$$

By definition $V\left(k+1,\pi^k\right)$ is the minimum renewal period cost, therefore $V\left(k+1,\pi^k\right) \le C + kC_I + V\left(0,\pi^0\right)$, so term 1 is not positive. We have proven (see the appendix) that $V\left(k,\pi^k\right)$ is non-decreasing in $\left(k,\pi^k\right)$. Then:

$$V\left(k+1,\pi^k\right) \le V\left(k+1,\pi^{k+1}(\theta)\right)$$

$$\Pr\left(\theta \mid k,\pi^k\right)V\left(k+1,\pi^k\right) \le \Pr\left(\theta \mid k,\pi^k\right)V\left(k+1,\pi^{k+1}(\theta)\right)$$

$$\sum_{\theta=1}^{M}\Pr\left(\theta \mid k,\pi^k\right)V\left(k+1,\pi^k\right) \le \sum_{\theta=1}^{M}\Pr\left(\theta \mid k,\pi^k\right)V\left(k+1,\pi^{k+1}(\theta)\right)$$

$$V\left(k+1,\pi^k\right) \le \sum_{\theta=1}^{M}\Pr\left(\theta \mid k,\pi^k\right)V\left(k+1,\pi^{k+1}(\theta)\right)$$

where θ is the indicator observed at $k+1^{\text{st}}$ inspection point. In this case, term 2 is not positive. Terms $\overline{R}\left(k,\pi^k,\Delta\right)$ and $\left[1-\overline{R}\left(k,\pi^k,\Delta\right)\right]$ are not negative by definition, so:

$$V\left(k+1,\pi^k\right) - V\left(k,\pi^k\right) < -K\left[1-\overline{R}\left(k,\pi^k,\Delta\right)\right] + g\overline{\tau}\left(k,\pi^k,\Delta\right)$$

Since we are considering the case where $K\left[1-\overline{R}\left(k,\pi^k,\Delta\right)\right] \ge g\overline{\tau}\left(k,\pi^k,\Delta\right)$:

$\therefore V\left(k+1,\pi^k\right) - V\left(k,\pi^k\right) < 0$.

On the other hand, we have proven that $V\left(k,\pi^k\right)$ is non-decreasing in $\left(k,\pi^k\right)$ (see the appendix) so $V\left(k+1,\pi^k\right) - V\left(k,\pi^k\right) \ge 0$, which is a contradiction. This means that the optimal decision is to replace the system immediately (Decision=0).

The optimal decision criteria can be briefly written as:

$$a\left(k,\pi^k\right) = \begin{cases} \infty & \text{if} \quad K\left[1-\overline{R}\left(k,\pi^k,\Delta\right)\right] < g\overline{\tau}\left(k,\pi^k,\Delta\right) \\ 0 & \text{if} \quad K\left[1-\overline{R}\left(k,\pi^k,\Delta\right)\right] \ge g\overline{\tau}\left(k,\pi^k,\Delta\right) \end{cases} \tag{11}$$

where $a\left(k,\pi^k\right)$ indicates the decision at period k while the CPDS is π^k. The optimal decision is dependent on the long run expected maintenance cost per unit time of the replacement system (excluding inspection cost) i.e. g. The *optimal decision*, based on minimum long-run average cost g^*, can be calculated based on (11) by replacing g by g^*.

$$a^*\left(k,\pi^k\right) = \begin{cases} \infty & \text{if} \quad K\left[1-\overline{R}\left(k,\pi^k,\Delta\right)\right] < g^*\overline{\tau}\left(k,\pi^k,\Delta\right) \\ 0 & \text{if} \quad K\left[1-\overline{R}\left(k,\pi^k,\Delta\right)\right] \ge g^*\overline{\tau}\left(k,\pi^k,\Delta\right) \end{cases} \tag{12}$$

The detailed calculation of g^* is given in the next section.

Figure 3 visualizes the replacement criterion for equipment with two possible states. The horizontal axis shows all possible values of π_2^k, which is the probability of being in state 2 at the k-th inspection point, the vertical axis represents the value of $\left[1-\overline{R}\left(k,\pi^k,\Delta\right)\right]/\overline{\tau}\left(k,\pi^k,\Delta\right)$, and finally the horizontal line indicates the value of

g^* / K. At any inspection point k, the value of π_2^k is calculated based on the latest observed indicator's value. If the regarding point is under the horizontal line the system will not be replaced preventively, otherwise, preventive replacement will be performed.

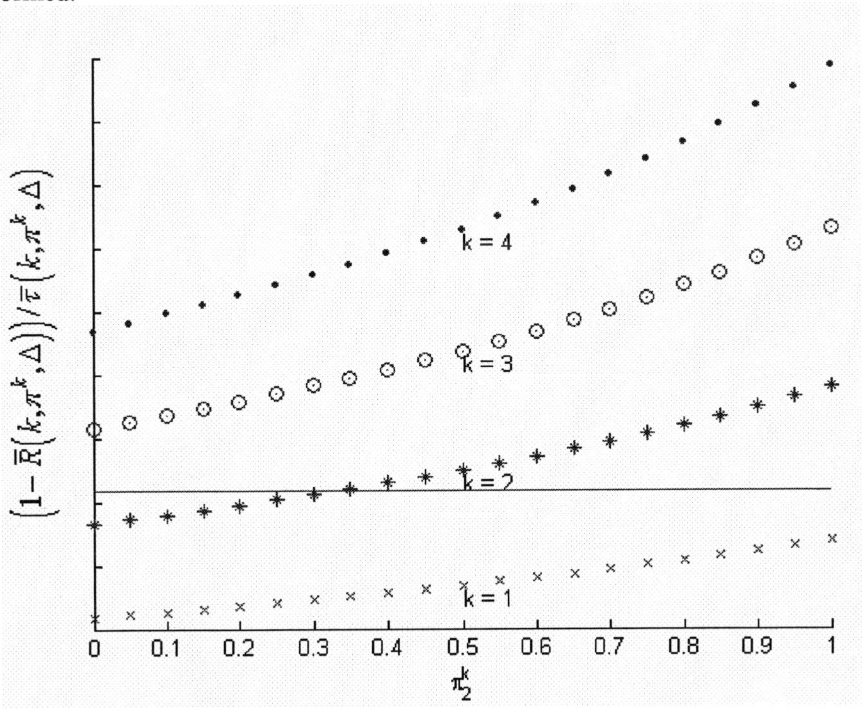

Figure 3 - Decision Criterion behavior for an imaginary problem

This means that in the example of Figure 3, the system will never be replaced preventively at first inspection point $k = 1$. At $k = 2$, Preventive Replacement will be the optimal decision, if the calculated amount of π_2^k is bigger than 0.35. At $k = 3$, since the curve is above the horizontal line, the system will be replaced at 3rd inspection point regardless of the observed value of the indicator.

We note that, if the inspection period can be changed, on one hand, there is a constant cost that is paid at every inspection epoch, i.e. more frequent inspections cost more. On the other hand, more frequent inspections provide more information that can lead to a more cost effective replacement policy. This means that the optimal inspection period can be selected between several feasible inspection periods. The criterion that helps us to select the optimal inspection period is the *minimum total long-run average maintenance cost,* G^*, which is the long-run average cost of replacements and inspections. In the next section, we calculate this measure as well.

LONG-RUN AVERAGE COST AND TOTAL LONG-RUN AVERAGE COST

In this section, we introduce a method to calculate the minimum long-run average cost g^* and the minimum total long-run average cost G^*. The iterative method introduced by [Ghasemi *et al.* 2007] can be used to calculate the optimal value of g, i.e. g^*. The minimum long run average cost per unit of time, where stopping-time is T_g, is the unique solution of $g = \dfrac{C + K\overline{\Pr}(T_g > T)}{\overline{E}_{\min}(T_g, T)}$ where T is the time to failure, $\overline{\Pr}(T_g > T)$ is the probability of a FR, and $\overline{E}_{\min}(T_g, T)$ is the expected average length of a replacement cycle. Stopping time is defined as $T_g = \Delta.\inf\left\{n \geq 0 \mid K\left[1-\overline{R}(n,\pi^n,\Delta)\right] \geq g\overline{\tau}(n,\pi^n,\Delta)\right\}$ [Ghasemi 2005].

Let $C^{T_{g^*}}$ and $P^{T_{g^*}}$ represent the expected cost and expected length over the renewal period associated with a replacement policy in which the optimal time to replacement is T_{g^*}. g^* represents the minimum long-run average cost of replacement (excluding the inspection cost). The minimum total long-run average cost per unit of time is:

$$G^* = \frac{C^{T_{g^*}}}{P^{T_{g^*}}} + \frac{C_I}{\Delta} = \frac{C\left[1-\overline{\Pr}\left(T_{g^*} > T\right)\right] + (C+K)\overline{\Pr}\left(T_{g^*} > T\right)}{\overline{E}_{\min}\left(T_{g^*}, T\right)} + \frac{C_I}{\Delta}$$

$$G^* = \frac{C + K\overline{\Pr}\left(T_{g^*} > T\right)}{\overline{E}_{\min}\left(T_{g^*}, T\right)} + \frac{C_I}{\Delta}$$

where C, K and C_I are the replacement cost, failure cost and inspection cost, respectively.

The following equations can be used to calculate $\overline{\Pr}\left(T_{g^*} > T\right)$ and $\overline{E}_{\min}\left(T_{g^*}, T\right)$. $\overline{\Pr}\left(T_g > T\right) = Q\left(0,\pi^0\right)$ and $\overline{E}_{\min}(T_g, T) = W\left(0,\pi^0\right)$ where:

$$W\left(j,\pi^j\right) = \begin{cases} 0 & j \geq k \\ \displaystyle\int_0^{t_g\left(\pi^j\right)-j\Delta} \overline{R}\left(j,\pi^j,s\right)ds & j = k-1 \\ \displaystyle\int_0^{\Delta} \overline{R}\left(j,\pi^j,s\right)ds + \sum_{\theta=1}^{M} W\left(j+1,\pi^{j+1}(\theta)\right)\overline{R}\left(j,\pi^j,\Delta\right)\Pr\left(\theta \mid j,\pi^j\right) & j < k-1 \end{cases}$$

also

$$Q(j,\pi^j)=\begin{cases} 0 & j\geq k \\ 1-\overline{R}(j,\pi^j,t_g(\pi^j)-j\Delta) & j=k-1 \\ 1-\overline{R}(j,\pi^j,\Delta)+\sum_{\theta=1}^{M} Q(j+1,\pi^{j+1}(\theta))\Pr(\theta\mid j,\pi^j)\overline{R}(j,\pi^j,\Delta) & j<k-1 \end{cases}$$

where $t_g(\pi)=\Delta\{r\in R^+ \mid K[1-\overline{R}(r,\pi,\Delta)]=g\overline{\tau}(r,\pi,\Delta)\}$. For more details, please refer to [Ghasemi *et al.* 2007].

OPTIMAL INSPECTION PERIOD

We assume that the optimal inspection period is to be chosen from a finite set of L possible inspection periods $\Delta_i, i=1,2,\ldots,L$. The optimal inspection period is the one with the minimum total long-run average cost G^*, calculated in the previous section. First, the optimal inspection period is found, and then based on the result obtained in the section entitled "Optimal Policy", the optimal replacement policy is found.

In the following section, we solve a replacement in an example where we assume that the inspection cost is negligible. The inspection period is predefined. Later, we assume that the inspections are costly. We compare two possible inspection periods, and we find the optimal inspection period among them. The long-run average cost and the corresponding optimal replacement criteria are also calculated.

THE APPLICATION OF THE OPTIMAL INSPECTION POLICY

In this section, a summary algorithm to find the optimal inspection interval and to apply corresponding replacement policy is shown. We assume that all the parameters of the model are pre-known. The algorithm consists of the following steps:

1. For all the possible inspection periods $\Delta_l, l=1,2,\ldots,$ calculate g_l^* which is the unique solution of $g=\dfrac{C+K\overline{\Pr}(T_g>T)}{\overline{E}_{\min}(T_g,T)}$.
2. Calculate G_l^* for all possible inspection periods $\Delta_l, l=1,2,\ldots$ by using $G_l^*=\dfrac{C+K\overline{\Pr}\left(T_{g_l^*}>T\right)}{\overline{E}_{\min}\left(T_{g_l^*},T\right)}+\dfrac{C_I}{\Delta}$. The smallest value of $G_l^*, l=1,2,\ldots$, say G_m^*, identifies the optimal inspection interval, Δ_m.
3. At any inspection point, considering the specific observed value of the indicator θ, update the CPDS by using Eq. (3).
4. Considering the Eq. (12) and using Δ_m and g_m^*, decide whether to perform PR or to Do-Nothing.

At any time, if the failure occurred, replace the system (FR), set $k=0$, reset CPDS to π^0 and continue from step 3.

NUMERICAL EXAMPLE

We use the example presented by [Ghasemi *et al.* 2007] and adapt it to our case of costly inspections. In this example, it is assumed that the system has a two parameter Weibull-like behaviour with a baseline distribution hazard function having the following parameters: $h_0(t)=\frac{\beta t^{\beta-1}}{\alpha^\beta}, t\geq 0,\ \alpha=1, \beta=2$ and $\psi(X_t)=e^{0.5(X_t-1)}$. The system's two possible degradation states are {1,2} with the transition probability matrix $P_1=\begin{bmatrix}0.4 & 0.6\\ 0 & 1\end{bmatrix}$ when the inspection period is $\Delta_1=0.5$. θ, the observed value of the system's indicator, can take three possible values. The indicator value and the system's degradation state are related by the probability distribution $Q=\begin{bmatrix}0.6 & 0.3 & 0.1\\ 0.2 & 0.4 & 0.4\end{bmatrix}$. $C=5$ and $K=2$ represent the replacement cost and the failure cost of the system consecutively. The long-run average cost of replacement, based on the provided method, is found to be $g_1^*=8.67$ and the optimum stopping time of the system is $T_{g_1^*}=\inf\left\{k\geq 0; 2\times\left[1-\overline{R}\left(k,\pi^k,0.5\right)\right]\geq 8.67\times\overline{\tau}\left(k,\pi^k,0.5\right)\right\}$.

Now we assume that inspection cost $C_I=1$ applies for each inspection We also assume that there is another possible inspection period, $\Delta_2=0.6$, with corresponding degradation state transition matrix $P_2=\begin{bmatrix}0.3 & 0.7\\ 0 & 1\end{bmatrix}$. We are interested in finding the optimal replacement interval among these two options, and the corresponding replacement criteria. The following table shows the final result of this method applied to the data.

TABLE 1. Numerical example

l	Δ_l	g_l^*	G_l^*	**Stopping Time (Replacement Policy)**
1	0.5	8.67	10.67	$T_{g_1^*}=\inf\left\{k\geq 0; 2\times\left[1-\overline{R}\left(k,\pi^k,0.5\right)\right]\geq 8.67\times\overline{\tau}\left(k,\pi^k,0.5\right)\right\}$
2	0.6	8.73	10.39	$T_{g_1^*}=\inf\left\{k\geq 0; 2\times\left[1-\overline{R}\left(k,\pi^k,0.6\right)\right]\geq 8.73\times\overline{\tau}\left(k,\pi^k,0.6\right)\right\}$

While the long-run average maintenance cost of replacement $g_1^*=8.67$ for the shorter inspection period $\Delta_1=0.5$, is smaller, the total long-run average maintenance cost $G_2^* = 10.39$, corresponding to $\Delta_2 = 0.6$, is the optimal one. This means that overall, costs will be reduced if we observe the system at an inspection period equal to 0.6 and apply the corresponding stopping-time.

CONCLUSION

For a system which is subject to a CBM program, inspections are performed to obtain proper indicators concerning the system's degradation state and to decide on an optimal replacement policy. In many practical cases, the inspections do not reveal the exact degradation state. In this work, we have relaxed the assumption of non-costly inspection, and determined the optimal replacement policy and the total long-run average cost of the system replacement and inspections. More frequent inspections, while may cost more, can lead to an overall less costly inspection and replacement policy due to the higher frequency of information provided regarding the system's degradation state. An optimal inspection period minimizing the total long-run average cost of the system can be identified. The numerical example shows the application of the model.

REFERENCES

1. Banjevic, D.; Jardine, A. K. S.; Makis, V., and Ennis, M. A control-limit policy and software for condition-based maintenance optimization. INFOR Journal. 2001; 39(1):pp 32-50.
2. Chelbi, Anis and Ait-Kadi, Daoud. An optimal inspection strategy for randomly failing equipment . Reliability Engineering and System Safety . 1999; 63:pp127-131.
3. Christer, A. H.; Wang, W., and Sharp, J. A state space condition monitoring model for furnace erosion prediction and replacement. European Journal of Operational Research. 1997; 101:1-14.
4. Cox, David Roxbee. Regression models and life tables (with discussion). Journal of the Royal Statistical Society - Series B. 1972; 26 pp 187-220.
5. Cox, David Roxbee. Renewal Theory. London: Chapman and Hall; 1962.
6. Ghasemi, Alireza. Condition Based Maintenance Using the Proportional Hazards Model with Imperfect Information: Masters Thesis, department of mathematics and industrial engineering, Ecole Polytechnique de Montreal; 2005.
7. Ghasemi, Alireza; Yacout S. , and Ouali, M. S. Optimal condition based maintenance with imperfect information and the proportional hazards model. International Journal of Production Research. 2007 Feb; 45(4):pp 989-1012.
8. Grall, A.; Berenguer, C., and Dieulle L. A condition-based maintenance policy for stochastically deteriorating systems. Reliability Engineering and System Safety. 2002; 76:pp 167-180.
9. Hontelez, A. M.; Burger, Helen H., and Wijnmalen, Diederik J. D. Optimum condition-based maintenance policies for deteriorating systems with partial information. Reliability Engineering and System Safety. 1996; 51:pp267-274.
10. Jardine, A. K. S.; Anderson, P. M., and Mann D. S. Application of the Weibull Proportional Hazards Model to Aircraft and Marine Engine Failure Data. Quality and Reliability Engineering International. 1987; 3:pp 77-82.
11. Lam, Teresa C. and Yen, R. H. Optimal Maintenance-Policies For Deteriorating Systems Under Various Maintenance Strategies. IEEE Transactions on Reliability . 1994; 43(3):pp 423-430.
12. Leemis, L. M. Reliability- Probabilistic Models and Strategical Methods. New Jersey: Prentice all, Inc., Engelwood Cliffs; 1995.

13. Lin, D.; Banjevic, D., and Jardine, A. K. S. Using principal components in a proportional hazards model with application in condition-based maintenance. Journal of the Operational Research Society. 2005; pp 1-10.
14. Maillart, L. M. Optimal condition-Monitoring Schedules for Multi-State Deterioration Systems with Obvious Failure. 2004; Technical Memorandum Number 778.
15. Makis, V. and Jardine, A. K. S. Computation of optimal policies in replacement models. IMA Journal of Mathematics Applied in Business & Industry. 1992; 3 pp 169-175.
16. Makis, V. and Jardine, A. K. S. Optimal replacement in the proportional hazards model. INFOR Journal . 1992a; 30(2):172-183.
17. Ohnishi, M.; Kawai, H., and Mine, H. An optimal inspection and replacement policy under incomplete state information. European Journal of Operational Research. 1986; 27: pp 117-128.

APPENDIX

Monotone behavior

In this section, the condition under which the cost function introduced in the Eq. (4) exhibits monotone behavior is established. Some definitions and propositions presented by [Ohnishi *et al.* 1994], [Rosenfield 1975] and [Kurano 1985] are adopted.

Definition 1: [Ohnishi 1994] An N-dimensional vector x, is said to be *stochastically less* than an N-dimensional vector y, if and only if $\sum_{i=k}^{N} x_i \leq \sum_{i=k}^{N} y_i$ for any $k; 1 \leq k \leq N$, and is denoted by $x \underset{ST}{\leq} y$.

Definition 2: [Ohnishi 1994] An N-dimensional vector x, is said to be *less* than an N-dimensional vector y in *Likelihood ratio,* if and only if $\begin{vmatrix} x_i & x_j \\ y_i & y_j \end{vmatrix} \geq 0$ for $1 \leq \mathrm{i} \leq \mathrm{j} \leq \mathrm{N}$ or equivalently $x_i y_j \geq x_j y_i$ for $1 \leq \mathrm{i} \leq \mathrm{j} \leq \mathrm{N}$ and is denoted by $x \underset{LR}{\leq} y$.

Definition 3: [Rosenfield 1995] An N-dimensional probability transition matrix P is said to be *Increasing Failure Rate* (IFR) if its rows are stochastically increasing i.e. $\sum_{j=k}^{N} p_{ij} \leq \sum_{j=k}^{N} p_{\hat{i}j}$ and $1 \leq k \leq N$ and $\hat{i} > i$. In other words, the $\sum_{j=k}^{N} p_{ij}$ is non-decreasing in i for $1 \leq k \leq N$.

Definition 4: [Rosenfield 1975] An N-dimensional probability transition matrix P, is said to be *Totally Positive of Order* 2 (TP2), if its rows are increasing in likelihood ratio, i.e. [Ohnishi 1994] $\begin{vmatrix} p_{im} & p_{in} \\ p_{jm} & p_{jn} \end{vmatrix} \geq 0$ for $\begin{cases} 1 \leq \mathrm{i} \leq \mathrm{j} \leq \mathrm{N} \\ 1 \leq \mathrm{m} \leq \mathrm{n} \leq \mathrm{N} \end{cases}$.

In this research, it is supposed that $P = \left[p_{ij} \right]$ $i, j \in S$, the Markovian probability transition matrix, and $Q = \left[q_{j\theta} \right]$ $j \in S, \theta \in \Theta$ are *TP2*.

Proposition 1: [Ohnishi 1994] Having $[a_i, 1 \leq i \leq N]$, an N-dimensional vector with non-decreasing elements, if $x \underset{ST}{\leq} y$ then $\sum_{i=k}^{N} a_i x_i \leq \sum_{i=k}^{N} a_i y_i$ for $1 \leq \mathrm{k} \leq \mathrm{N}$.

Proposition 2: [Rosenfield 1976] If P is *TP2* then P is *IFR*.

Proposition 3: [Ohnishi 1994] $\pi \underset{LR}{\le} \hat{\pi} \Rightarrow \pi \underset{ST}{\le} \hat{\pi}$, where π, $\hat{\pi}$ are two N-dimensional vectors.

The following lemmas are adopted without the proofs from [Ghasemi *et al.* 2007] where π^k and $\hat{\pi}^k$ are two hypothetical CPDSs at the k-th period.

Lemma 1: If $\pi^k \underset{ST}{\le} \hat{\pi}^k$ then for any a, $\overline{R}(k,\pi^k,a) \ge \overline{R}(k,\hat{\pi}^k,a)$ where $\overline{R}(k,\pi^k,a)$ is the probability that the system is still working at $k\Delta + a$ while the CPDS at period k is π^k

Lemma 2: $\overline{R}(k,\pi^k,a)$ is non-increasing in k for any a.

Lemma 3: If $\pi^k \underset{ST}{\le} \hat{\pi}^k$, then $\overline{\tau}(k,\pi^k,a) \ge \overline{\tau}(k,\hat{\pi}^k,a)$ for any a where $\overline{\tau}(k,\pi^k,a)$ is the mean sojourn time of the system between $k\Delta$ and $k\Delta + a$ when the CPDS at the k-th period is π^k.

Lemma 4: $\overline{\tau}(k,\pi^k,a)$ is non-increasing in k for any a.

Theorem 1: Assuming that assumptions 1 through 5 as stated in [Makis and Jardine 1992] are satisfied, function V introduced by Eq. (6) defined on $\overline{S}$, where $\overline{S}$ is the set of all possible variations of the pair (k,π^k), with a constant $g \ge 0$, is a bounded measurable *non-decreasing* function.

Proof:

We consider the restricted action space to be defined as $A_\varepsilon = \{\varepsilon, \infty\}$ where ε means taking the action in a short time. [Ghasemi *et al.* 2007] have shown that under this condition, there is a non-negative real valued function $v_\varepsilon(k,\pi^k) \in D$ for the restricted action space A_ε, such that $v_\varepsilon(k,\pi^k) = U^\varepsilon v_\varepsilon(k,\pi^k)$ where D is a Borel subset of $\overline{S}$. The map U^ε is defined as $U^\varepsilon u(k,\pi^k) = \min\{kC_I + C + U(0,\pi^0,+\infty,u), U(k,\pi^k,+\infty,u)\}$ $u \in D$. By letting $k=0$, since $C>0$, $U^\varepsilon u(0,\pi^0) = \min\{kC_I + C + U(0,\pi^0,+\infty,u), U(0,\pi^0,+\infty,u)\} = U(0,\pi^0,+\infty,u)$ and then we can write:

$$U^\varepsilon u(k,\pi^k) = \min\{ C + U^\varepsilon u(0,\pi^k),\ U(k,\pi^k,+\infty,u)\} \tag{13}$$

where we define:

$$U(k,\pi^k,a,u) = [kC_I + K + C + u(0,\pi^0)] - g\overline{\tau}(k,\pi^k,a)$$
$$+\sum_{\theta=1}^{M}\Big[\big[u(k+1,\pi^{k+1}(\theta)) - (kC_I + K + C + u(0,\pi^0))\big]\Pr(\theta \mid k,\pi^k)\Big]\overline{R}(k,\pi^k,a)$$

$$\begin{aligned} U(k,\pi^k,a,u) &= [kC_I + K + C + u(0,\pi^0)](1-\overline{R}(k,\pi^k,a)) - g\overline{\tau}(k,\pi^k,a) \\ &+\sum_{\theta=1}^{M}\big[u(k+1,\pi^{k+1}(\theta))\Pr(\theta \mid k,\pi^k)\big]\overline{R}(k,\pi^k,a) \end{aligned} \tag{14}$$

for each $u \in D$ and any constant $g > 0$.

Corollary1:

For any non-decreasing function $u(k,\pi)$, where $u(k,\pi) \le kC_I + K + C + u(0,\pi^0)$ for any (k,π), $U(k,\pi,\Delta,u)$ is non-decreasing in k i.e. $U(k,\pi,\Delta,u) \ge U(k',\pi,\Delta,u)$ where $k \le k'$.

Proof:

Using Eq. (16), we can write:

$$\frac{\delta U(k,\pi,\Delta,u)}{\delta k} = C_I\left(1-\overline{R}(k,\pi,\Delta)\right) - \left[kC_I + K + C + u\left(0,\pi^0\right)\right]\frac{\delta \overline{R}(k,\pi,\Delta)}{\delta k} - g\frac{\delta \overline{\tau}(k,\pi,\Delta)}{\delta k}$$
$$+\left[\sum_{\theta=1}^{M} u\left(k+1,\pi(\theta)\right)\Pr(\theta \mid k,\pi)\right]\frac{\delta \overline{R}(k,\pi,\Delta)}{\delta k}$$
$$+\frac{\delta\left[\sum_{\theta=1}^{M} u\left(k+1,\pi(\theta)\right)\Pr(\theta \mid k,\pi)\right]}{\delta k}\overline{R}(k,\pi,\Delta)$$

$$\frac{\delta U(k,\pi,\Delta,u)}{\delta k} = \underbrace{\left[\sum_{\theta=1}^{M} u\left(k+1,\pi(\theta)\right)\Pr(\theta \mid k,\pi) - \left(kC_I + K + C + u\left(0,\pi^0\right)\right)\right]\frac{\delta \overline{R}(k,\pi,\Delta)}{\delta k}}_{1}$$
$$-g\frac{\delta \overline{\tau}(k,\pi,\Delta)}{\delta k} + C_I\left(1-\overline{R}(k,\pi,\Delta)\right)$$
$$\underbrace{+\frac{\delta\left[\sum_{\theta=1}^{M} u\left(k+1,\pi(\theta)\right)\Pr(\theta \mid k,\pi)\right]}{\delta k}}_{2}\overline{R}(k,\pi,\Delta) \tag{15}$$

Now we show that term 1 is negative. Following the assumption of the corollary which states that $u\left(k+1,\pi^{k+1}\right) < kC_I + K + C + u\left(0,\pi^0\right)$ by multiplying both sides with $\Pr\left(\theta \mid k,\pi^k\right)$ and summing up all possible amounts of θ,we can write:

$$u\left(k+1,\pi^{k+1}\right)\Pr\left(\theta \mid k,\pi^k\right) < \left[kC_I + K + C + u\left(0,\pi^0\right)\right]\Pr\left(\theta \mid k,\pi^k\right)$$
$$\sum_{\theta=1}^{M} u\left(k+1,\pi^{k+1}\right)\Pr\left(\theta \mid k,\pi^k\right) < \sum_{\theta=1}^{M}\left[kC_I + K + C + u\left(0,\pi^0\right)\right]\Pr\left(\theta \mid k,\pi^k\right)$$
$$\sum_{\theta=1}^{M} u\left(k+1,\pi^{k+1}\right)\Pr\left(\theta \mid k,\pi^k\right) < kC_I + K + C + u\left(0,\pi^0\right)$$

which means that $\sum_{\theta=1}^{M} u\left(k+1,\pi^{k+1}\right)\Pr\left(\theta \mid k,\pi^k\right) - kC_I - K - C - u\left(0,\pi^0\right) < 0$

Now we consider term 2:

$$\frac{\delta\left[\sum_{\theta=1}^{M} u\left(k+1,\pi(\theta)\right)\Pr(\theta\mid k,\pi)\right]}{\delta k}=\frac{\sum_{\theta=1}^{M}\delta\left[u\left(k+1,\pi(\theta)\right)\Pr(\theta\mid k,\pi)\right]}{\delta k}$$

$$=\frac{\sum_{\theta=1}^{M}\Pr(\theta\mid k,\pi)\,\delta u\left(k+1,\pi(\theta)\right)}{\delta k}$$

$$=\sum_{\theta=1}^{M}\Pr(\theta\mid k,\pi)\frac{\delta u\left(k+1,\pi(\theta)\right)}{\delta k}$$

Since $u(k,\pi)$ is non-decreasing in k the summation will be positive or zero.

From lemma 2, it follows that $\frac{\delta\overline{R}(k,\pi,\Delta)}{\delta k}\le 0$, and from lemma 4, it follows that $\frac{\delta\overline{\tau}(k,\pi,\Delta)}{\delta k}\le 0$ and also the term $C_I\left(1-\overline{R}(k,\pi,\Delta)\right)$ is always positive. So that in EQ. (17), $\frac{\delta U(k,\pi,\Delta,u)}{\delta k}\ge 0$ i.e. $U(k,\pi,\Delta,u)$ defined by Eq. (16) is non-decreasing in k.

Corollary 2:

For any non-decreasing function $u(k,\pi)$, where $u(k,\pi)\le kC_I+K+C+u(0,\pi^0)$, for any (k,π), $U(k,\pi^k,\Delta,u)$ is non-decreasing in π, i.e. $U(k,\pi,\Delta,u)\ge U(k,\pi',\Delta,u)$ if $\pi\underset{LR}{\le}\pi'$.

Proof:

$$\frac{\delta U(k,\pi,\Delta,u)}{\delta\pi}=-\left[kC_I+K+C+u\left(0,\pi^0\right)\right]\frac{\delta\overline{R}(k,\pi,\Delta)}{\delta\pi}-g\frac{\delta\overline{\tau}(k,\pi,\Delta)}{\delta\pi}$$
$$+\left[\sum_{\theta=1}^{M}u\left(k+1,\pi(\theta)\right)\Pr(\theta\mid k,\pi)\right]\frac{\delta\overline{R}(k,\pi,\Delta)}{\delta\pi}$$
$$+\frac{\delta\left[\sum_{\theta=1}^{M}u\left(k+1,\pi(\theta)\right)\Pr(\theta\mid k,\pi)\right]}{\delta\pi}\overline{R}(k,\pi,\Delta)$$

$$\frac{\delta U(k,\pi,\Delta,u)}{\delta\pi}=\left[\sum_{\theta=1}^{M}u\left(k+1,\pi(\theta)\right)\Pr(\theta\mid k,\pi)-kC_I-K+C+u\left(0,\pi^0\right)\right]\frac{\delta\overline{R}(k,\pi,\Delta)}{\delta\pi}$$
$$-g\frac{\delta\overline{\tau}(k,\pi,\Delta)}{\delta\pi}+\frac{\delta\left[\sum_{\theta=1}^{M}u\left(k+1,\pi(\theta)\right)\Pr(\theta\mid k,\pi)\right]}{\delta\pi}\overline{R}(k,\pi,\Delta)$$

By using lemmas 1 and 3, $\frac{\delta U(k,\pi,\Delta,u)}{\delta\pi}\ge 0$ i.e. $U(k,\pi,\Delta,u)$ defined by Eq. (16) is non-decreasing in π.

Suppose $u_0(k,\pi^k)=0$ for any (k,π^k) in (15), so u_0 is non-decreasing in (k,π^k). By corollaries 1 and 2, and also by considering that $u_n = U^{\varepsilon} u_{n-1}$, as given by [Kurano 1985], then u_1 is non-decreasing. By induction, $u_n(k,\pi^k)$ is non-decreasing in (k,π^k) for any n.

Since $v_n \to v_\varepsilon$ when $n \to \infty$ [Kurano 1985], v_ε is non-decreasing as well. It can be seen that $v_\varepsilon(k,\pi^k) \le v_{\overline{\varepsilon}}(k,\pi^k)$ if $\varepsilon \le \overline{\varepsilon}$. Suppose $v(k,\pi^k) = \lim_{\varepsilon \to 0} v_\varepsilon(k,\pi^k)$ for any (k,π^k). This implies that $v(k,\pi^k)$ is non-decreasing in (k,π^k). We note that $A_\varepsilon \to A$ while $\varepsilon \to 0$. Since map U is monotone [Kurano 1985], and by the monotone convergence theorem [Capinski 2004] we get:

$$\lim_{\varepsilon \to 0} U(k,\pi^k,+\infty,v_\varepsilon) = U(k,\pi^k,+\infty,v) \tag{16}$$

From equations (15) and (16), when $n \to \infty$ we get:

$$\lim_{n\to\infty} U^{\varepsilon} u_n(k,\pi^k) = \min\left\{ \lim_{n\to\infty}\left(kC_I + C + U^{\varepsilon} u_n(0,\pi^k)\right),\ \lim_{n\to\infty} U(k,\pi^k,+\infty,u_n)\right\}$$

$$U^{\varepsilon} v_\varepsilon(k,\pi^k) = \min\left\{ kC_I + C + U^{\varepsilon} v_\varepsilon(0,\pi^k),\ U(k,\pi^k,+\infty,v_\varepsilon)\right\}$$

and when $\varepsilon \to 0$ we get:

$$\lim_{\varepsilon\to\infty} U^{\varepsilon} v_\varepsilon(k,\pi^k) = \min\left\{\lim_{\varepsilon\to\infty}\left(kC_I + C + U^{\varepsilon} v_\varepsilon(0,\pi^k)\right), \lim_{\varepsilon\to\infty} U(k,\pi^k,+\infty,v_\varepsilon)\right\}$$

$$v(k,\pi^k) = \min\left\{kC_I + C + v(0,\pi^k), U(k,\pi^k,+\infty,v)\right\} \tag{17}$$

U and v defined by (18) and (19) respectively, represent W and V defined in (6) and (7) respectively which finishes the proof of theorem 1.

A Survey On Management Of Software Engineering In Japan

Yasuo Kadono[a], Hiroe Tsubaki[b] and Seishiro Tsuruho[c]

[a] *Management Science Institute Inc., 5-31-6 Taishido, Setagaya-ku, Tokyo 154-0004, JAPAN (e-mail: kadono@msi21.co.jp).*
[b] *Risk Analysis Research Center, the Institute of Statistical Mathematics.*
[c] *Software Engineering Center, Information-Technology Promotion Agency, Japan.*

Abstract. The purpose of this study is to clarify the mechanism of how software engineering capabilities relate to the business performance of IT vendors in Japan. To do this, we developed a structural model using factors related to software engineering, business performance and competitive environment. By analyzing the data collected from 78 major IT vendors in Japan, we found that superior deliverables and business performance were correlated with the effort expended particularly on human resource development, quality assurance, research and development and process improvement.

Keywords: business performance, competitive environment, software engineering, statistical analysis
PACS: 89.20.Bb, 89.20.Ff, 89.20.Kk, 89.70.-a.

INTRODUCTION

Many companies that use enterprise software in Japan have not been fully satisfied with the quality, cost, speed and productivity of software that IT vendors deliver. At the same time, IT vendors in Japan are facing drastic changes in their business environment, such as technology innovations and new entrants from China and India. In order for the IT industry in Japan to meet these challenges, an important step is to understand the extent to which software engineering is a core competence for achieving medium- and long-term success. To do so, we designed the research on software engineering capabilities and conducted it in SEE2005 [7].

The objectives of the study were:

- to assess the achievement of software engineering by IT vendors in Japan, and
- to better understand the mechanism of how software engineering capabilities relate to the business performance of IT vendors.

To achieve these objectives, we developed a new measurement tool called "Software Engineering Excellence (SEE)", which can evaluate the overall software engineering capabilities of IT vendors from the viewpoint of deliverables, project management, quality assurance, process improvement, research and development, human development, and contact with customers. Also, we introduced other indicators: business performance and competitive environment. The competitive environment complements the relationship between SEE and the business performance of the software vendors.

CP1007, *Current Themes in Engineering Technologies*, edited by S.-I. Ao, M. A. Amouzegar, and S.-S. Chen

In the SEE2005 survey, we analyzed the relationship among software engineering excellence (SEE), business performance and competitive environment based on the data collected from 55 major IT vendors in Japan. As we confirmed using the path analysis, we found that software engineering excellence (SEE) has a direct positive impact on business performance and that the competitive environment directly as well as indirectly (i.e., via SEE) affects business performance [7].

In the SEE2006 survey, we modified the measurement model used in SEE 2005 and increased the number of surveyed Japanese IT vendors from 55 to 78 in order to more deeply investigate more the impact of software engineering on business performance, as well as the competitive environment. In particular, in this study we focus on the relationships among factors of SEE, the competitive environment, and business performance as measured by operating profit ratio [8].

RESEARCH METHOD

Structural Model

We assume the following research questions on the relationships among the three primary indicators (i.e., software engineering, business performance, and competitive environment) in **FIGURE 1**.

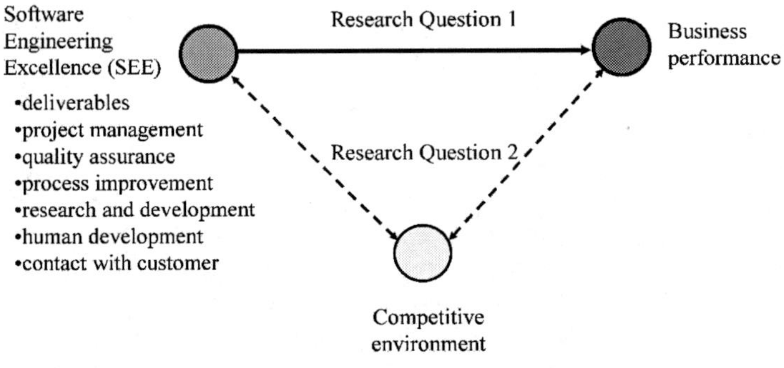

FIGURE 1. Structural model.

Research question 1: the "software engineering excellence (SEE)" has a positive impact on "business performance".

Research question 2: the "competitive environment" directly and indirectly (i.e., via SEE) affects "business performance".

Here, "software engineering excellence (SEE)" expresses the extent to which IT vendors can put software engineering into practice. SEE is a measurement tool for evaluating the overall capabilities of the software engineering of IT vendors from the viewpoints of deliverables, project management, quality assurance, process improvement, research and development, human development, and contact with customers.

"Business performance" expresses the overall business performance of individual IT vendors, such as profitability, growth, and management efficiency. "Competitive environment" expresses the company profile of IT vendors as well as the business environment in which the vendors work, e.g., number of software engineers, business model, and average age of employees.

Measurement Model

Our measurement model of software engineering excellence (SEE) was developed through interviews with over 20 experts in the industry and literary searches [1,3,4,9,10]. The SEE measurement model has a hierarchical structure with three layers: observed responses, seven detailed concepts, and SEE as a primary indicator, as we developed in our research on IT management effectiveness [5, 6].

The observed responses such as readiness for state-of-the-art technology, moral support, and clarification of user specification are newly added to the measurement model as a result of feedback from the respondents to SEE2005 and interviews with the experts.

- Software engineering excellence (SEE)
    - deliverables: achievement ratio of quality, cost, speed, and productivity
    - project management: project monitoring, assistance to project managers, project planning capability, ratio of PMP(Project Management Professional)
    - quality assurance: organization, method, review, testing, guideline, management of outsourcers
    - process improvement: data collection, improvement of estimation, assessment method, CMM/CMMI [2]
    - research and development: strategy, organization, sharing technological skills, learning organization, development methodology, intellectual assets, commoditized software, readiness to state-of-the-art technology,
    - human development: training hours, skill development systems, incentive schemes, measure of human development, moral support
    - contact with customers: ratio of prime contracts, scope of service offered, direct communication with customer's top management, deficit prevention, clarification of user specification

In addition to SEE, we assume two primary indicators, business performance and competitive environment as follows.

- Business performance
    - profitability: operating profit ratio
    - growth: annual sales growth
    - management efficiency: return on equity
- Competitive environment
    - number of software engineers including programmers
    - average age of employees
    - business model: ratio of customized development, ratio of development based on

mainframe computer, ratio of prime contractors

- aspirations of senior managers: sales growth orientation, profitability orientation, business model innovation orientation, spirit of challenge
- corporate culture: information sharing and agility

SURVEY ON SOFTWARE ENGINEERING EXCELLENCE

To research the questions in the previous section, we conducted a new survey on Software Engineering Excellence (SEE2006). In this survey, we designed a questionnaire on the practice of software engineering and the nature of the company. This questionnaire was sent to the CEOs of 537 major Japanese IT vendors with over 300 employees, and was then distributed to the departments in charge of software engineering. Responses were received from 86 companies and valid responses totaled 78 at SEE2006 (response rate of15%), while valid responses numbered 55 at SEE2005 (response rate of 23%). Profitability as measured by operating profit ratio of 72 (36) vendors was available at SEE2006 (SEE2005).

The measurement model in the previous section is fitted to the data by confirmatory factor analyses to estimate the scores of software engineering excellence (SEE) in the same way we developed it at SEE2005 [7].

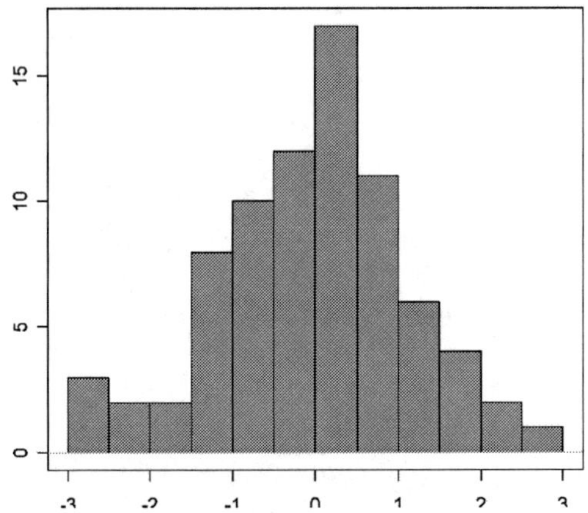

FIGURE 2. Histogram of deviations of SEE.

FIGURE 2 shows a histogram of the deviations of software engineering excellence (SEE) of 78 IT vendors. We consider that the result of SEE analysis is appropriate for further analyses since some scores of SEE are reasonable in light of the results of the interviews with the individual respondents and since the histogram has an almost unimodal distribution.

FIGURE 3 contains a box-and-whisker plots which show that the median software engineering excellence (SEE) of vendors who were originally users of IT (2) is higher than that of vendors who were originally makers of IT (1). Also, the median SEE of vendors who were originally vendors of IT (1) is higher than that of independent vendors (3). However, the maximum SEE of 3 is higher than that of 2. This tendency in SEE206 is the same as that in SEE2005 except the slight order reversal of the median SEE between 1 and 2.

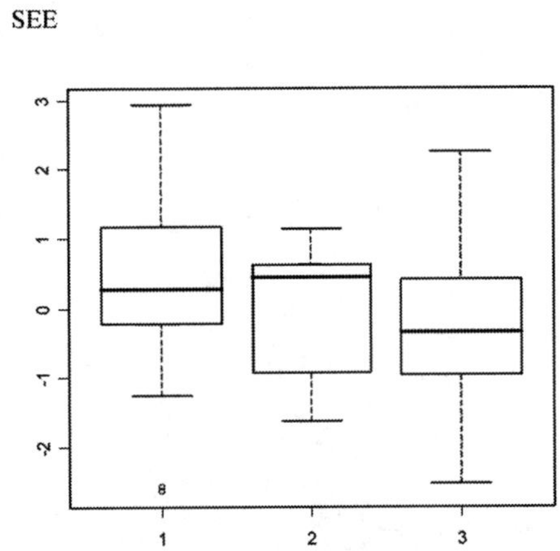

FIGURE 3. SEE by origin of vendors (1: makers, 2: users, 3: independent).

The median achievement ratios of quality, cost, and delivery are higher than 70% (**FIGURE 4**). The achievement level of delivery is the highest of quality, cost, and delivery. This tendency was also observed in the previous study at SEE2005 [7].

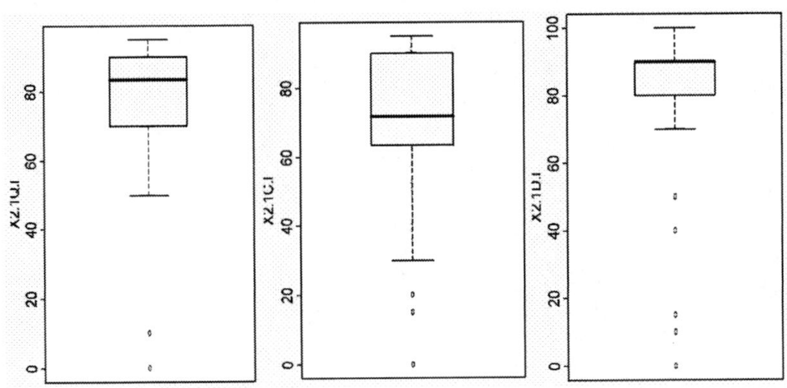

FIGURE 4. Achievement ratio of quality, cost, and delivery.

Regarding training hours, the average hours for new recruits is almost 500 hours per year, while the average hours for software engineers except new recruits is 40 hours per year (**FIGURE 5**).

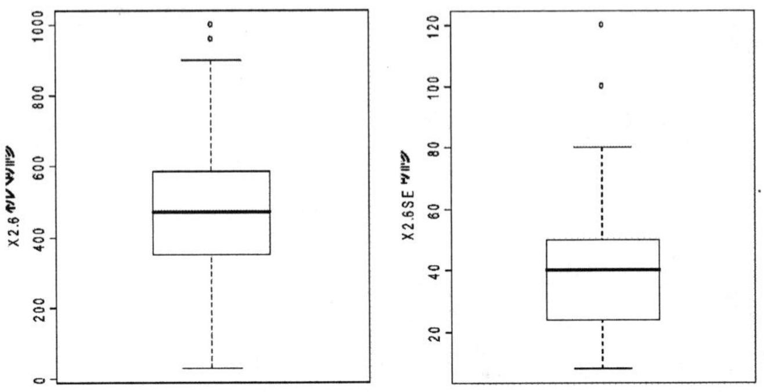

FIGURE 5. Training hours.
(left side: new recruits, right side: except new recruits)

PRELIMINARY ANALYSES

To construct a path model analysis on the research questions in the next section, we conducted the following preliminary analyses of relationships among indicators such as SEE, business performance, and competitive environment.

FIGURE 6 shows that vendors who have a larger number of software engineers tend to score higher for software engineering excellence (SEE). This tendency was similar to that of the previous study at SEE2005 [7].

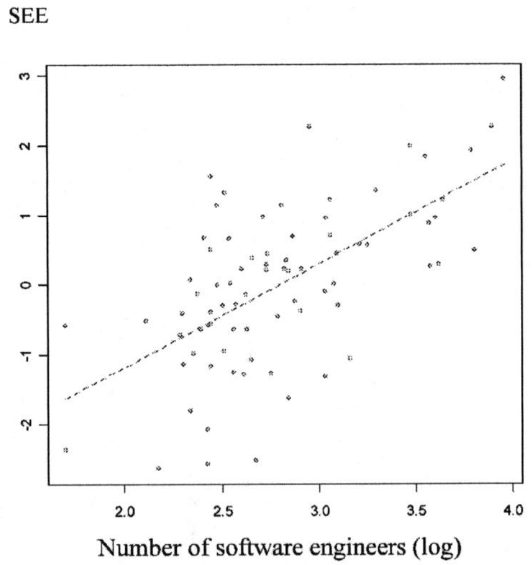

FIGURE 6. Relationship between number of software engineers and software engineering excellence.

FIGURE 7 shows that vendors whose employees are older tend to score higher for software engineering excellence (SEE).

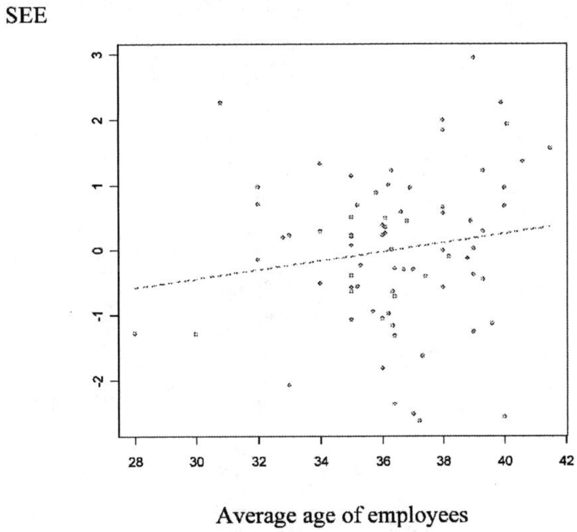

FIGURE 7. Relationship between average age of employees and SEE.

FIGURE 8 shows that vendors who develop more custom software tend to score lower for software engineering excellence (SEE).

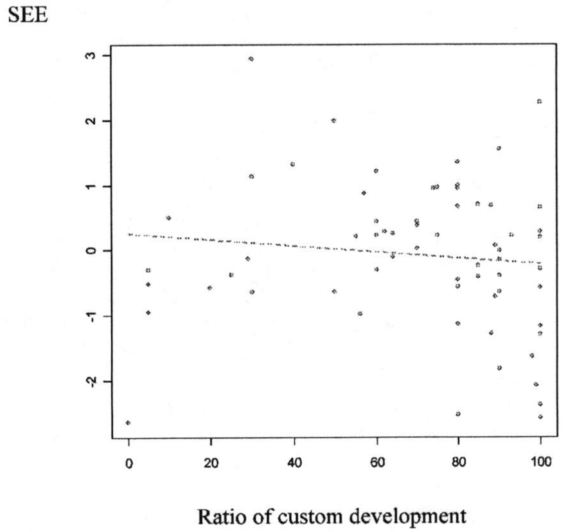

FIGURE 8. Relationship between ratio of custom development and SEE.

FIGURE 9 shows that vendors whose employees are older tend to be less profitable. This relationship is weaker than that of the previous study at SEE2005. The reason for this change is thought to be recovery of revenue as well as rejuvenation of the software industry in Japan. However, we need further investigation of the reasons.

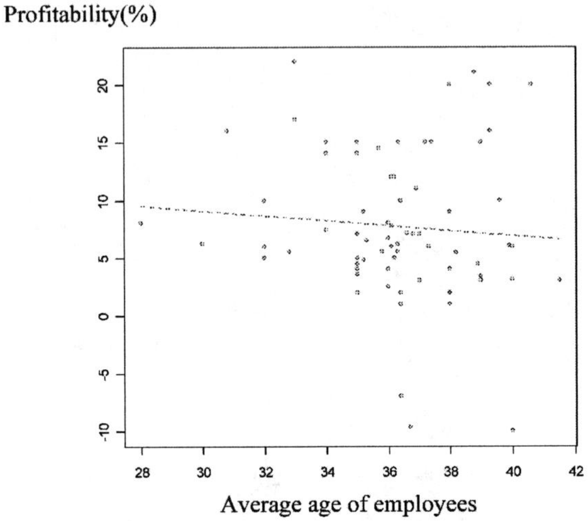

FIGURE 9. Relationship between average age of employees and profitability.

FIGURE 10 shows that vendors who have a higher software engineering excellence (SEE) tend to be slightly more profitable (i.e., higher operating profit ratio) based on the 72 vendors whose data of business performance is available.

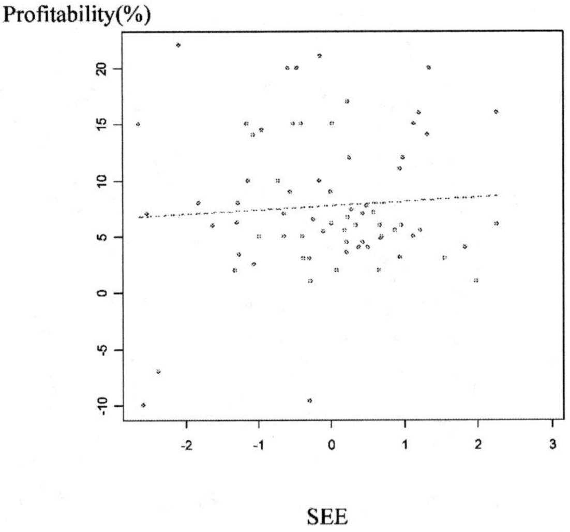

FIGURE 10. Relationship between SEE and profitability.

PATH MODEL ANALYSIS

On the basis of the results of the preliminary analyses, we constructed a path model that consists of all seven factors of software engineering excellence (SEE) as well as the selected factors for business performance and competitive environment in section 2, owing to the constraints of sample size, statistical significance, and realistic interpretation, e.g., average age of employees, number of software engineers, ratio of development based on mainframe computer, ratio of prime contractors, ratio of manufacturer customers, and ratio of operating profit.

By a trial and error method, we succeeded in constructing a well-fitted path model (CFI = 1.0), where all the existing path coefficients are significant at the 5% level (FIGURE 11 and FIGURE 12).

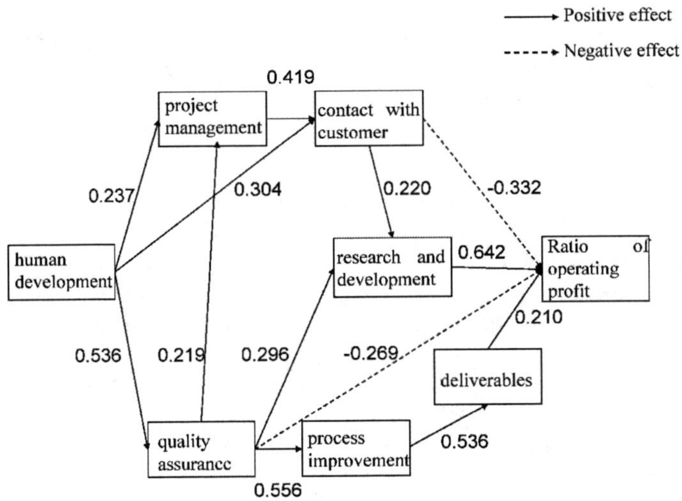

FIGURE 11. Result of path analysis among factors of SEE and business performance.

Regarding research question 1 in section 2, i.e., the relationship among factors of SEE and business performance measured by operating profit ratio, we found the following (FIGURE 11):

- Within the SEE factors, human development is positioned in the uppermost stream.
- Human development has a positive impact on quality assurance, project management and customer contact.
- Quality assurance has a direct negative impact on the operating profit ratio. This suggests that the cost of quality assurance does not pay off.
- However, indirectly quality assurance has positive impact on the ratio of operating profit via the positive influence of process improvement, deliverables, and research and development.
- Research and development has a positive impact on the ratio of operating profit directly. This tendency is similar to the result of the previous study at SEE2005

[7].

- Process improvement has a positive impact on the ratio of operating profit via deliverables. This effect of process improvement is the same as that of the previous study.

Similarly, in terms of the research question 2 in section 2, i.e., the relationships among factors of competitive environment, factors of SEE, and business performance measured by the ratio of operating profit, by using a trial and error method, we succeeded in constructing a well-fitted path model (CFI = 1.0), where all the existing path coefficients are significant at the 5% level. We found the following direct influences (FIGURE.12):

- The number of software engineers positively affects human development, research and development, quality assurance, and project management, while the number of software engineers negatively influences the ratio of operating profit.
- The ratio of customized development has a positive impact on deliverables, while it has a negative impact on human development and research and development.
- The ratio of prime contractors has a positive impact on customer contact, quality assurance, and research and development.
- The average age of employees has a positive impact on research and development, while it has a negative impact on deliverables.
- The ratio of development based on mainframe computer has a positive impact on process improvement and project management, while it has a negative impact on deliverables, the ratio of operating profit, and customer contact.

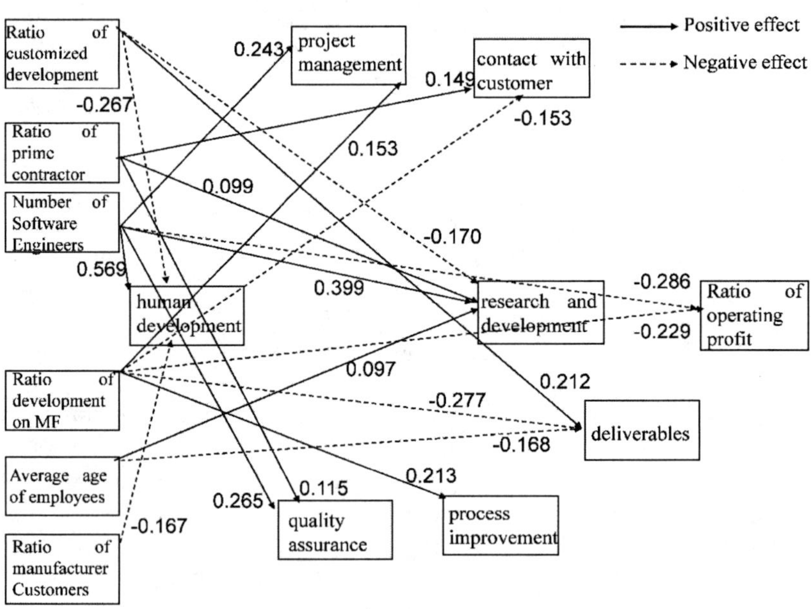

FIGURE 12. Result of path analysis among factors of competitive environment, factors of SEE and business performance.

The above observations suggest that improving software engineering excellence (SEE) is a significant method of improving the profitability of IT software vendors in Japan. Also, improving SEE depends on the competitive environment, as well as the nature of the individual software vendors, such as the number of the vendor's software engineers, the ratio of prime contractors, and the ratio of customized development.

CONCLUSION

In this study, we analyzed the relationship among software engineering excellence (SEE), business performance and competitive environment based on the data collected from 78 major IT vendors in Japan. As we confirmed using the path model, we found that human development, research and development, and process improvement mostly improve deliverables and the profitability of the software vendors. In addition, the competitive environment, such as the number of software engineers, the average age of employees, and the business model should be considered for the further growth of the software industry in Japan.

To better understand the reality and issues facing Japan's software industry in the medium- and long-term, we suggest that future studies be conducted as follows:

- further refinement of the measurement model and analysis of the cause-and-effect relationships in more detail,
- data collection over a wider range of IT vendors,
- global benchmarking, and time series analysis.

ACKNOWLEDGMENTS

The author gratefully acknowledges the valuable suggestions and support of the Ministry of Economy, Trade and Industry, and Software Engineering Center, Information-Technology Promotion Agency, Japan.

REFERENCES

1. Erik Brynjolfsson, (2004) *Intangible Asset*, Diamond.
2. Carnegie Mellon University, Software Engineering Institute, http://www.sei.cmu.edu/cmmi/.
3. Micheal A. Cusumano, (2004) *The Business of Software*, Free Press.
4. Takahiro Fujimoto, (2003) *Nouryoku Kouchiku Kyousou*, Chuohkouronshinsya (in Japanese).
5. Yasuo Kadono, (2004) *Evolution of IT Management Creating Business Value*, Nikkagiren.
6. Yasuo Kadono, Hiroe Tsubaki, (2006) "Development of IT Management Effectiveness and Analyses of Japanese Companies," Journal of Japanese Association for Management Information, Vol.14, No.4, pp.69-83.
7. Kadono, Y., H. Tsubaki, and S. Tsuruho (2006). "A Study on the Reality and Economy of Software Development in Japan," The sixth Asia Pacific Industrial Engineering Management Systems Conference, Bangkok, Thailand, pp.1425-1433.
8. Kadono, Y., H. Tsubaki, and S. Tsuruho (2007). "A Survey on Management of Software Development in Japan," World Congress on Engineering and Computer Science, San Francisco, USA pp.983-988.
9. Yoshihiro Matsumoto, (2005) *Application of SWEBOK to Software Development*, Ohmsha.
10. Ministry of Economy, Trade and Industry, (2006) *Report on Software Industry*.

AUTHOR INDEX

S

T

V

W

Y